P9-AFZ-599

METHODS IN
# MICROBIOLOGY

# METHODS IN
# MICROBIOLOGY

## Volume 24
## Techniques for the Study of Mycorrhiza

Edited by

### J. R. NORRIS

*Reading, UK*

### D. J. READ

*Department of Animal & Plant Sciences,*
*University of Sheffield, UK*

### A. K. VARMA

*School of Life Sciences, Jawaharlal Nehru University*
*New Delhi, India*

UNIVERSITY OF THE PACIFIC

ACADEMIC PRESS
*Harcourt Brace Jovanovich, Publishers*
London   San Diego   New York   Boston
Sydney   Tokyo   Toronto

ACADEMIC PRESS LIMITED
24–28 Oval Road
London NW1

*United States Edition published by*
ACADEMIC PRESS INC.
San Diego, CA 92101

Copyright © 1992 by
ACADEMIC PRESS LIMITED

Chapter 20 © 1992 US Government in the jurisdictional territory of the USA

*All Rights Reserved*
No part of this book may be reproduced in any form, by photostat, microfilm or any other means,
without written permission from the publishers

*British Library Cataloguing in Publication Data is available*

*ISBN 0-12-521524-X*

Editorial and production services by Fisher Duncan
10 Barley Mow Passage, London W4 4PH, UK
Printed in Great Britain by St Edmundsbury Press,
Bury St Edmunds, Suffolk

# CONTRIBUTORS

**L. K. Abbott**  Soil Science and Plant Nutrition, School of Agriculture, The University of Western Australia, Nedlands, WA 6009, Australia

**R. Azcon**  Departamento de Microbiología, Estación Experimental del Zaidín, Profesor Albareda 1, 18008 Granada, Spain

**C. Azcon-Aguilar**  Departamento de Microbiología, Estación Experimental del Zaidín, Profesor Albareda 1, 18008 Granada, Spain

**D. J. Bagyaraj**  Department of Agricultural Microbiology, University of Agricultural Sciences, GKVK Campus, Hebbal, Bangalore 560056, India

**J. M. Barea**  Departamento de Microbiología, Estación Experimental del Zaidín, Profesor Albareda 1, 18008 Granada, Spain

**G. Bécard**  United States Department of Agriculture, Agricultural Research Service, Eastern Regional Research Center, 600 East Mermaid Lane, Philadelphia, PA 19118, USA

**G. J. Bethlenfalvay**  United States Department of Agriculture, Agricultural Research Service, Western Regional Research Center, Albany, CA 94710, USA

**H. Beyrle**  Department of Botany, Technische Universität of München at Weihenstephan, W-8050 Freising 12, Germany

**M. Bodmer**  Swiss Federal Research Station, CH-8820 Wädenswil, Switzerland

**P. Bonfante-Fasolo**  Dipartimento di Biologia Vegetale, Università di Torino, Viale Mattioli 25, 10125 Torino, Italy

**G. W. Butcher**  Monoclonal Antibody Centre, AFRC Institute of Animal Physiology and Genetics Research, Babraham, Cambridge CB2 4AT,UK

**E. Cervantes**  IRNA-CSIC, Apartado 257, 37071 Salamanca, Spain

**J. Creighton Miller, Jr**  Department of Horticultural Sciences, Texas A&M University, College Station, TX 77843, USA

**K. Dressel**  Department of Botany, Technische Universität of München at Weihenstephan, W-8050 Freising 12, Germany

**F. Feldmann**  Institut für Angewandte Botanik, Universität Hamburg, Marseiller Strasse, Postfach 302762, 2000 Hamburg 36, Germany

**C. Gazey**  Soil Science and Plant Nutrition, School of Agriculture, The University of Western Australia, Nedlands, WA 6009, Australia

**S. Gianinazzi**  Laboratoire de Phytoparasitologie, INRA-CNRS, Station de Génétique et d'Amélioration des Plantes, INRA, BV 1540, 21034 Dijon Cédex, France

**V. Gianinazzi-Pearson**  Laboratoire de Phytoparasitologie, INRA-CNRS, Station de Génétique et d'Amélioration des Plantes, INRA, BV 1540, 21034 Dijon Cédex, France

**M. Habte**  Department of Agronomy and Soil Science, Hawaii Institute of Tropical and Human Resources, University of Hawaii, Honolulu, HI 96822, USA

**B. Hock** Department of Botany, Technische Universität of München at Weihenstephan, W-8050 Freising 12, Germany

**E. Idczak** Institut für Angewandte Botanik, Universität Hamburg, Marseiller Strasse, Postfach 302762, 2000 Hamburg 36, Germany

**S. Liebmann** Department of Botany, Technische Universität of München at Weihenstephan, W-8050 Freising 12, Germany

**F. Malavasi** Department of Medical Genetics, Biology and Medical Chemistry, University of Turin, Centre of Immunogenetics and Histocompatibility, via Santena 19, 10126 Torino, Italy

**P. A. McGee** School of Biological Sciences, A12, University of Sydney, 2006 NSW, Australia

**D. D. Miller** Department of Horticulture, Ohio State University, 2001 Fyffe Court, Columbus, OH 43210, USA

**J.-E. Nylund** Department of Forest Mycology and Pathology, Swedish University of Agricultural Sciences, PO Box 7026, S-750 07 Uppsala, Sweden

**G. Pacioni** Dipartimento di Scienze Ambientali, Università, I-67100 L'Aquila, Italy

**S. Perotto** John Innes Institute, John Innes Centre for Plant Research, Colney Lane, Norwich NR4 7UH, UK

**Y. Piché**, Département des Sciences Forestières, Centre de Recherche en Biologie Forestière, Faculté de Foresterie et de Géomatique, Université Laval, Ste-Foy, G1K 7P4 Québec, Canada

**S. Rajapakse** Department of Horticulture, Clemson University, Clemson, SC 29634, USA

**A. D. Robson** Soil Science and Plant Nutrition, School of Agriculture, The University of Western Australia, Nedlands, WA 6009, Australia

**C. Rodríguez-Barrueco** IRNA-CSIC, Apartado 257, 37071 Salamanca, Spain

**S. Rosendahl** Institute for Sporeplanter, University of Copenhagen, Øster Farimagsgade 2 D, DK-1353 Copenhagen K, Denmark

**H. Schüepp** Swiss Federal Reserach Station, CH-8820 Wädenswil, Switzerland

**R. Sen** Department of Botany, University of Helsinki, Unioninkatu 44, SF-00170 Helsinki, Finland

**S. E. Smith** Department of Soil Science, Waite Agricultural Research Institute, University of Adelaide, Glen Osmond, 5064 SA, Australia

**P. Spanu** Abteilung Pflanzenphysiologie Botanisches Institut der Universität Basel, Hebelstrasse 1, CH-4056 Basel, Switzerland

**D. M. Sylvia** Soil Science Department, University of Florida, Gainesville, FL 32611-0151, USA

**I. C. Tommerup** Division of Forestry, CSIRO, Private Bag, Wembley, WA 6014, Australia

**R. Toth** Department of Biological Sciences, Northern Illinois University, De Kalb, IL 60115-2861, USA

**H. Wallander** Department of Forest Mycology and Pathology, Swedish University of Agricultural Sciences, PO Box 7026, S-750 07 Uppsala, Sweden

**P. G. Williams** School of Biological Science, University of New South Wales, PO Box 1, Kensington, NSW 2033, Australia

# PREFACE

It is now almost ten years since the appearance of the only previous volume devoted to the methodologies used in mycorrhizal research: *Methods and Principles of Mycorrhizal Research* (N. C. Schenck, ed) American Phytopathological Society, St Pauls, Minnesota, 1982. These have been momentous years which have seen not only an explosive increase of interest in mycorrhiza as evidenced by the surge in number of published papers, but also, and most importantly, the welcome development of a greater awareness amongst plant physiologists, ecologists and microbiologists that the mycorrhizal condition is the norm rather than the exception in nature. Through the same period sophisticated new technologies (e.g. NMR, RFLP, DNA mediated transformation, immunocytochemistry), which are readily applicable to mycorrhizal research, have become widely available.

These changes have given rise to an urgent need for a comprehensive and up-to-date source book of techniques which can be routinely used in laboratories involved in mycorrhizal research to facilitate, and increase the effectiveness of, their research. The present volumes are designed to meet this need.

The volumes are divided so that the major emphasis of the first (Vol. 23) is placed upon a description of techniques applicable to ectomycorrhizal and ericoid systems, while that of the second (Vol. 24) is upon vesicular–arbuscular systems. Each contains a mix of papers, written by the leading authorities in their fields, which emphasise techniques, but which also include up-to-date reviews of particular topics e.g. carbon metabolism, nitrogen metabolism and applied aspects. The methods whereby, and the extent to which, mycorrhizal inoculation can be used to facilitate revegetation, as well as increases of agricultural or forest productivity are also described.

We envisage that over the next decade these volumes will provide essential information for all those initiating projects in this important subject area, and that at the same time they will facilitate considerable increases in effectiveness in established programmes of mycorrhizal research.

Inevitably in multi-authored volumes such as these the terminology used in the original submissions differed greatly between papers. Not the least of the difficulties was posed by the word 'mycorrhiza' itself. This appeared in various guises including 'mycorrhiza', 'mycorrhizas' and 'mycorrhizae'. Faced with such problems, and with the need for internal consistency foremost in their minds, the editors elected to employ the word 'mycorrhiza' throughout, since it can be used to refer to both the singular and plural state and to describe 'the mycorrhizal condition'. On etymological grounds, 'mycorrhiza' is preferable to 'mycorrhizae' because the latter involves the combination of two elements of Greek origin, 'myco' (fungus) and 'rhiza' (root), with a Latin 'ae' ending.

We hope that the need for internal consistency in the use of terminology will be recognised. For their part, the editors have striven to ensure that any minor changes of style arising from alteration of terminology are made in such a way as to have minimal impact on the essential meaning of the text.

*October 1991*                                                    *J. R. Norris*
                                                                 *D. J. Read*
                                                                 *A. K. Varma*

# CONTENTS

# 1

# Selection of Inoculant Vesicular-arbuscular Mycorrhizal Fungi

L. K. ABBOTT, A. D. ROBSON and C. GAZEY

*Soil Science and Plant Nutrition, School of Agriculture, The University of Western Australia, Nedlands, WA 6009, Australia*

## I. Introduction

Procedures for selecting inoculant vesicular-arbuscular mycorrhizal fungi need to be considered in the broader context of whether inoculation is indeed the appropriate management option (Abbott and Robson, 1991). First, after identifying the limitations to plant growth or soil structure, evidence is needed to predict whether vesicular-arbuscular mycorrhizal fungi would be able to overcome these limitations.

Inoculation with selected vesicular-arbuscular mycorrhizal fungi is clearly necessary in phosphate-deficient soils that have no indigenous populations of these symbiotic fungi. Such situations will include eroded

METHODS IN MICROBIOLOGY
VOLUME 24   ISBN 0-12-521524-X

Copyright © 1992 by Academic Press Limited
All rights of reproduction in any form reserved

sites (Hall and Armstrong, 1979), severely disturbed areas including those from which topsoil has been removed or stored (e.g. after mining (Jasper *et al.*, 1989) or roadworks), soils that have been fallowed (Thompson, 1987) and fumigated soils used for horticulture (both field and nursery soils).

Inoculation may be an option in soils that have either high or low levels of "ineffective" vesicular-arbuscular mycorrhizal fungi (Abbott and Robson, 1985a). Although there is evidence that species of vesicular-arbuscular mycorrhizal fungi differ in their ability to increase phosphate uptake and plant growth, little is known of their effectiveness under field conditions from various quantities of inoculum. There is limited knowledge of the potential for inoculant vesicular-arbuscular mycorrhizal fungi to displace indigenous species that occur at high levels in field soils, but which may be ineffective. Evidence from other areas of soil microbiology (e.g. Holland, 1970) indicate a poor chance of success in displacing the background population of vesicular-arbuscular mycorrhizal fungi with inoculant species.

In all other situations, the assumption is that effective vesicular-arbuscular mycorrhizal fungi already occur in the soil and the need for inoculation depends on the abundance of these fungi. If the abundance of fungi is high, there is no potential benefit of inoculation. If effective fungi are present, but in low abundance, then inoculation with more of the same fungi or with other species can be considered. An alternative option in this case is to change the agricultural practices to increase the abundance of the mycorrhizal fungi already present.

In this chapter we present procedures for selecting inoculant vesicular-arbuscular mycorrhizal fungi based on an understanding of the criteria that are essential for their successful use. It is not possible to be prescriptive in this approach. Procedures for selection need to be designed for each situation. All of the techniques that are necessary for characterizing species of vesicular-arbuscular mycorrhizal fungi and for assessing mycorrhiza formation and functioning are described elsewhere in this volume.

## II. Definition of selection criteria

### A. What is "effectiveness"

"Effectiveness", "efficiency", "efficacy" and "effectivity" are terms applied to species of vesicular-arbuscular mycorrhizal fungi, but what do they mean? In practice the difinition of these words needs to be tailored for a particular situation. Species of vesicular-arbuscular mycorrhizal fungi are required that can either directly increase plant growth or indirectly increase plant growth by improving soil conditions. Direct benefits are usually related to the enhancement of phosphate uptake into the plant, but in some soils enhanced uptake of zinc, copper and ammonium may also be important (Stribley, 1987). Indirect benefits may include increased soil aggregation (Tisdall and Oades, 1982) or stabilization of soil (Koske and Polson, 1984) associated with hyphae formed in soil. Other potential benefits are either poorly defined (e.g. hormonal response) or associated with alleviation of phosphorus deficiency (e.g. drought tolerance, disease tolerance).

Processes that will be important in enhancing phosphate uptake are: hyphal growth in soil outside the zone of depletion of phosphorus around roots; absorption of phosphorus by the hyphae; transport of phosphorus towards the root; and transfer of phosphorus to the root (Smith and Gianinazzi-Pearson, 1988). These authors pointed out that although variations in symbiotic response have usually been measured in terms of the growth of the host plant, methods are now available to assess the responses in terms of physiological processes. Although effectiveness will be related to physiological processes, it is also essential that the fungi are present within roots and soil in sufficient quantities and are active at an appropriate time so that their "intrinsic" abilities lead to benefits for the plant. Thus, a fungus may be effective according to the physiological processes assessed (Smith and Gianinazzi-Pearson, 1988) without increasing plant growth under field conditions. A fungus may be effective in one situation but may be ineffective in another, if soil management affects either the functioning of the symbiosis or the maintenance of sufficient mycorrhizal roots with an appropriate level of active hyphae. The effectiveness of a fungus has practical significance only if the fungus can benefit the plant. Therefore a definition of effectiveness that includes a measure of plant growth will take into account features of the fungus that enable it to colonize roots and soil extensively at appropriate stages of plant growth as well as factors that relate to the exchange of phosphorus and carbon between the two symbionts (Harley, 1984). The selection of inoculant fungi requires a knowledge of the intrinsic characteristics of fungi in the context of a range of soil and environmental conditions relevant to those under which the inoculant fungi will be used.

Effective fungi for large scale agriculture may be those that can enhance plant growth where phosphate is limiting. In this case a measure of the benefits in terms of seed or forage yield is relevant. Furthermore, the definition of effectiveness in this context may include the maintenance of benefits into subsequent seasons. Therefore, characteristics of the fungi that relate to persistence of high levels of inoculum in the soil will be relevant to effectiveness in such circumstances. Yield may be an appropriate measure of effectiveness for comparing the potential of fungi for most horticultural crops, but the effect on time taken to reach important stages of plant development may also be important. For example, time to flowering may be relevant for plants that are sold in flower. When re-establishing natural ecosystems after major soil disturbance, total biomass of plants in the community may not be as important as plant diversity. On disturbed sites, vesicular-arbuscular mycorrhizas may assist in the establishment of plant communities that include species that are highly mycorrhiza-dependent. In this case a measure of effectiveness should take into account the biomass of specific species within the community.

In other situations, mycorrhizal fungi may be most beneficial for plant growth through their capacity to assist in the process of soil stabilization (Tisdall and Oades, 1979, 1982). Some characteristics of fungi that are essential for their capacity to enhance plant growth in a phosphate-deficient soil by alleviating phosphate deficiency may not be important in the development of hyphal networks that play a key role in aggregate stability of soil. These benefits may require a longer-term view of the potential of vesicular-arbuscular mycorrhizal fungi than can be assessed within one season.

The term "effectiveness" defined in relation to the ability of a fungus to increase plant growth in a phosphate-deficient soil (Abbott and Robson, 1981a) is not an attribute that can be used as a universal label for a species or strain of vesicular-arbuscular mycorrhizal fungus. This is because the effectiveness defined in this way will vary according to the effect of the soil environment on characteristics of the fungus that relate to the formation and functioning of the hyphae in soil and within the root. Each characteristic of a fungus can be described in terms of whether it contributes to effectiveness at achieving a required benefit in plant growth. However, soil and environmental conditions and the quantity of infective propagules (Daniels *et al.*, 1981; Haas and Krikun, 1985) of the fungus will have an overriding influence on whether or not the mycorrhiza benefit the plant. The use of "effective" (or of similar adjectives) to describe vesicular-arbuscular mycorrhizal fungi (Powell, 1982) therefore has little practical significance out of the context of a specific soil environment.

The definition of effectiveness of vesicular-arbuscular mycorrhizal fungi to be used in the context of selecting inoculants is a general one: "effectiveness is the ability of fungus to benefit plant growth". The nature of the benefit will depend on each situation and will include the time-span for which benefits are required. Conditions that limit the quantity of fungus in the soil/root will also affect effectiveness.

## B. Selection criteria

Criteria for the selection of inoculant vesicular-arbuscular mycorrhizal fungi depend on characteristics of the fungi that relate to effectiveness (as defined above) and include their suitability for large-scale inoculum production. The criteria for selection of fungi will depend on details of the local environment, soil conditions and host plant to be used.

The criteria for selection of inoculants to benefit the plant directly by alleviating phosphorus deficiency are the ability to: (1) colonize roots rapidly following inoculation; (2) absorb phosphate from soil; (3) transfer phosphorus to the plant; (4) increase plant growth; (5) persist in soil as required; (6) re-establish mycorrhiza in following seasons (if required); (7) form a large number of propagules; and (8) form propagules able to withstand manipulation during and after inoculum production (Table I).

## 1. Selection for phosphate uptake

*(a) Hyphae.* The growth of hyphae of vesicular-arbuscular mycorrhizal fungi in soil is fundamental to the effective functioning of the symbiosis as well as to the colonization of roots (Sylvia, 1990). These two roles of the hyphae in soil are of primary importance and are relatively poorly understood, especially for a large number of fungal species. Species of fungi can be screened for their ability to form hyphae in soils. Differences vary in relation to soil chemical properties, for example salinity (Juniper, 1990) and pH (see review by Robson and Abbott, 1989).

*(b) Transfer of phosphorus to host.* Although little is known of the relative abilities of fungi to transfer phosphorus to host plants (separate processes of uptake, translocation and transfer of phosphorus all being involved), this capacity can generally be assessed indirectly in terms of the host response. Therefore in the absence of a "complete" under-standing of these processes, selection of fungi can still be made. However, characterizing fungi in terms of their ability to absorb and transfer phosphorus will enable predictions to be made about the potential value of particular species. Species of fungi differ in their

**TABLE I**

Characteristics of fungi important for the selection of inoculant vesicular-arbuscular mycorrhizal fungi that can be quantified.

| Activity required | Relevant characteristics of fungi |
|---|---|
| Ability to absorb and transport phosphate from soil into plant | External hyphae<br>(1) quantity in relation to mycorrhizal development<br>(2) distribution in relation to mycorrhizal development<br>(3) P absorption/transfer<br>(4) role in colonization along and between roots |
| | Capacity to colonize roots in relation to<br>(1) P supply<br>(2) presence of other vesicular-arbuscular mycorrhizal fungi<br>(3) soil properties<br>(4) host<br>(5) other soil micro-organisms and soil fauna |
| Ability to increase plant growth | Capacity to transfer P to host relative to requirement for host C<br>Capacity to colonize roots (see above) |
| Ability to persist in soil (if required) | Capacity to produce extensive persistent propagules in roots and soil |
| Ability to re-establish mycorrhiza in following season (annuals or perennials) | Capacity of propagules to re-grow and infect |
| Ability to withstand large-scale inoculum production procedures | Propagule number<br>Propagule resistance to desiccation<br>Propagule survival |

ability to increase plant growth when grown in phosphate-deficient soils. In many cases, the relative effectiveness ($C_m/C_{nm}$) of fungi is generally independent of phosphate supply (Abbott and Robson, 1984a). However, for *Scutellospora calospora* this was not the case for mycorrhiza formed on subterranean clover (Thomson *et al.*, 1986). Shoot growth was increased by this fungus when phosphate deficiency was severe, but at a level of phosphorus just below that required for maximum growth of non-mycorrhizal plants, the growth of the shoots was decreased. Therefore, selection of inoculum fungi for their capacity to increase phosphate uptake and plant growth cannot be carried out wisely using only a single application of fertilizer. A range of levels of application of phosphate that clearly defines the extent of phosphorus deficiency needs to be used.

*(c) Mycorrhiza formation.* In spite of the limitations identified below, it appears most appropriate to select fungi on their ability to colonize roots rapidly and extensively. In general, the benefits of vesicular-arbuscular mycorrhizal fungi relate to the rate and extent of mycorrhizal formation (e.g. Abbott and Robson, 1981a), although this is not always clear (Medina *et al.*, 1987). In many studies there has been no clear assessment of the *rate* of colonization of roots in relation to plant growth. Hence the absence of a relationship between the extent of mycorrhiza formation (assessed at one or two times) and the enhancement of plant growth may merely reflect differences in the rate of mycorrhiza formation (Abbott and Robson, 1982; McGonigle, 1988). Consideration also needs to be given to the observation that some fungi cannot be clearly stained within roots using current techniques and benefits of inoculation have been recorded in the absence of any observed mycorrhiza development (Morton, 1985). In several studies, there are indications that certain fungi can enhance plant growth without extensive mycorrhiza formation (e.g. Sylvia and Burks, 1988) and further research to examine whether these effects are consistent for particular species or strains of fungi are required to characterize these fungi.

Another aspect of the relationship between mycorrhiza formation and plant growth is that of the general lack of information about the activity of hyphae. An assessment of the viability of hyphae within the root and in the soil (Schubert *et al.*, 1987; Sylvia, 1988) may help to overcome the limited value of measurements of length of mycorrhizal roots in relation to plant growth. Such comparisons of vesicular-arbuscular mycorrhizal fungi will only be useful for the selection of inoculants if the processes of mycorrhiza formation and plant growth are followed through time. Furthermore, when there is a likelihood of loss of fine mycorrhizal roots during sampling, estimates of the relationship between mycorrhizal root length (or percentage root length) and plant growth, for different species of fungi, will essentially be only crude.

Species of fungi have been shown to differ in their ability to form hyphae within roots (Wilson, 1984), although this attribute is likely to vary with factors that change the quantity of infective hyphae relative to root density. In experiments where equal numbers of infective propagules were used, the extent and rate of colonization of roots was not always the same for each fungus (Haas and Krikun, 1985). Descriptions of the rate of colonization and quantification of fungal structures formed within roots are often not recorded, but would be useful for selection of inoculant fungi because these parameters may be more closely related to mycorrhizal function.

*(d) Choice of propagules.* An understanding of the relative importance of different propagules of vesicular-arbuscular mycorrhizal fungi for the formation of mycorrhiza in field soils is essential for assessing their ability to increase plant growth. The propagules in inocula may not be the same as those which are responsible for the colonization of roots under field conditions following inoculation. A knowledge of the types of propagules of vesicular-arbuscular mycorrhizal fungi and the role of these propagules in a particular field situation is necessary to ensure that appropriate propagules are used for screening vesicular-arbuscular mycorrhizal fungi for their effects on plant growth.

There have been many comparisons of the effect of inoculation with various species (and isolates) of vesicular-arbuscular mycorrhizal fungi for their ability to increase plant growth. Difficulties arise in both the choice of propagules and in the quantity of inoculum used in such comparisons. However, for the reasons outlined above, it is not possible to use these studies to generalize about the suitability of species as inoculants. Selection of fungi needs to be made in relation to particular soils and climates.

## 2. *Selection for persistence*

Persistence of fungi following inoculation will be essential in many sites. Reference to the literature on selection of rhizobial inoculants is very relevant here (Chatel *et al.*, 1968). The characteristics of fungi related to propagule formation and their tolerance of conditions need to be known. The relative importance of propagules varies according to fungal species, and may also differ for the same fungus when growing in a range of environments. The fungi need to be able to colonize soil and plant roots away from points of inoculation so that they will be present in an extensive volume of soil to initiate colonization of roots that appear in the following season (Abbott and Robson, 1982).

*(a) Mycorrhiza formation.* The ability of fungi to colonize roots and form extensive hyphae away from roots needs to be assessed in the presence of other vesicular-arbuscular mycorrhizal fungi and the general soil population of micro-organisms and soil fauna (see Linderman,

1988). Most comparisons of fungi are made by inoculating single species into separate pots. Some evidence of interactions among fungi has been presented. Competitive ability does not appear to be a fixed attribute of species of vesicular-arbuscular mycorrhizal fungi, as factors that influence the quantity of infective hyphae at the root surface appear to dictate the outcome of joint colonization (Abbott and Robson, 1984b).

Several studies have demonstrated seasonal changes in the extent of colonization by different types of vesicular-arbuscular mycorrhizal fungi (Jakobsen and Nielson, 1983; Dodd and Jeffries, 1986; Rosendahl, 1989). The possibility of a succession of fungi colonizing roots during a growing season highlights the potential for considering fungi in terms of their capacity to remain infective (even if most roots are colonized by another species) until the early colonizers have lost their infectivity. As more new roots grow, the original colonizer may no longer be infective. This will depend on the species of fungi. In field soils in southwestern Australia, *Glomus* spp. dominated in the colonization of roots of pasture species (Scheltema *et al.*, unpubl. res.). However, occasionally there was a change in the relative abundance of fine endophyte and *Scutellospora* spp. during the season. The reasons for a change in the abundance of species of fungi could relate to differences in soil or other factors that altered the competitive advantage of a species. The contribution of hosts to changes in the abundance of species of vesicular-arbuscular mycorrhizal fungi (Dodd *et al.*, 1990b; McGonigle and Fitter, 1990) also needs to be considered in selecting inoculant fungi.

*(b) Infectivity of propagules.* Persistence is only a characteristic required where inoculation is not practical each season. The propagules that persist need the ability to germinate or re-grow rapidly as soon as conditions are suitable and roots are present. Tolerance of heavy metals may be an important criterion for persistence in some mined areas (Gildon and Tinker, 1983). Sieverding and Toro (1988) described a method for screening species of fungi for tolerance to water stress. The ability of fungi to survive drought may also be a selection criterion for persistence.

*3. Selection for inoculum production*

Inoculum production will impose additional selection requirements for potential inoculant fungi. Fungi that are unable to tolerate conditions imposed during or after inoculum preparation, or that lose other important characteristics during the process, such as their ability to infect rapidly, will be unacceptable as inoculants. Extensive propagule

formation and resistance of propagules to desiccation could be of
primary importance, but not necessarily (Sylvia and Burks, 1988).

The selection of inoculant vesicular-arbuscular mycorrhizal fungi cannot be considered in isolation from a knowledge of the ecology of
naturally occurring species within the same region. Reasons for any low
abundance of fungi at sites where inoculation is planned are crucial to
decision making. Why are the fungi currently at low levels in the soil?
Perhaps their current levels are already as high as is possible in the
circumstances. Furthermore, what is the basis for any ineffectiveness in
naturally occurring fungi? The outcome of possible interactions between
the resident and inoculant vesicular-arbuscular mycorrhizal fungi, and
the influence of the other resident soil microflora and fauna also need to
be anticipated. Attempts to answer ecological questions about indigenous vesicular-arbuscular mycorrhizal fungi are essential in the selection
programme.

### III.   Characterizing and comparing fungi

The selection of inoculant vesicular-arbuscular mycorrhizal fungi needs
to be made using (1) knowledge of the conditions where the fungi will
be introduced and (2) knowledge of the response of fungi to these
conditions. It is not possible to make an informed selection of inoculant
fungi without a thorough understanding of the biology of the fungi for
the soil, host and environmental conditions likely to be encountered.
Steps involved in developing and using a collection of fungi to select
inoculants are: (1) collection and maintenance of isolates; (2) identification; (3) characterization (detailed assessment of fungi in relation to soil
and environment variables); and (4) inoculum preparation (Table II).

### A.   Locating a source of fungi

Although more than 120 species of vesicular-arbuscular mycorrhizal
fungi have been described, the mycorrhizal status of many of them is
not known (Morton, 1988). It has been suggested that inoculant fungi
ought to be selected from among those that occur naturally within an
area (Abbott and Robson, 1982; Adelman and Morton, 1986; Sylvia and
Burks, 1988). Suitable fungi may also be sought from locations with
similar climates and soil conditions for plant growth—such fungi are
most likely to be best adapted to the required conditions. It is also
possible that certain fungi collected elsewhere will have appropriate
characteristics for a particular situation (bringing in attributes not

**TABLE II**

Specific steps involved in selection of inoculant vesicular-arbuscular mycorrhizal fungi.

| | |
|---|---|
| Collection and maintenance: | Maintain fungi as separate isolates |
| Identification: | Identify where possible |
| Characterization: | Identify major type of propagule for range of environments |
| | Assess infectivity in relation to (1) soil properties (chemical, physical), (2) propagule number, (3) presence of other species of fungi, (4) range of host plants |
| | Assess effectiveness in relation to infectivity in the field |
| Inoculum preparation: | Identify most suitable method for large-scale inoculum preparation for field use |

present in the local population that has developed under different selection pressures, such as no soil disturbance).

The establishment of pot cultures of vesicular-arbuscular mycorrhizal fungi from field collections is generally a tedious process and one often avoided by mycorrhiza researchers. The extent to which fungi that are relatively easy to maintain in pot culture are representative of species that form the majority of mycorrhizal root length in field soils is not generally known. In most cases the abundance of species in field soils is assessed using spore numbers or infectivity measurements (bioassays) that may not reflect the mycorrhizal population in field roots. If the ease of establishing pot cultures of fungi limits the pool of fungi available for detailed study, this will also have serious consequences for the selection of inoculants.

## B. Understanding the biology of fungi

A broad data-base of biological characteristics of each potential inoculant fungus is a prerequisite for selection (Table I). This data can be sought from field surveys and experiments conducted in field soils as well as in glasshouse and laboratory studies. Similar knowledge of indigenous fungi present at sites where inoculation is planned would also be crucial in assessing the likely success of inoculation with selected species. An initial source of useful knowledge can be obtained from details of the site where each fungus is collected (Table III). These data, particularly when compiled to include information for the same or similar species from different locations, can provide a basis for hypotheses about the soils and situations where they would be most suitable as inoculants.

As an integral part of a selection programme there is a need for research aimed at assessing the extent to which generalizations among isolates and species of vesicular-arbuscular mycorrhizal fungi can be made. Are all species of *Acaulospora* similar in the manner in which they respond to a range of soil conditions? Do all species of *Glomus* colonize roots in similar ways from similar quantities and forms of inocula? If groups of species within genera (e.g. those that are most similar morphologically; Morton and Benny, 1990) are similar in characteristics relevant to selection of inoculants, then the number of fungi that need to be screened for these attributes can be reduced. Information such as this, collected at a limited number of institutions, could be used as a basis for narrowing the pool of fungi from which selection will be made at a particular location. At present this information is scanty and not collated.

An example of the type of information that is fundamental to selecting inoculant vesicular-arbuscular mycorrhizal fungi is presented for three species in Table IV. These data have been accumulated mainly from glasshouse experiments over many years and provide a basis for identifying the suitability of each fungus as an inoculant. Similar compilations of data for other species that have been studied extensively, such as *Glomus mosseae*, would be of value in identifying the limitations to current knowledge relating to selection of inoculants.

**TABLE III**

Field data that should be recorded when collecting possible inoculant vesicular-arbuscular mycorrhizal fungi.

|  | Data required |
|---|---|
| Site | Exact location |
| Environment | Climatic and vegetation type/density (root growth) |
| Soil properties | pH, extractable soil P, organic carbon, exchangeable potassium, field texture, electrical conductivity |
| Management history | Previous cropping history and agricultural practices used at site (pesticides, fertilizer, soil amendments, tillage) |

**TABLE IV**

Example of information that is useful as a basis for selecting inoculant fungi.

| Characteristic of fungi | Glomus sp. isolate WUM 10 (1)[a] | Acaulospora laevis isolates WUM 11 (1,4)[a] | Scutellospora calospora isolates WUM 12 (2,3)[a] |
|---|---|---|---|
| Ease of isolation | Sporocarps hard to obtain, best isolated using bait pots (this fungus was an accidental isolation) | Spores relatively easy to obtain, %germination of field spores usually not high | Spores relatively easy to obtain but occur in small numbers in the field |
| Ease of maintenance of pot cultures | Easy to maintain in pot culture, infective from fresh root pieces | Easy to maintain in pot culture, but must ensure spores are mature before pot cultures are dried and stored. Spore dormancy 6 months [6] | Relatively low level of propagule formation in pot culture under conditions used so far, not infective from fresh root pieces |
| Form of propagules in dry pot culture soil | Spores and infected root pieces [1,2][b] | Spores only [2] | Spores and infected root pieces [2] |
| Number of propagules in dry pot culture soil | Number of spores or sporocarps in pot culture soil not known. Germination characteristics of spores not known | Large numbers of spores in dry pot culture soil (up to 80 spores g⁻¹ soil). Sporulation number and timing can be manipulated by increasing initial inoculum level in pot culture [8]. 70–80% of quiescent spores usually germinate within 5 weeks. Spores require 6 months dormancy period [6] | Moderate number or spores in pot culture soil (maximum of 5 spores g⁻¹ soil). Spore dormancy period 6–12 weeks. 70% of quiescent spores usually germinate within 5 weeks |

| Characteristic of fungi | Glomus sp. isolate WUM 10 (1)[a] | Acaulospora laevis isolates WUM 11 (1,4)[a] | Scutellospora calospora isolates WUM 12 (2,3)[a] |
|---|---|---|---|
| Persistence in dry soil | > 5 years | > 8 years | Up to 4 years |
| Infectivity from propagules in dry pot culture soil | Rapid and extensive from high levels of inoculum only. Infection usually has a high intensity [3] | Infection development often slow from spores stored in dry soil, related to level of inoculum. Infection intensity increases with time and inoculum level | Rapid and extensive from high levels of inoculum only. Infection intensity often low |
| Infectivity from fresh mycorrhizal root pieces and in relation to sporulation | Infective from fresh root pieces of any age [1] | Only infective from young (< 5 weeks old) infected root pieces excised from living plants if sporulation has not occurred extensively [1] | Not infective from root pieces of any age. Infectivity of hyphae attached to mycorrhizal roots appears to stop in association with sporulation |
| External hyphae: amount and distribution | Low length of external hyphae per unit of infected root [3] | Moderate abundance of external hyphae formed per unit root length infected [3], well distributed away from the root [9] | High abundance of external hyphae formed per unit of root length infected [3], poorly distributed away from the root [9] |
| pH tolerance | Wide tolerance to pH (infects well for range pH 5.3–7.5 in 1:5 0.01 M $CaCl_2$) [5] | Tolerant of acidic soils up to pH 6.2 (1:5 0.01 M $CaCl_2$)—decrease in high pH soils. Generally absent from alkaline soils [11] | pH range not known. Forms mycorrhiza at pH 5.3 (1:5 0.01 M $CaCl_2$) |

| | | |
|---|---|---|
| P tolerance | Extensively infects plants supplied with 60% of P required for maximum yield of shoots of subterranean clover [7] | More sensitive to P than *Glomus* sp. (WUM 10). Decreases growth when P sufficient for 80% of maximum growth [7] |
| Competitiveness with other vesicular-arbuscular mycorrhizal fungi | Competitiveness depends on relative inoculum infectivity (quantity) at any time [4,11] | |
| Effectiveness at increasing plant growth | Effective at high levels of inoculum in the absence of competition. Related to rate and extent of infection | Effectiveness related to the P status of the plant. Decreases growth when P sufficient for 80% of maximum growth [7] |

[a] Isolate numbers for fungi in pot culture at The University of Western Australia.
[b] Reference indicated by numbers in square brackets as follows: 1, Abbott and Robson (1981b); 2, Tommerup and Abbott (1981); 3, Abbott and Robson (1985b); 4, Abbott and Robson (1984b); 5, Abbott and Robson (1985c); 6, Tommerup (1983); 7, Thomson *et al.* (1986); 8, Gazey *et al.* (1990); 9, Jakobsen *et al.* (1990); 10, Pearson (unpubl. data); 11, Porter *et al.* (1987a,b).

Although the data compiled in Table IV are not complete (especially in terms of field evaluation), several points relevant to the selection of inoculants are apparent. First, the suitability of a fungus as an inoculant will generally depend on the supply of a high quantity of inoculum in a form suitable for infecting roots rapidly and extensively. If indigenous vesicular-arbuscular mycorrhizal fungi are present, the competitive advantage of inoculants can probably only be maintained if a high inoculum level is used. Some fungi (e.g. *Scutellospora calospora*) may be ineffective at increasing plant growth, although they may play an important role in improving soil structure by forming copious quantities of hyphae in soil. *S. calospora* may therefore be a suitable inoculant in sites of re-vegetation, but it may not function effectively in agricultural soils where more efficient use of phosphate may be the goal of the inoculation programme.

A data-base of characteristics of isolates of fungi could be established independently of a selection programme for a particular location. However, the enormous amount of work involved may mean that information not relevant to the selection of inoculants is accumulated at the expense of data more relevant for particular sites. The assemblage of these data in association with requirements for specific sites would ensure that limited resources are directed towards collecting the most useful information.

## C.  Assessment of inoculation success

In a selection programme it is essential to monitor the success of introduction of the fungi. This can be done either by assessing the total level of mycorrhizal formation or by identifying the individual inoculant fungi. The relatively simple procedures for assessing total colonization or roots contrasts with the more difficult procedures that can be used to measure the amount of root colonized by individual fungi. These include the use of (1) morphological criteria (Abbott *et al.*, 1983); (2) electrophoretic separation of proteins or isoenzymes (Rosendahl *et al.*, 1989); (3) serology (Aldwell *et al.*, 1985; Wright *et al.*, 1987); and (4) DNA fingerprinting (Morton, 1988). Assessing the success of inoculation into soils containing fungi that are closely related to the inoculants is likely to be difficult. For all techniques the greater the taxonomic differences between the inoculant and indigenous species, the greater the chance of assessing their level of colonization.

When field-testing inoculants, soils should be characterized as in Table III. Additionally, the abundance and types of naturally occurring vesicular-arbuscular mycorrhizal fungi should be recorded after isolating spores and using bioassays.

Success of inoculation may need to be judged in terms of time periods longer than one year. Benefits of field inoculation in subsequent years (Dodd *et al.*, 1990a,b) could result from either rapid initial mycorrhiza formation following inoculation or a slow build-up of mycorrhiza to levels that benefit plants in the next season. In this context, the opportunity for inoculation of plant species to boost the population of mycorrhizal fungi, in order to obtain benefits in the following year, may be worth considering when selecting inoculant fungi. Limited success of introductions of inoculant fungi into field soils may reflect (1) selection of unsuitable fungi or (2) a potential for delayed benefits that was not assessed.

## IV.  Suggested protocol for selection

The following procedures are suggested for establishing a programme to select inoculant vesicular-arbuscular mycorrhizal fungi:

1. Collect detailed information for the site where inoculation is proposed (Table III).
2. Determine the current pool of fungi from which selection could be made; collect detailed information on the fungi available (Tables II, IV).
3. If suitable fungi are not available, identify possible sources of appropriate fungi; collect and characterize these fungi as in step 2 above.
4. Characterize and compare fungal attrributes in relation to the selection criteria. For this, useful information will come from laboratory, glasshouse and field studies. Emphasis is required on field evaluation for final selection.

This outline represents an ambitious programme for selection of inoculant fungi. The alternative is to choose fungi on the basis of limited information. The suggested protocol for selecting inoculant fungi emphasizes the link between criteria for selection of inoculant fungi and conditions for use of the inoculants. The alternative, using fungi for which limited knowledge is available, could be valuable in identifying the suitability of fungi, but may be costly in terms of failures of field-scale inoculation. There are no simple short-cuts for selecting inoculant fungi except in narrowing the initial pool of fungi from which to select. This could be done by selecting from among fungi that occur in soils or locations similar to that for which the inoculant fungi are required. In particular, initial comparisons could be made among fungi which are well adapted to a key soil characteristic such as pH. Our conclusion is that currently there is generally a lack of information upon which to base selection of inoculants.

## V.    Examples of selection programmes

A comprehensive programme of selection of vesicular-arbuscular mycorrhizal fungi has been carried out for tropical agriculture in Colombia (Howeler *et al.*, 1987). Fungi were collected from a range of edaphic and climatic zones. Soil properties were recorded for sites of origin of the fungi. They were evaluated using cassava for their infectivity and effectiveness at increasing plant growth first in pasteurized soil, and then in similar soil containing the original large population of indigenous vesicular-arbuscular mycorrhizal fungi. The next step included similar trials using different soils and a wider range of plant species. Fungi were characterized in particular for their response to soil pH. The final step was field evaluation.

Another example of a selection programme is that of Sylvia and Burks (1988) in which the aim was to select vesicular-arbuscular mycorrhizal fungi that would enhance growth of beach grasses and lead to the stabilization of sand dunes. In this study, five *Glomus* spp. (three isolates from sea oats in Florida; two isolates from other sources) were assessed.

## VI.   Conclusion

The main characteristics of the fungi that are relevant to selection are:

- the ability to colonize roots rapidly
- the ability to absorb phosphate and transport it to the host
- the ability to increase plant growth
- the ability to persist in soil
- the ability to re-establish mycorrhiza in following seasons
- the ability to form a large number of propagules
- the ability to form propagules able to withstand manipulation during and after inoculum production.

It is important to note that fungi that are effective in one environment may not be effective elsewhere. A data-base of knowledge of a wide range of vesicular-arbuscular mycorrhizal fungi is required as a key resource for an efficient selection programme. The major component of a selection programme is the collection and maintenance of fungi, and their identification and characterization. This component is also essential for maintaining inoculant fungi at suitable levels in field soils after their introduction so that they continue to be effective.

## Acknowledgements

We are grateful to the Australian Wool Research and Development Fund, and Alcoa of Australia Ltd, AMC Mineral Sands Ltd, Comalco Aluminium Ltd, Consolidated Rutile Ltd, Western Colleries Ltd and Westralian Sands Ltd (through the Australian Minerals Industries Research Association) for their continued support of our mycorrhiza research programme. David Jasper and Mark Brundett made useful suggestions on this manuscript.

## References

Abbott, L. K. and Robson, A. D. (1981a). *Austral. J. Agric. Res.* **32**, 631–639.

Abbott, L. K. and Robson, A. D. (1981b). *Austral. J. Agric. Res.* **32**, 621–630.

Abbott, L. K. and Robson, A. D. (1982). *Austral. J. Agric. Sci.* **33**, 1049–1059.

Abbott, L. K. and Robson, A. D. (1984a). In *VA Mycorrhizae* (C. L. Powell and D. J. Bagyaraj, eds), pp. 113–130. CRC Press, Boca Raton, FL.

Abbott, L. K. and Robson, A. D. (1984b). *New Phytol.* **96**, 275–281.

Abbott, L. K. and Robson, A. D. (1985a). In *Proceedings of the 6th North American Conference on Mycorrhizae* (R. Molina, ed.), pp. 76–79. Forest Research Laboratory, Oregon State University, Corvallis, OR.

Abbott, L. K. and Robson, A. D. (1985b). *New Phytol.* **99**, 245–255.

Abbott, L. K. and Robson, A. D. (1985c). *Austral. J. Soil Res.* **23**, 253–261.

Abbott, L. K. and Robson, A. D. (1991). In *The Rhizosphere and Plant Growth* (D. L. Keister and P. B. Gregan, eds), pp. 353–362. Kluwer Academic Publishers, Dordrecht.

Abbott, L. K., Robson, A. D. and Hall, I. R. (1983). *Austral. J. Agric. Res.* **34**, 741–749.

Adelman, M. J. and Morton, J. B. (1986). *Soil Biol. Biochem.* **18**, 7–83.

Aldwell, F. E. D., Hall, I. R. and Smith, J. M. B. (1985). *Trans. Br. Mycol. Soc.* **84**, 399–412.

Chatel, D. L., Greenwood, R. M. and Parker, C. A. (1968). *Trans. 9th Intern. Congr. Soil Sci.* **II**, 65–73.

Daniels, B. A., McCool, P. M. and Menge, J. A. (1981). *New Phytol.* **89**, 385–391.

Dodd, J. C. and Jeffries, P. (1986). *Soil Biol. Biochem.* **18**, 149–155.

Dodd, J. C., Arias, I., Koomen, I. and Hayman, D. S. (1990a). *Plant Soil* **122**, 229–240.

Dodd, J. C., Arias, I., Koomen, I. and Hayman, D. S. (1990b). *Plant Soil* **122**, 241–247.

Gazey, C., Abbott, L. K. and Robson, A. D. (1990). In *Proceedings of the 8th North American Conference on Mycorrhizae* (M. F. Allen and S. E. Williams, Compilers), p. 115. Agricultural Experimental Station, University of Wyoming.

Gildon, A. and Tinker P. B. (1983). *New Phytol.* **95**, 247–263.

Haas, J. H. and Krikun, J. (1985). *New Phytol.* **100**, 613–621.

Hall, I. R. and Armstrong, P. (1979). *N. Z. J. Agric Res.* **22**, 479–484.

Harley, J. L. (1984). In *Cellular Interactions* (A. Pirson and M. H. Zimmermann, eds.) pp. 148–186. Springer-Verlag, New York.

Holland, A. A. (1970). *Plant Soil* **32**, 293–302.

Howeler, R. H., Sieverding, E. and Saif, S. (1987). *Plant Soil* **100**, 249–283.

Jakobsen, I. and Nielson, N. E. (1983). *New Phytol.* **93**, 401–413.

Jakobsen, I., Abbott, L. K. and Robson, A. D. (1990). In *Proceedings of the 8th North American Conference on Mycorrhizae* (M. F. Allen and S. E. Williams, Compilers), p. 152. Agricultural Experimental Station, University of Wyoming.

Jasper, D. A., Abbott, L. K. and Robson, A. D. (1989). *Austral. J. Bot.* **37**, 33–42.

Juniper, S. (1990). Hons Thesis, The University of Western Australia.

Koske, R. E. and Polson, W. R. (1984). *Bioscience* **34**, 420–425.

Linderman, R. G. (1988). *Phytopathology* **78**, 366–371.

McGonigle, T. P. (1988). *Func. Ecol.* **2**, 473–478.

McGonigle, T. P. and Fitter, A. H. (1990). *Mycol. Res.* **94**, 120–122.

Medina-Gonzales, O. A., Sylvia, D. M. and Kretschmer A. E., Jr. (1987). *Tropical Grasslands* **21**, 24–27.

Morton, J. B. (1985). *Soil Biol. Biochem.* **17**, 383–384.

Morton, J. B. (1988). *Mycotaxon* **32**, 267–324.

Morton, J. B. and Benny, G. L. (1990). *Mycotaxon* **37**, 471–491.

Porter, W. M., Abbott, L. K. and Robson, A. D. (1987a). *J. Appl. Ecol.* **24**, 659–662.

Porter, W. M. Abbott, L. K. and Robson, A. D. (1987b). *J. Appl. Ecol.* **24**, 663–672.

Powell, C. L. (1982). *Plant Soil* **68**, 3–9.

Robson, A. D. and Abbott, L. K. (1989). In *Soil Acidity and Plant Growth* (A. D. Robson, ed.), pp. 139–165. Academic Press, Sydney.

Rosendahl, S., (1989). PhD Thesis, University of Copenhagen.

Rosendahl, S., Sen, R., Hepper, C. M. and Azcon-Aguilar, C. (1989). *Soil Biol. Biochem.* **21**, 519–522.

Schubert, A., Marzachi, C., Mazzitelli, M., Cravero, M. C. and Bonfante-Fasolo, P. (1987). *New Phytol.* **107**, 183–90.

Sieverding, E. and Toro, S. (1988). *J. Agron. Crop Sci.* **161**, 322–332.

Smith, S. E. and Gianinazzi-Pearson, V. (1988). *Ann. Rev. Plant Physiol. Mol. Biol.* **39**, 221–244.

Stribley, D. P. (1987). In *Ecophysiology of VA Mycorrhizal Plants* (G. R. Safir, ed.), pp. 59–70. CRC Press, Boca Raton, FL.

Sylvia, D. M. (1988). *Soil Bio. Biochem.* **20**, 39–43.

Sylvia, D. M. (1990). *Agric. Ecosyst. Environ.* **31**, 253–261.
Sylvia, D. M. and Burks, J. N. (1988). *Mycologia* **80**, 565–568.
Thompson, J. P. (1987). *Austral. J. Agric. Res.* **38**, 847–867.
Thomson, B. D., Robson, A. D. and Abbott, L. K. (1986). *New Phytol.* **103**, 751–765.
Tisdall, J. M, and Oades, J. M. (1979). *Austral. J. Soil Res.* **17**, 429–441.
Tisdall, J. M. and Oades, J. M. (1982). *J. Soil Sci.* **33**, 141–163.
Tommerup, I. C. (1983). *Trans. Br. Mycol. Soc.* **81**, 751–765.
Tommerup, I. C. and Abbott, L. K. (1981). *Soil Biol. Biochem.* **13**, 431–433.
Wilson, J. M. (1984). *New Phytol.* **97**, 413–426.
Wright, S. F., Morton, J. B., and Suorobuk, J. E. (1987). *Appl. Environ. Microbiol.* **53**, 2222–2225.

# 2

# Methods for the Study of the Population Biology of Vesicular-arbuscular Mycorrhizal Fungi

I. C. TOMMERUP

*Division of Forestry, CSIRO, Private Bag, Wembley, WA 6014, Australia*

## I.  Introduction

For vesicular-arbuscular mycorrhizal fungi to effect a role contributing to the production of crops, plants on revegetation sites or to species

METHODS IN MICROBIOLOGY
VOLUME 24  ISBN 0-12-521524-X

Copyright © 1992 by Academic Press Limited
All rights of reproduction in any form reserved

abundance and speciation of natural plant populations, they must influence the survival, growth or reproductive output of their hosts. The effects of mycorrhiza on plant development are greatly influenced by the timing and amount of mycorrhiza formation. The ecologically important question that arises in relation to mycorrhiza formation is at any instant during plant development, what is the source of inoculum and its composition in terms of speciation and propagule type, amount and colonization potential? Expressed in another way, what is known about the population biology of the resident or inoculant fungi?

A population is defined here as a group of individuals of the same species occupying a particular space at the same time. Populations are the components from which communities are constructed. The study of populations deals with the life-conditions of a species as shown by the numbers of "survivors", "births" and "deaths", life-history stage, physiological state, disease and other factors. It also deals with immigrants and emigrants which for vesicular-arbuscular mycorrhizal fungi can take place with wind and water erosion, animal movement and man's deliberate inoculation. The study of populations of vesicular-arbuscular mycorrhizal fungi is complicated by the levels at which it can be approached. The fungi have dynamic phases at several stages in their life-cycles, some host dependent and some host independent. This is further complicated by the task of distinguishing an individual. This arises from the widespread asexual propagation by mycelium in soil, by spores and by hyphal fusion which results in two or more initially discrete individuals becoming physiologically one. Hence instead of juxtaposed individuals competing, they fuse and co-operate. In the extreme, a population becomes an individual. This type of behaviour, although common amongst fungi, is the converse of that seen in most plants and animals.

At present an individual mycelium arising from a discrete propagule such as a spore can be distinguished at an early stage in colonization of roots. At later stages, all the colonization by a fungus in a root or roots of a plant would have to be regarded as a unit if there are no markers to distinguish the mycelium. The same reasoning applies to extraradical (extramatrical) hyphae. This is a necessary simplification of differentiating individuals for some life-history stages of fungi for the present. It is likely to change in the future as methods using molecular biological tools are developed for differentiating mycelia in field samples.

To begin to perceive the true complexity of communities of vesicular-arbuscular mycorrhizal fungi *in situ* and to have some hope of understanding the many aspects of their ecology, we must have some way of measuring populations and understanding their ecophysiological processes. During the past 30 years by observing the fungi in their natural habitats and controlled environments, the critical importance of the habitats in determining the nature of the community has been recognized. Simultaneously, the life-histories of some populations have been unravelled. For some species the behaviour of a few populations is known and an outline of the plasticity of the species is becoming apparent. Although the particular life-history reflects the habitat from which the fungi were isolated, many of the same characteristics occur in the same species from different habitats, indicating that they are a species characteristic. Functional diversity associated with persistence may dominate the biology and hence the population development of many species.

This chapter will emphasize methods for measuring populations and understanding the behaviour of vesicular-arbuscular mycorrhizal fungi growing in association with plants and soil in field-based, glasshouse or laboratory studies. The methods are a basis for developing procedures for each set of circumstances.

## II.  Propagule phase

Population biology requires an understanding of the numbers of organisms and the consequences of those numbers, such as their relative competitive ability, as they become established in the community. The question of counting arises. What should be counted? The total number of propagules regardless of their capacity to initiate colonization of plants when they are counted, or only the propagules capable of initiating colonies at a particular time? To understand the dynamics of the population, both measurements are necessary.

The question also arises, what is a propagule? Currently five major types of propagules have been recognized. They are asexual spores, zygospores, hyphae in dead root-fragments, hyphae ($\pm$ vesicles) in living roots and extraradical hyphae. The relative importance of each of them to the persistence of a population depends on the interaction of the fungus with the climatic and soil environments and the plant community, and in many circumstances the way in which man intervenes to produce crops or graze animals, or, for example, to revegetate a mine site. In arid regions with annual plants, the fungi may persist as asexual spores or as hyphae associated with dead, dry roots (Tommerup, 1985c). In moist temperate ecosystems having perennial vegetation, the fungi may persist largely in association with living roots, spreading from root to root via extraradical mycelium and seldom producing spores or other propagules that are a host-independent phase (Baylis, 1969). The methods for enumerating propagules for population biology overlap with those used for quantifying inoculum, whether the inoculum is being introduced to or already exists in a soil.

## A. Experimental design and sampling techniques

Soil is a heterogeneous physical, chemical and microbiological environment and vesicular-arbuscular mycorrhizal fungi are one microbiological component. The fungi are biotrophs and, like other biotrophic fungi, several phases of their life-cycle are dependent on host plants for development. Root function and architecture are variable. Both soil and plants affect the function and distribution of the fungi in time and space. This chapter is primarily concerned with methods for collecting data which need to be analysed rather than with experimental design, analysis in time and space, and modelling. The reader's attention is drawn to the many statistical texts which discuss, for example, the initial steps in planning experiments; sampling procedures in inherently heterogenous environments which need to be assessed in the vertical and horizontal profiles; methods for increasing the accuracy of experiments such as determining sampling numbers to reduce errors; and factors involved in choosing time intervals between samplings versus the numbers of samples that can be processed by an operator. Recent texts and papers on population biology of animals, plants or microbes and plant disease epidemiology provide an additional, valuable background for solving technical problems as well as discussing relevant statistical methods for data analysis (e.g. Campbell and Madden, 1990). Some of the techniques that are now in use for sampling mycorrhizal fungi in soil are still being refined and this is the case for sampling for soil micro-organisms in general.

## B. Methods for measuring propagule numbers

Two facets of enumerating propagules need to be considered. They are those dealing with the numerical estimates and those concerned with whether the propagules are viable. In this chapter viable propagules are considered to be those which have the ability to germinate or regrow and therefore still have some potential to colonize a host or develop another form of replacement propagule. This definition includes spores that are innately dormant or environmentally dormant (quiescent: Tommerup, 1983a) at the time the estimate is made, however additional measurements are needed to distinguish them from non-viable spores (Section II.F).

Numerical estimation of propagules involves either separating propagules from soil and counting or using a biological assay for detecting propagules. In many combinations of fungi and soils more than one method of estimating propagules will need to be used to provide a valid test of propagule numbers. Selection of those methods will depend on the types of propagules formed by the fungi. The challenge is to devise methods estimating a constant proportion of the actual number of propagules. Due to characteristics of soil and/or propagule types it is often practically impossible to extract and identify all propagules. The five main reasons for this are the problem of separating spores from soil or from within roots; the variable tendency among sporocarps to release spores while they are being separated from soil; the variability in viability among spores; the difficulty in distinguishing non-spore forms of propagules as discrete propagules; and/or destruction of their viability or increase in the number of functional units during the separation process.

In most studies questions concerning propagule units will have to be resolved. For example, should a sporocarp or root fragment each having three spores or a root fragment with three discrete infection units be scored as one or three propagules, and will the propagule unit of three disintegrate during the assay procedures such as during dilutions for the most probable number method? A related question is that of size of the propagule units. In experiments to test some hypotheses it may be appropriate to cut root inocula into a particular size. However, in other experiments disturbance may not be appropriate because that will affect propagule viability or spatial relationships. In many circumstances a series of experiments will be required to provide sufficient understanding of the fungal and host biology and of environmental factors at the descriptive level, before research extends to examining the dynamics of population biology.

## 1. Asexual spores

Several methods have been developed for enumerating spores, most of which depend on wet sieving and decanting (Gerdemann and Nicolson, 1963), either alone or in conjunction with other separation techniques. In many soils it is almost impossible to separate all the spores from other soil particles so that they can be positively identified and counted. A balance must be struck between the time required for extracting spores from each soil sample and the gain in precision obtained.

Treatment of soils with dispersants such as sodium hexametaphosphate may increase the numbers of spores separated, but destroys spore viability. The method has value only for aspects of population biology that do not rely on testing viable spores. Many commonly used procedures have been described by Daniels and Skipper (1982). Some are summarized here with an emphasis on technical accuracy and recent modifications. Similar methods (see Pacioni, Chapter 16, this volume) are used for the production of spore inocula, but the technical emphasis is different. For population biology the emphasis is on extracting all spores from a soil sample, whereas for inoculum production the emphasis is on extracting and concentrating large numbers of spores. For both activities spore viability must be retained.

Separation of propagules of vesicular-arbuscular mycorrhizal fungi from the soil matrix usually involves a long process of wet sieving, flotation and sedimentation. All methods for enumerating separated propagules use wet sieving. This is a disadvantage where the propagules are in a dry state but at a practical level there are few methods available for separating fine organic matter from the dry soil matrix. Propagules can be concentrated in dry soil and complete separation achieved by a short flotation (Tommerup, 1982).

*(a) Procedures for extracting spores: methods based on wet sieving*

(1) Sieve soil through a series of meshes of decreasing pore size to 1–2 mm to remove stones and large undegraded pieces of organic matter and retain or pulverize clods.

(2) Mix a known volume or weight of soil (maximum of 250 g, larger amounts are inefficient to handle and 50 g is often a more efficient sample size) with approximately four times the volume of water, stir, allow heavier particles to settle and decant through a series of sieves from approximately 1 mm to 38 $\mu$m. In some circumstances smaller meshes may be necessary. (They can be constructed from nylon meshes held between two rings of perspex.) Repeat the washing/decanting operation until a statistically constant number of spores is removed from the sample. The number of repetitions will depend on the soil type and volume of the sample and needs to be experimentally determined. Wash the sieves to ensure that all fine particles have passed through the meshes.

Verification that all spores have been removed can only be achieved by indirect means. Plant baiting will indicate what fungi have germinable propagules remaining in the soil after spores have been extracted.

However, whether the propagules were likely to be spores or other types can only be surmised on the basis of information about the propagule types produced by the fungi and the likelihood that such propagules would be present in the soil. The experiments need to take the form of a time-course study of sufficient duration to enable those fungi capable of producing spores to do so. The hyphal types need to be identified to further indicate the range of fungi present in the soil. Baiting experiments are discussed in detail below (Section II.B.1(c)).

(3) Wash the material from the meshes into a beaker. In practice, sievings collected on 500 $\mu$m or larger meshes often have sporocarps and few spores, so whether all sievings need to be examined will depend on the species of fungi in the soil.

(4) Sievings can be handled in a number of ways to locate spores:

  (a) Transfer small amounts to a Petri dish or nematode-counting dish (Hayman and Stovold, 1979), examine under a dissecting microscope and identify and count spores; remove spores with a capillary or microsuction pipette and measure viability.

For all further separation techniques given below each step needs to be assessed to verify that all spores are being retained. While the density gradient and flotation bubbling methods concentrate spores and separate them from debris, they often do not retain all the spores. The suitability of the method varies with the soil type and the size class of the spores to be separated. Spores less than 80 $\mu$m in diameter are more difficult to separate from soils than larger ones and hence quantitative measurements require technical modifications and frequently greater replication.

  (b) Density gradient—prepare sucrose density layers as pairs or multiple layers of 20–70% in clear tubes and then pipette an aliquot of sievings equal in volume to that of each sucrose layer (Menge in Daniels and Skipper, 1982; Ohms, 1957; Ross and Harper, 1970), centrifuge at about 1000 $g$ for 3 min, remove spores usually with some debris at each interface, identify, count and measure viability; radiopaque media can be used to reduce osmotic shock (Furlan et al., 1980); Ficoll 10–30% is a more expensive alternative to sucrose (Daniels in Daniels and Skipper, 1982).

  (c) Density gradient—large-scale sucrose density gradient (Mertz et al., 1979). This method is useful for separating spores but it is very inefficient for quantitative analysis (I. C. Tommerup, unpubl. data);

    (d) Sucrose centrifugation—add 45 ml of sievings with 2 g of pow-
dered kaolinite, which forms a compact layer over the pellet, in
a heavy duty 50 ml centrifuge tube. Centrifuge at 700 $g$ (equivalent separation is obtained for 15 ml tubes at 400–700 $g$) for
5 min, decant supernatant and suspend pellet in 20–50% sucrose depending on the spore size. Recentrifuge for 45 s, decant
supernatant onto a fast flow filter paper in a Buchner funnel
under vacuum, wash spores with water onto a glass fibre filter
from which they can be rapidly removed with a fine needle or
pipette (Tommerup and Kidby, 1979, modification of Jenkins,
1964 and J. C. Sutton, pers. commun.; Tommerup, 1982).
(Extraction of spores using Steps 1–3 and sucrose centrifugation
was reliable ($p = 0.001$) for sandy soils with fungi having spores
of 70–250 $\mu$m (Tommerup, 1982 and unpubl. data).)

    (e) Flotation bubbling method—this method is useful for extracting
spores (Furlan and Fortin, 1975) but inefficient for quantitative
analysis. It was tested by adding a known number of spores to
sievings and re-extracting them (I. C. Tommerup, unpubl. data;
see also Daniels and Skipper, 1982).

*(b) Procedures for extracting spores: methods without sieving*

    (i) *Concentration of propagules in dry soil.* The principle of this
method is to use airstreams of known velocity moving upwards through
a bed of soil to transport particles from the soil and deposit them onto a
filter. Soil particles including spores, and other propagules such as viable
hyphae in dead roots, are sorted according to their sedimentation
velocities. By adjusting the airstream velocity, graded separation of the
soil is achieved and the weight of each fraction separated can be
controlled (Tommerup, 1982; Tommerup and Carter, 1982). The simple
equipment can be built by laboratory workshops. For quantitative
studies this method is useful for sandy soils but it can be inefficient for
highly aggregated soils which are difficult to comminute.

    (1) Comminute dry soil and place 25 g sample in the elutriator
chamber, separate the smallest soil particles and propagules using
the lowest of a series of suction velocities. Transfer each sample,
usually about 1 g, to a dry tube. Increase the airstream velocity,
separate the next largest fraction and so forth.

    (2) Separate spores from each fraction, including the heavy fraction
remaining in the elutriator after all airstream fractions have been
removed, by flotation in sucrose, centrifugation for 30 s and

decanting the supernatant (modification of sucrose centrifugation, Step 4(d) above).

(ii) *Eelworm-counting slide.* This method is useful for dealing with large soil samples. It depends on an ability to separate all spores from the soil and to suspend them in water so that uniform subsampling is achieved (J. Menge, pers. commun. and described in Daniels and Skipper, 1982). The method is more efficient with sandy soils than with heavier or aggregated ones. It is extremely inefficient when spore numbers are very low.

(1) Stir a known weight of soil into five times the weight of water and continue stirring at a constant rate.
(2) Withdraw 1 ml of solution and place in an eelworm-counting slide. This is marked similarly to a haemocytometer and uses the same principles for calculating the numbers of spores. Count the spores in a number of segments of the slide and calculate the number per ml.

(iii) *Plate method.* This method is useful when spore numbers in the soil are above $10 \text{ g}^{-1}$ and the amount of soil available for sampling is small (Smith and Skipper, 1979). Aggregated soils need to be comminuted.

(1) Add 1 g soil to 9 ml water, shake vigorously and pipette supernatant onto a filter paper. Repeat the operation until all spores have been removed from the soil. Pipetting the supernatant onto filter paper in a Buchner funnel under suction increases the efficiency of handling samples.
(2) Count spores on the filter or in a nematode-counting dish.

(iv) *Most probable number method.* This method has been used for estimating spore numbers and compared with other enumeration techniques (Porter, 1979; Daniels *et al.*, 1981; Wilson and Trinick, 1982). The success of the method has been shown to vary from good to negligible. Such variability is explained by the effects of several physical, chemical and biological factors. Probabilities but not actual numbers are measured. The assumptions and limitations of the method are discussed below (Section II.B.6(a)). Similar limitations apply to other dilution methods which have been used to estimate propagule numbers.

*(c) Procedures for extracting spores: viability test.* Two types of methods have been used to determine the viability of spores, one being a baiting

assay, the other measuring the proportion of germinating spores. Viability of a population of spores is often higher when incubated in soil than in aseptic conditions (Tommerup, 1985a). However soil factors and the physical conditions of incubation can induce environmental dormancy in spores and reduce or eliminate germination of otherwise germinable spores. To develop a reliable system enabling all quiescent spores from a community to germinate, a series of experiments is usually necessary to define the precise conditions for a standard test.

(i) *Spore germination*

(1) Incubate spores between pairs of cellulose membrane filters (e.g. Millipore 0.2–0.4 $\mu$m pore size, 47 mm diameter) in pasteurized soil wet to approximately 60% of field capacity and at 15–30 °C, depending on the soil temperature range in the ecosystem from which the fungus was removed (Tommerup and Kidby, 1979). This soil should have the same pH as that from which the spores were removed. Due to the effects of various soil chemical factors on germination, a sandy soil with low organic matter may give more uniform results, by avoiding germination suppression, than one with a high organic matter component (Tommerup, 1985a,b,c). Care should be taken not to cause suppression of germination due to microbial decontamination of soils, e.g. by autoclaving (Wilson, 1984a; McGee, 1987).

(2) Remove membrane pairs from soil, open the pair and stain with trypan blue in lactic-glycerol (as in staining mycelium in roots), remove excess stain by pipetting about 2 ml of lactic-glycerol onto the membranes and examine for germinated spores using magnifications up to 400×. Hyphal length from each spore or the total for the replicate pair of filters can be measured directly or by using a gridline-intersect method such as that of Newman (1966). Hyphal length is a measure of vigour of germinated spores and is a more sensitive indicator of the effects of suboptimal environmental conditions on spore germination than is germination (Tommerup, 1982).

(3) The duration of the incubation may need to be a time-course with periods of 2–3 weeks for non-dormant spores. When dormant spores are present some samples may have to be retained for up to a year (Tommerup 1983a).

(ii) *Plant baiting assays.* It is usually more accurate to measure spore germination directly as described above than to use a plant baiting assay. Estimates of spore germination are likely to be underestimated by plant baiting assays because they measure the probability that a prop-agule will intercept a susceptible portion of a root and form a colony within a certain period of time. For a germinated spore, any factor which affects root and/or hyphal architecture, growth rate and function may alter the probability of contact between a root and a germinated spore and the probability of a colony being initiated (see Sections III and IV).

### 2. Zygospores

Zygospores are presently known for one species of vesicular-arbuscular mycorrhizal fungus, *Gigaspora decipiens* (Tommerup, 1988; Tommerup and Sivasithamparam, 1990). In the field and in pot cultures, zygospores form from extraradical mycelium and from hyphae of germinated asexual spores. Colonies in roots and asexual spores of this fungus can be secondary homothallic phenotypes (two compatible mating types in the same thallus) so that a germinated spore or colony forms zygos-pores, or may be heterothallic (one mating type in a spore or colony) so that two mating-type compatible mycelia interact to produce zygospores. Zygospores attached to mycelium of colonies or asexual spores are easily recognized. The mature teleomorph (zygospore in a closely appressed sac) is usually 15–20 $\mu$m, brown, often coated in fine soil particles and difficult to distinguish from these particles in material separated from soil.

Enumeration of zygospores poses the same problems as enumeration of other small propagules. It is not feasible to detect them microscopic-ally, except in a pure coarse white sand. Recovery rates were 1–60% for replicates having a known number added to soils (I. C. Tommerup, unpubl. data). Zygospores will form a component of the propagules capable of colonizing roots measured by baiting techniques.

The presence of zygospores in soil can be verified by using gentle techniques to separate them in association with mycelium. Techniques are adaptations of those described for separating asexual spores such as gently mixing 5 g soil in 40 ml 50% sucrose, centrifuging at 700 $g$ for 30 s and decanting onto filter paper under suction. Likewise colonized roots with extramatrical mycelium can be floated free of soil particles in a continuous slow-flow water bath. If the mycelium needs to be stained it should be by flotation with addition and removal of solutions by a pipette. The viability of zygospores can be measured by incubating them between a pair of membranes in soil as described above (Section II.B.1(c)) and examining them for germination. Zygospores are innately dormant so that, as for dormant asexual spores, long periods of incubation of 5–12 months may be necessary.


### 3. *Hyphae in living roots*

Hyphae in living roots are both a prominent propagule and a developmental precursor for other propagules such as spores and free hyphal propagules. As propagules in association with extraradical hyphae, they may be the dominant means of spread of the fungus in root systems even in environments where living fine roots are absent for many months due to seasonal droughts (Tommerup and Abbott, 1981; Tommerup, 1985c; Jasper *et al.*, 1989a,b). Almost every article on the effects of mycorrhizal fungi on plant growth and many on the assessment of propagule performance and production deal with the question of measuring the amount of colonization in roots. Microscopic and chemical methods have been developed.

Microscopic measurements provide an assessment of the relative abundance of mycelium in roots expressed as either the total mycorrhizal root length, the density of hyphae within roots, the length per entry point and the number of entry points, and more recently estimates of whether the mycelium is alive or dead (early methods reviewed by Kormanick and McGraw, 1982; current procedures reviewed by Gianinazzi and Gianinazzi-Pearson, Chapter 7 and Schüepp *et al.*, Chapter 4, this volume). When mycelium has distinct anatomical characteristics, measurements for each of two or three fungi which have colonized the roots can be made (Abbott, 1982). To develop an understanding of relationships such as between the amount of colonization and the propagule capacity of a root, quantity of extraradical hyphae and sporulation, and to describe them, either one type of assessment in a detailed time-course study or more than one type of assessment are the necessary minimum. In the future, such relationships may or may not be found to be a causal ones.

The two main chemical methods to determine the extent of colonization of root tissue are assays for chitin and ergosterol content. Chitin, a major constituent of fungal cell walls, is assayed by hydrolysing chitin to glucosamine, which is then measured colorimetrically with a gas chromatograph or an amino acid analyser (Hepper, 1977; Whipps and Lewis, 1980). Chitin assays are time-consuming, have a low sensitivity and are subject to interference by root components and soil particles (Hepper, 1977; Plassard *et al.*, 1983). The ergosterol assay is presently the best chemical method. Ergosterol is a constituent of membranes, the prominent sterol in most living fungal cells, and is either absent or a minor component of higher plants.

Neither chemical method distinguishes colonization by vesicular-arbuscular mycorrhizal fungi from that of other fungi or differentiates amongst vesicular-arbuscular mycorrhizal fungi. For population biology studies where more than one mycorrhizal fungus is present or where there are other root-inhabiting fungi, as is very commonly the case in field samples, chemical methods would need to be used in combination with microscopic analysis. Methods for subsampling of replicates would be those used when chemical analysis and mycorrhizal fungal content are measured in the same replicate (Amijee *et al.*, 1989). The ergosterol method (see Nylund and Wallander, Chapter 5, this volume) would be particularly useful for the rapid estimation of large numbers of samples from experiments in controlled environments using soils decontaminated for root-colonizing fungi prior to inoculation.

## 4. Extraradical hyphae

Extraradical mycelium has a major role in the spread of mycorrhizal colonization within a plant and among adjacent plants (Wilson and Tommerup, 1991). This mycelium has not been quantitatively measured as a propagule. Its development has been measured as a total amount and as the proportion of viable hyphae in pot experiments (Abbot *et al.*, 1984; Schubert *et al.*, 1987). Some hyphae remain viable in dry soil which is not disturbed and may be important propagules of some fungi in environments with long dry periods during which the fine roots die (Jasper *et al.*, 1989a,b). Extraradical hyphae have not been quantified in the field, the main problem being differentiating mycorrhizal from non-mycorrhizal mycelium. At present it is difficult to envisage how this will be achieved in natural or agricultural ecosystems unless chemical methods for quantification and identification of the fungi are developed. The methods used for pot experiments could be adapted for field plots into which identified fungi are introduced (see Sylvia, Chapter 3 and Schüepp *et al.*, Chapter 4, this volume).

## 5. Hyphae in dead root fragments

Hyphae in dry dead root fragments are an important propagule of some species of vesicular-arbuscular mycorrhizal fungi in climates having long dry seasons which lead to the death of all fine roots of perennial plants and the death of annuals (Tommerup and Abbot, 1981; Tommerup, 1982; McGee, 1987). Viable hyphae may or may not be associated with intraradical vesicles. The presence of these propagules can be verified by sieving out dried root fragments or by fractionating soil and separating them by flotation using the techniques described above for spores and testing their viability. Viability can be measured by incubating the hyphae between membranes in soil and scoring the proportion producing new mycelium (Section II.B.1(c); Tommerup and Abbott, 1981).

Enumeration of these propagules can be achieved in a relative sense by using a baiting technique, such as the most probable number method, to give a relative estimate of the colonizing ability of the propagules. One problem not already addressed is the possibility that root pieces might fragment during dilution and therefore the number of propagules would increase. Some propagule units would be smaller; since small propagules produce less hyphal growth than large ones, this would affect the space and time components of the test. The density of propagules would be increased at lower dilutions which would also affect the test. Another approach is to take a large number of small samples of the soil which should be collected with minimal disturbance for the reasons described. Place these in a small pot, sow seedlings and measure the number of entry points at daily intervals. Use the numbers of entry points detected at the earliest times to provide an assessment of primary colonization and therefore relative numbers of propagules. As in the most probable number method, a statistical assumption of the estimates is that the propagules are individual reproductive units and therefore clumping of propagules will lead to an underestimate of numbers. The statistical effects of spatial relationships between roots and propagules, the experimental conditions and the polycyclical colonization habit of these fungi on estimates of propagule density are discussed below in relation to the most probable number method and in articles on modelling estimates of propagule numbers (Wilson and Trinick, 1982; Tommerup, 1985b,c; Gilligan, 1990). The constraints apply to all baiting methods.

### 6.  Total number of propagules

In field communities usually the form of the mycorrhizal propagule is not known and experience has shown that it is unlikely that spores will be the only propagules for all the mycorrhizal fungi. Even if the propagules are spores, they may not be easy to isolate quantitatively. Estimates of total propagule density are made using plant baiting assays which have the limitation of estimating only those propagules which germinate or regrow, intercept a root and initiate an identifiable colony during the experiment. Estimates are affected by all the variables that change plant or fungal growth. Some of the assays are those used to estimate "inoculum potential" in mycorrhizal studies.

"Inoculum potential" defines the fungal energy for growth per unit area of host surface (Garrett, 1956) and is a conceptual term that is non-quantifiable. For pathogenic fungi, Gilligan (1983) has used the term "infection efficiency", which is the proportion of the soil inoculum capable of initiating infection of hosts. The number of propagules must be known independently of the bioassay measurements to quantify the infection efficiency *sensu* Gilligan. For mycorrhizal fungi, the term "colonization capacity" seems more appropriate. Colonization capacity defines the amount of root colonized in a certain time or the time taken to colonize a certain amount of root by a measured dose of inoculum. It is a relative measure for the reasons explained above.

*(a) Most probable number method.* The most probable number method has been used for estimating the numbers of infective propagules of soil-borne plant pathogens or symbiotic organisms (Maloy and Alexander 1958; Porter, 1979; Wilson and Trinick, 1982; Daniels *et al.*, 1981). It provides a relative measure of the density of propagules capable of colonizing roots. Four main assumptions of the method are: (1) that the propagules are randomly distributed in the soil; (2) that propagules are single and not aggregates; (3) that dilution is proportional to the numbers of propagules; (4) and that if one organism is present it will be detected by the assay method. Estimates derived from plant colonization tests rarely detect all the propagules present. Wilson and Trinick (1982) have pointed out that although the method can provide very useful data on the colonizing capacity of the inocula of vesicular-arbuscular mycorrhizal fungi, estimates of the propagule numbers are highly dependent on the conditions used and the values are relative, not absolute. They draw attention to the factors which affect the estimation of the most probable number. If these factors are not taken into account then results are likely to be unreliable. The factors include the rate of root growth and pattern of branching; environmental variables that are optimal for root and fungal growth and close control of such variables; soil nutrient levels; selection of a time for the test that has a high probability of detecting all propagules independently of their size, since small propagules produce less regrowth than large ones. To reduce variability due to the probability that an organism may not reach a root, a standardized system must be used for all most probable number estimates. The procedure is as follows:

(1) Make 10-fold to 2-fold series dilutions of the propagules in soil

from which mycorrhizal propagules have been removed. Small dilution series values and large replication decrease the variability. Field experiments are likely to require a preliminary experiment to enable appropriate choices of these two variables in experimental design.

(2) Place an inoculum core of 10 g in the centre of a pot containing sufficient soil for unimpeded plant growth. Plant germinated seed above the inoculum, grow under standardized conditions for 4–8 weeks, and assess for the presence of colonization.

(3) Values for the most probable number are calculated from the number of pots having colonized plants by an approximation of the maximum likelihood method (Fischer and Yates, 1963)

*(b) Other plant baiting methods.* Several methods based on plant-baiting have been developed for measuring the colonizing capacity of inoculum (see Abbott *et al.*, Chapter 1, this volume). The purpose of some of these methods is not to quantify inoculum exactly, but rather to measure a constant portion of the population so as to have a relative comparison of the colonization capacity of inoculum. These methods are relevant to experiments seeking to differentiate among inocula or explain aspects of fungal behaviour, but cannot be used for enumeration of propagules.

## C.   Use of single spores and single colonies

Studies using single propagules can provide explanatory information, for example, about spread of colonization in a root system and potential for propagule production by an individual. Such studies may provide more limited genetic variability than bulk inoculation of an isolate. However, because spores and single colonies are highly multinucleate and likely to be heterokaryons for some characters, single sporing does not have the genetic advantages that occur for uninucleate or homokaryotic propagules (Tommerup, 1988). Various methods have been tested for maximizing the success of maintaining contact between propagules and roots in soil. The most successful methods are those using germinated spores on a supporting substrate. These include placing a small block of agar or a millipore filter with an aseptic germinated spore against a seedling root (San Antonio, 1988; I. C. Tommerup, unpubl. res.); or growing a root between a pair of membrane filters or stainless steel or nylon mesh disks with a germinated spore (as in Tommerup, 1985a). Alternatively, a spore can be placed on a seedling root which is then transplanted.

## D. Estimating propagule size

Size affects the colonizing capacity of a propagule. Estimates of relative sizes of propagules may take the form of comparing the time to produce a certain amount of colonization in a host under a given set of environmental conditions or the amount of colonization produced in a certain time. Because the fungi vary in the way in which they colonize roots (Wilson and Trinick, 1983), quantification of colonization will need to take this into account if more than one fungus is being studied. At known propagule densities, statistical and descriptive biological solutions of size are possible (e.g. Gilligan, 1990 and references therein).

## E. Propagule longevity

Propagules of vesicular-arbuscular mycorrhizal fungi can survive for more than 20 years in dry soils stored in laboratories and for shorter periods in moist soils in cold storage. However, these are not the types of conditions likely to occur in the field (Loree and Williams, 1987). The duration of survival is affected by the physiological state of the propagule and by environmental factors (Tommerup, 1985a,b). Propagule dormancy is one factor which may contribute to longevity, particularly if the dormant period coincides with a period adverse for survival after germination. Survival in populations having no input of new propagules can be measured by time-course studies of viability that take into account the duration of dormancy and the effects of factors that relieve dormancy.

## F. Propagule dormancy

Innate dormancy has been identified in several species of vesicular-arbuscular mycorrhizal fungi. Spore dormancy may be one reason for the lack of correlation between spore numbers in inocula and colonization of roots. The duration of dormancy is probably species rather than isolate dependent (Tommerup, 1983a, 1988). A dormant spore is one that does not germinate although it is exposed to physical and chemical conditions that will support germination of apparently identical, but non-dormant spores of the same isolate. In contrast, a quiescent spore does not germinate unless the above physical and chemical conditions are fulfilled, and germination of a quiescent spore is prevented only because the environment is unsuitable (also termed environmental dormancy). Asexual spore dormancy is variable amongst species ranging from a few weeks to six months. Dormancy can be relieved in some species by, for example, controlled drying. In others no method for relieving dormancy is known and storage until the germination block is physiologically eliminated is the only solution.

The methods for distinguishing dormant and quiescent spores from other viable spores in the population are the same as those used to test germination (Section II.B.1(c)). To test for quiescent spores, the physical, chemical and microbiological aspects of the incubation system need to be considered. To do this, major soil factors may need to be altered. Possible changes include use of another soil, use of a range of soil-sterilizing treatments or of soil pasteurization rather than sterilization, and checks that the soil pH and moisture levels are appropriate. The soil in which spores form may inhibit germination, and this germination suppression may be relieved by, for example, planting a new crop or by pasteurization (Tommerup, 1985a). The methods employed to extract spores from soil may need to be altered, for example washing spores to ensure removal of contaminating soil or other germination inhibitors prior to incubation.

It can be difficult to show dormancy in a population of mixed ages because some spores may be losing viability while others are becoming non-dormant, and new spores which are innately dormant are forming. In some habitats spore formation occurs more or less continuously while in others, due to climate or cropping practices, spore production is episodic. Hence populations of field spores can have ages ranging from years to several months. The design of the experiments to determine the proportion of dormant spores in these soils needs to be a time-course one with long incubation periods for some samples to allow time for a change from dormant to quiescent or germination. Alternatively, a large soil sample stored in a controlled environment can be sampled over a long period.

## G.  Propagule dispersal

Emigration and immigration of propagules from and to populations occurs. It is likely to take place when there is soil movement such as by wind and water erosion, by large animals and by soil fauna. Immigration is easier to demonstrate than emigration, for example when propagules are transported into an ecosystem with few types of fungi or no means of supporting the fungi such as a devegetated mine site (McMahon and Warner, 1984). Detection of immigration is dependent upon methods (Section V) to demonstrate the presence and identity of the fungi existing in a community and to distinguish new arivals from them.

## H. Propagule production

In ecosystems having a discrete period for fine root production and activity, in experimental field plots and pots in controlled environments, the production of new propagules can be distinguished and the numbers quantified in time-course experiments. Environmental and biological factors affecting production can be identified and their effects on propagule production measured. It is necessary to identify the products of propagule regrowth and to determine whether a propagule produces another asexual propagule such as a smaller asexual spore or a sexual spore. The relationships among single environmental factors or factor interactions and host plant growth and the extent of colonization in roots, the amount of extraradical mycelium, and the rate of propagule production must all be determined. Spore production is related to root colonization and percentage of colonization is proportional to the logarithm of the number of new spores produced (Furlan and Fortin, 1973; Wilson and Trinick, 1982). Other relationships also need to be quantified such as that between living mycelium in roots and the length of living extraradical mycelium, or the relationship between the number of propagules surviving in dried roots or as extraradical hyphae from dried roots and living mycelium in living roots. This type of information will need to be obtained before population and community biology can be understood at descriptive and mathematical levels.

## III. Propagule germination or regrowth

Propagules have two major roles in the life-history of a fungus: one is survival, the other to pass its genes onto the next generation. To achieve this the propagule must regrow and colonize a root. Soil chemical and physical factors influence the rate of propagule regrowth and the rate and amount of mycelium produced. These factors directly affect propagule capacity as inoculum and the same factors also affect root growth, but the interactions have not been critically examined in controlled environment or field experiments. However, it is reasonable to speculate that under suboptimal conditions the effect on the interaction will be greater than on the individual components. Vesicular-arbuscular mycorrhizal fungi, like plants, vary at the strain as well as at the species level in their responses to a level of a factor. At present there is practically no information about the effects of interactions among factors, yet in the field they are often probably dominant over single factor variations.

Investigations of the effects of factors on regrowth involve selecting the parameters to describe regrowth with respect to time and selecting the duration of the test period. The lag period prior to the onset of germination, the maximum proportion of spores that are capable of germinating, and the time taken to reach the maximum germination, the rate of elongation and the maximum elongation of the hyphae are necessary to describe the effects of temperature, soil water potential and soil biological factors (Koske, 1981; Tommerup, 1983b, 1984a, 1985a,b,c). It has been shown that the presence of plants does not change the effects of physical factors on spore germination but that it did influence other biological factors (Daniels and Trappe, 1980; Koske, 1981; Tommerup, 1983b, 1984a, 1985a). Spores are used as the example in the procedures, but the methods are appropriate for other types of propagules.

## 1.  Experimental design

A time-course study is usually necessary. When there is no previous information about an isolate a choice of many treatment levels and times of assessing germination, using few replicates, would give more information than using fewer treatment levels and greater replication. To achieve a consistent interpretation of germination and hyphal development among experiments, germination should be defined in terms of germ tube length (e.g. 5–20 $\mu$m) and the remainder of the mycelium defined as hyphae.

## 2.  Temperature

To investigate the effect of temperature, spores can be incubated directly in soil. It is easier to recover them and retain the mycelium attached to each spore if they are incubated between membrane filters as described in Section II.B.1(c). Soil with spores is incubated in containers, which maintain water content and allow gas exchange, in incubators with thermostats or variable temperature blocks.

## 3.  Water potential

Water potential in the field or in undisturbed cores can be measured by the filter paper method (Hamblin, 1981; Tommerup, 1984a). Germinated and ungerminated spores in cores of field soil or from cores incubated under controlled conditions are counted after dispersing 2 g samples of soil in water or 0.1% lactic acid with trypan blue to separate spores from soil.

Constant water potential can be maintained by incubating soils on suction plates or pressure membrane plates. The water potential of soil columns 0.5–1.5 cm in height is stabilized, spores are introduced into the column which is quickly replaced in the suction/pressure apparatus, so that restabilization occurs. After incubation, spores are separated and examined for germination.

### 4.  Soil pH

Soil pH affects spore germination and can be varied by adding acids or bases to soil (Siqueira et al., 1984; Porter et al., 1987). A pH buffering curve for the soil should be prepared. The effects on germination and growth of changing pH may not arise only through changes in hydrogen ion concentration—a number of other ions may change in soil solution concentration. Additionally, changes in pH modify the activities of other soil microbes, which may in turn affect propagule growth. Rhizosphere changes in soil pH associated with plant growth have been elegantly measured (Marschner and Römheld, 1983) and these techniques could be adapted for examining propagule behaviour in the presence and absence of roots.

### 5.  Mineral nutrients

Mineral nutrient levels in many soils probably have only a minor role in propagule regrowth. Spore germination in soil appears not to be affected by phosphorus status, or probably by that of nitrogen, even when the changing amounts of these elements can lead to reduced root colonization of the host in the same soil (I. C. Tommerup, unpubl. data). The effects of other mineral nutrients and heavy metals on propagule regrowth in soil have not been measured. To measure the effects of metals and mineral nutrients, polycarbonate membranes and stainless steel or nylon meshes that do not bind them should be used to support propagules. Alternatively, propagules should be incubated directly in soil.

### 6.  Soil texture

Soil texture affects physical, chemical and microbial characteristics of soils. Effects on germination of combinations of factors, but not texture alone, have been measured as discussed above.

## 7. Microbial factors

Microbial factors are probably involved in preventing and promoting spore germination in natural soils (Tommerup, 1985a). Organisms specifically involved in such effects have not been identified. The factors may involve microbial degradation of organic matter and interactions with living roots as well as interactions with soil type. Specific techniques have yet to be developed.

## 8. Propagule viability

Germinated spores and regrown hyphae from fragments of dead roots can persist for many weeks as viable propagules in field soils without plants or with plants which are genetically or environmentally not susceptible. The presence of such propagules can be shown by separating them from soil (Section II.B.2). Their colonizing capacity can be demonstrated by baiting assays which do not disturb the soil and disrupt the mycelium (Tommerup, 1984b,c, unpubl. res.). Experimentally, the propagules can be introduced into field soils by incubating them between membranes, disks of nylon or stainless steel mesh, or as soil cores in cylinders of mesh.

# IV.  Interception of roots and colony formation

## A.  Interception of roots

### 1.  Pre-penetration and penetration structures

Any environmental factors that affect root growth may influence the probability of fungus–root interception. The outcome of interception will depend on many factors including the degree of susceptibility of the root; whether or not it has been previously colonized by a mycorrhizal or other soil organism; and the inherent capacity of the propagule to initiate a colony. The types of environmental factors likely to change plant, fungus and interaction behaviour are soil physicochemical and microbial properties and climate. Other factors such as shading are important, because of their effects on root growth and physiology.

The main way of measuring interception between roots and vesicular-arbuscular mycorrhizal hyphae in soil is to count the numbers of pre-penetration and penetration structures. Many plant pathogenic fungi form appressoria prior to penetration. However, after contact with roots vesicular-arbuscular mycorrhizal hyphae often appear, at the light microscope level, to have unmodified tips. They do adhere to roots prior to penetration in compatible interactions. In incompatible interactions such as those due to high soil phosphorus or nitrogen levels or unsusceptible host species, hyphae may attach to roots and may or may not begin to penetrate but fail to initiate a colony successfully (Tommerup, 1984c; Amijee et al., 1989). If adhesion does not occur there is no sign that interception took place. Even in susceptible portions of roots in highly compatible interactions, adhesion does not always follow interception.

The portion of a root tip which has the same level of susceptibility varies amongst fungal isolates and species for a given host (Hepper et al., 1988). Species differ in the number and clustering of penetration structures and in the number per discrete colony unit, and in the rate of development of secondary penetration sites (Wilson and Trinick, 1983; Walker and Smith, 1988). Biological and mathematical descriptions need to take these into account when experiments are designed to use pre-penetration and/or penetration structures as a measure of population change.

### 2. Fungus–host genetical factors

Recent information indicates that specificity amongst mycorrhizal fungi and host plants may function at the level of host variety–fungal strain and therefore be explained by the concept of the gene-for-gene relationship. To what extent does the host plant select the fungi and the fungi select the host? The effect of mycorrhizal fungi on the diversity of natural communities is largely unknown. The value of mycorrhizal fungi during juvenile phases of plant development may be of critical importance to some plant species, particularly during episodes of re-colonization and immigration. The same may equally be true for the fungi but as yet there is very little evidence to support or refute the idea. Evidence indicates that certain plants may depend on mycorrhizal fungi for survival, but the fungi probably survive in the absence of these plants due to their ability to use alternative hosts (Tommerup, 1988). The effectiveness of this strategy could be determined by "baiting" techniques (Section II.B.6) and the precision of quantification increased in the future by restriction fragment length-polymerase chain reaction techniques for identifying fungi. Distinction between levels of compatibility/incompatibility interactions can be quantified by the numbers of pre-penetration and penetration events and successful colonies and their rates of formation.

## B.  Colony formation

To understand the fluctuations in, or build-up of, mycorrhiza, quantitative methods are necessary to measure the amount of colonization in roots and to determine whether the mycelium is alive or dead. To collect the data required for population studies, growth and changes in host root architecture, behaviour of the fungus, environmental factors and the increase of mycorrhizal root length have to be examined. Methods for measuring root length, the proportion of the root which is mycorrhizal, and estimating total and living mycelium lengths, are discussed above (Section II.B.3, 4). Where more than one fungus is present total colonization is often not affected. Differentiation between fungi is possible where individuals have been identified and grown singly in pot cultures. This is discussed in Section V.

## C.  Mathematical models of colonization

The foundation for much of the theory of epidemics of soil-borne plant pathogenic fungi is provided by mathematical models. Propagule production, propagule density, pre-penetration, penetration, and colonization phases are integral components of epidemics. The theory for analysis and modelling of these components applies also to mycorrhizal fungi. Models of mycorrhizal colonization require further development (Walker and Smith, 1984, 1988; Smith *et al.*, 1986). They provide a useful framework for assessing early colonization in pots. They do not yet cope with the complexities in the field nor the successive cycles of secondary colonization where the primary colony from inoculum in soil may become of negligible importance in the build-up of a mycelial network in roots. Selection and fitting of models for some of these aspects are discussed by Campbell and Madden (1990) and Gilligan (1990).

## V.  Distinguishing among fungi

One of the major problems of mycorrhizal research in the field or where more than one fungus is present is to identify and quantify them. Spores are distinguishable (Daft and Hogarth, 1983). Hyphae in roots and the extraradical mycelium are crucial to the mycorrhizal process and it is very important to identify them. In pot experiments where fungi have different morphological characteristics, they can be identified (Abbott, 1982; Wilson and Trinick, 1983). To differentiate among populations of closely related species, immunological methods such as the use of monoclonal antibodies, or chemical methods such as diagnostic isozymes or restriction fragment length-polymerase chain reaction techniques are required. Diagnostic isozymes have been successfully used in competition studies (Rosendahl *et al.*, 1989; see also Chapter 9, this volume).

## VI.  Genetic diversity

Genetic diversity in populations can at present be defined by the frequencies of isozyme phenotypes, and in a single species by detection of mating type phenotypes. Other markers such as restriction fragment length polymorphisms will become identified in the future and provide a basis for population analysis and for defining the effects of relative frequencies of certain genotypes and phenotypes.

## VII.  Competition among fungi

A central requirement in studies of mycorrhizal communities is to explain the coexistence of several species in the same root. Competitive interactions among vesicular-arbuscular mycorrhizal fungi and the interrelationships between factors are discussed by Wilson and Tommerup (1991). The possibility exists that the fungi might express "niche" differentiation by exploiting different portions of the same root axis. Where roots are not uniformly susceptible to colonization, the number of successful penetration sites will be limited and their positions will vary with root growth and distribution. This hypothesis has been tested to some extent by Hepper and colleagues (Hepper, 1985; Hepper *et al.*, 1988) using detailed time-course studies and spatially separated inocula.

Fungi differ in their capacities to colonize plants and in their abilities to compete with other fungi (Wilson and Trinick, 1983; Wilson, 1984b,c). To investigate these features equivalent inoculum levels are required. These have been defined as equivalent numbers of propagules, although propagule size also affects the amount of colonization (see Section II.D; Wilson and Tommerup, 1991). To examine competition, Wilson used a replacement series experiment using species differing in their colonizing habit, and plant monocultures. In these studies the total inoculum density remained the same, while the proportion of the interacting species was changed. Colonization of each species alone at the same (proportional) series of densities was also measured. The relative yields and crowding coefficients were then calculated (Wilson, 1984b). If the plant community is mixed, experiments need to take into account plant competition and the fact that hyphal links can occur among plant species.

## VIII.  Interactions in a community

The structure of communities is usually defined in terms of the number and relative abundance of species within taxonomic groups. Some communities are easier to study than others. For example, in agricultural soils it is relatively easy to gain a complete census by setting up many replicates to assess variation within and between seasons and crops. Such detailed data are virtually unattainable for very complex plant ecosystems. The rate of growth of a population can be matched against the theoretical ideal to determine whether a population is increasing periodically or continuously by examining the fit of the logistic equation

$$\frac{\mathrm{d}N}{\mathrm{d}t} = \frac{rN(K - N)}{K}$$

where $r$ defines the potential rate of increase of a population under specified environmental conditions; $K$ is the maximum attainable population size in the ecosystem; $N$ is the original size of the population; and $t$ is time.

Any stage of the life-history is counted as an individual. Certain stages will be restricted to particular resource patches, such as the functioning arbuscular phase occurring in living root cells. Mycorrhiza are usually confined to the finer lateral roots and are highly sensitive to processes of root production and turnover (St. John and Coleman, 1983; Sutton, 1973). Edaphic factors influence the plant community and in turn the fungal biomass, as well as directly influencing fungal species distributions (White et al., 1989). Techniques have been developed for studying root activity, both spatially and temporally, in the field. Spatial relationships of functioning roots in soil can be determined using radioactive tracers and non-radioactive tracers fed to plants or incorporated into soil (Nye and Foster, 1961; Baldwin and Tinker, 1972). Combinations of tracers give greater precision for field work (Caldwell et al., 1985; Fitter, 1986). The techniques are capable of being adapted for use in mycorrhizal studies in association with existing ones which involve two-dimensional sampling (Section IV.B; see also Tinker et al., Chapter 11, Methods in Microbiology Volume 23).

## IX. Conclusions and future developments

Field studies have shown a great deal that is new about vesicular-arbuscular mycorrhizal fungi, their populations and communities. To provide a complete picture field studies must be integrated with laboratory investigations which attempt to elucidate general principles from controlled environments. To achieve this a series of techniques is being applied including more advanced developments in microscopical techniques, radiotracers, microbial genetics, molecular and biochemical tools and mathematical models. A long recognised problem has been to achieve differentiation among many of the fungi and to quantify them at stages other than as spores. This is now possible, and future refinements are likely to improve the practicality of doing this with a wider range of field material. The complex issues involved in examination of the establishment processes and in the dynamic interplay of organisms introduced into new environments as well as determination of attributes which favour rapid ramification and/or long-term persistence within a simple or mixed plant community in natural or agricultural habitats can now be examined with an increasing range of techniques. This chapter has emphasized the quantitative approach to collecting biological information. Although it is difficult to collect data from the field and to model mathematically the dynamics of the populations, there is a considerable store of ecophysiological information and techniques are available with which to make meaningful analyses.

## References

Abbott, L. K. (1982). *Austral. J. Bot.* **30**, 485–490.

Abbott, L. K., Robson, A. D. and De Boer, G. (1984). *Austral. J. Agric. Res.* **34**, 741–749.

Amijee, F., Tinker, P. B. and Stribley, D. P. (1989). *New Phytol.* **111**, 435–446.

Baldwin, J. P. and Tinker, P. B. (1972). *Plant Soil* **37**, 209–213.

Baylis, G. T. S. (1969). *N. Z. J. Bot.* **7**, 173–174.

Caldwell, M. M., Eissenstat, D. M., Richards, J. H. and Allen, M. F. (1985). *Science* **299**, 384–386.

Campbell, C. L. and Madden, L. V. (1990). *Introduction to Plant Disease Epidemology*. John Wiley & Sons, New York.

Daft, M. J. and Hogarth, B. (1983). *Trans. Br. Mycol. Soc.* **80**, 945–952.

Daniels, B. A. and Skipper, H. D. (1982). In *Methods and Principles of Mycorrhizal Research* (N. C. Schenck, ed.), pp. 29–35. American Phytopathological Society, St Paul, MN.

Daniels, B. A. and Trappe, J. M. (1980). *Mycologia* **72**, 457–471.

Daniels, B. A., McCool, P. M. and Menge, J. A. (1981). *New Phytol.* **89**, 385–391.

Fischer, R. A. and Yates, F. (1963). *Statistical Tables for Agricultural, Biological and Medical Research*. Oliver and Boyd, Edinburgh.

Fitter, A. H. (1986). *Oecologia* **69**, 594–599.

Furlan, V., Bartschi, H. and Fortin, J.-A. (1980). *Trans. Br. Mycol. Soc.* **75**, 336–338.

Furlan, V. and Fortin, J.-A. (1973). *Nat. Can.* **100**, 467–477.

Furlan, V. and Fortin, J.-A. (1975). *Nat. Can.* **102**, 663–667.

Garrett, S. D. (1956). *Biology of Root-infecting Fungi*. Cambridge University Press, Cambridge.

Gerdemann, J. W. and Nicolson, T. W. (1963). *Trans. Br. Mycol. Soc.* **46**, 235–244.

Gilligan, C. A. (1983). *Ann. Rev. Phytopathol.* **21**, 45–64.

Gilligan, C. A. (1990). In *The Rhizosphere* (J. M. Lynch, ed.), pp. 207–232. John Wiley & Sons, Chichester.

Hamblin, A. P. (1981). *J. Hydrol.* **53**, 355–360.

Hayman, D. S. and Stovold, G. E. (1979). *Austral. J. Bot.* **27**, 227–233.

Hepper, C. M. (1977). *Soil Biol. Biochem.* **9**, 15–18.

Hepper, C. M. (1985). *New Phytol.* **101**, 685–693.

Hepper, C. M., Azcon-Aguilar, C., Rosendahl, S. and Sen, R. (1988). *New Phytol.* **110**, 207–215.

Jasper, D. A., Abbott, L. K. and Robson, A. D. (1989a). *Austral. J. Bot.* **37**, 33–42.

Jasper, D. A., Abbott, L. K. and Robson, A. D. (1989b). *New Phytol.* **112**, 101–107.

Jenkins, W. R. (1964). *Plant Dis. Rep.* **48**, 692.

Kormanick, P. P. and McGraw, A. -C. (1982). In *Methods and Principles of Mycorrhizal Research* (N. C. Schenck, ed.), pp. 37–45. American Phytopathological Society, St Paul, MN.

Koske, R. E. (1981). *Mycologia* **73**, 288–300.

Loree, M. A. J. and Williams, S. E. (1987). *New Phytol.* **106**, 735–744.

Maloy, O. C. and Alexander, M. (1958). *Phytopathology* **48**, 126–128.

Marschner, H. and Römheld, V. (1983). *Z. Pflanzenphysiol.* **111**, 241–251.

Mertz, S. M., Heithaus, S. J. and Bush, R. L. (1979). *Trans. Br. Mycol. Soc.* **72**, 167–169.

McGee, P. A. (1987). *Plant Soil* **101**, 227–233.

McMahon, J. A. and Warner, N. (1984). In *VAM Mycorrhiza and Reclamation of Arid and Semi-arid Lands* (S. E. Williams and M. F. Allen, eds), pp. 28–41. University of Wyoming Agricultural Experimental Station Report SA1261.

Newman, E. I. (1966). *J. Appl. Ecol.* **3**, 139–145.

Nye, P. H. and Foster, W. N. M. (1961). *J. Aric. Sci.* **56**, 299–306.

Ohms, R. E. (1957). *Phytopathology* **47**, 751–752.

Plassard, C., Coll, A. and Mousain, D. (1983). *C. R. Acad. Sci. III* **297**, 233–236.

Porter, W. M. (1979). *Austral. J. Soil Res.* **17**, 515–519.

Porter, W. M., Abbott, L. K. and Robson, A. D. (1987). *J. App. Ecol.* **24**, 663–672.

Rosendahl, S., Sen, R., Hepper, C. M. and Azcon-Aguilar, C. (1989). *Soil Biol. Biochem.* **21**, 519–522.

Ross, J. P. and Harper, J. A. (1970). *Phytopathology* **60**, 1552–1556.

San Antonio, J. P. (1989). *Mycologia* **81**, 658–662.
Schubert, A., Marzachi, C., Mazzitelli, M., Cravero, M. C. and Bonfante-Fasolo, P. (1987). *New Phytol.* **107**, 183–190.
Siqueira, J. O., Hubbell, D. H. and Mahmund, A. W. (1984). *Plant Soil* **76**, 115–124.
Smith, G. W. and Skipper, H. D. (1979). *Soil Sci. Soc. Am. J.* **43**, 722–725.
Smith, S. E., Walker, N. A. and Tester, M. (1986). *New Phytol.* **104**, 547–558.
St. John, T. V. and Coleman, D. C. (1983). *Can. J. Bot.* **61**, 1005–1014.
Sutton, J. C. (1973). *Can. J. Bot.* **51**, 2487–2493.
Tommerup, I. C. (1982). *Appl. Environ. Microbiol.* **44**, 533–539.
Tommerup, I. C. (1983a). *Trans. Br. Mycol. Soc.* **81**, 37–45.
Tommerup, I. C. (1983b). *Trans. Br. Mycol. Soc.* **81**, 381–387.
Tommerup, I. C. (1984a). *Trans. Br. Mycol. Soc.* **83**, 193–202.
Tommerup, I. C. (1984b). *Trans. Br. Mycol. Soc.* **82**, 275–282.
Tommerup, I. C. (1984c). *New Phytol.* **98**, 487–495.
Tommerup, I. C. (1985a). *Trans. Br. Mycol. Soc.* **85**, 267–278.
Tommerup, I. C. (1985b). In *Proceedings of the 6th North American Conference on Mycorrhizae* (R. Molina, ed.), pp. 87–88. Forest Research Laboratory, Corvallis, OR.
Tommerup, I. C. (1985c). In *Proceedings of the 6th North American Conference on Mycorrhize* (R. Molina, ed.), p. 331. Forest Research Laboratory, Corvallis, OR.
Tommerup, I. C. (1988). In *Advances in Plant Pathology* (G. S. Sidhu, ed.), Vol. 6, pp. 81–92. Academic Press, London.
Tommerup, I. C. and Abbott, L. K. (1981). *Soil Biol. Biochem.* **13**, 431–433.
Tommerup, I. C. and Carter, D. J. (1982). *Soil Biol. Biochem.* **14**, 69–71.
Tommerup, I. C. and Kidby, D. K. (1979). *Appl. Environ. Microbiol.* **37**, 831–835.
Tommerup, I. C. and Sivasithamparam, K. (1990). *Mycol. Res.* **94**, 897–900.
Walker, N. A. and Smith, S. E. (1984). *New Phytol.* **96**, 55–69.
Walker, N. A. and Smith, S. E. (1988). *Phytopathology* **78**, 253–255.
Whipps, J. M. and Lewis, D. H. (1980). *Trans. Br. Mycol. Soc.* **74**, 416–418.
White, J. A., Munn, L. C. and Williams, S. E. (1989). *Soil Sci. Soc. Am. J.* **53**, 86–90.
Wilson, J. M. (1984a). *Soil Biol. Biochem.* **16**, 433–435.
Wilson, J. M. (1984b). *New Phytol.* **97**, 427–435.
Wilson, J. M. (1984c). *New Phytol.* **97**, 413–426.
Wilson, J. M. and Tommerup, I. C. (1991). In *Mycorrhizal Functioning: An Integrative Plant–Fungus Process* (M. F. Allen, ed.). Chapman & Hall, London (in press).
Wilson, J. M. and Trinick, M. J. (1982). *Austral. J. Soil. Res.* **21**, 73–81.
Wilson, J. M. and Trinick, M. J. (1983). *New Phytol.* **93**, 543–553.

# 3

# Quantification of External Hyphae of Vesicular-arbuscular Mycorrhizal Fungi

DAVID M. SYLVIA

*Soil Science Department, University of Florida, Gainesville, FL 32611-0151, USA*

## I. Introduction

Root colonization by vesicular-arbuscular mycorrhizal fungi has both an internal and external phase. The development of structures within the root has been studied in detail (Smith and Walker, 1981; Sanders and Sheikh, 1983; Bonfante-Fasolo, 1984); however, the external phase of growth of vesicular-arbuscular mycorrhizal fungi has not received adequate study. Even though the hyphae that grow into the soil matrix

METHODS IN MICROBIOLOGY
VOLUME 24  ISBN 0-12-521524-X

Copyright © 1992 by Academic Press Limited
All rights of reproduction in any form reserved

from the root are the functional organs for nutrient uptake and translocation, few researchers have provided quantitative data on their growth and distribution. The objectives of this chapter are: (1) to present current knowledge on the function and distribution of the external hyphae of vesicular-arbuscular mycorrhizal fungi; and (2) to describe techniques for quantifying vesicular-arbuscular mycorrhizal hyphae in soil.

## II.  Function of hyphae in soil

### A.  Uptake and translocation of nutrients and water

Rhodes and Gerdemann (1975) demonstrated that external hyphae of vesicular-arbuscular mycorrhizal fungi take up phosphorus from the soil and translocate it some distance to the plant. However, this study was conducted in a model system using autoclaved soil. Fitter (1985) suggested that, under natural conditions, hyphal integrity may be compromised and intact hyphal lengths for transport may be greatly reduced. Owusu-Bennoah and Wild's (1979) work supports Fitter's contention; they found that depletion zones around mycorrhiza in sterile soil were greater than those around mycorrhiza in non-sterile soil. The mechanisms of nutrient uptake by hyphae have been reviewed by several authors (Tinker and Gildon, 1983; Cooper, 1984).

Nelsen (1987) has reviewed the water relations of vesicular-arbuscular mycorrhizal plants and concluded that most changes attributed to vesicular-arbuscular mycorrhizal fungi were secondary responses due to improved plant nutrition. Indeed, improved nutrient and water uptake by vesicular-arbuscular mycorrhizal plants are closely related phenomena, since nutrients are available to the plant only in the soil solution. Under conditions of water stress movement of nutrients in the root zone is restricted, limiting uptake of poorly-mobile ions (Viets, 1972). However, Hardie and Leyton (1981) demonstrated improved hydraulic conductivities of vesicular-arbuscular mycorrhizal plants versus non-mycorrhizal plants over a range of applied phosphorus concentrations. These authors concluded that improved conductivities were mainly due to hyphal translocation of water in the soil. In contrast, Graham and Syvertsen (1984) found that water flow to roots via hyphae could not account for greater water uptake by mycorrhizal roots. Further evaluation of this phenomenon, especially under field conditions, is needed.

Intra- and interspecific nutrient transfers via mycorrhizal hyphae are known to occur. For vesicular-arbuscular mycorrhizal systems, transfers of carbon (Hirrel and Gerdemann, 1979; Francis and Read, 1984), nitrogen (van Kessel *et al.*, 1985; Francis *et al.*, 1986), and phosphorus (Heap and Newman, 1980; Chiariello *et al.*, 1982; Whittingham and Read, 1982; Francis *et al.*, 1986) have been demonstrated. However, the extent and implications of these interconnections are not fully known (Newman, 1988; Eissenstat, 1990).

### B. Soil aggregation

Koske *et al.* (1975) found extensive hyphal networks binding sand to plant roots in sand dunes. Sutton and Sheppard (1976) followed up these field observations with an experiment in which benomyl was applied to some treatments to prevent mycorrhizal development. They reported that reduced mycorrhiza resulted in reduced aggregation of soil particles. Tisdall and Oades (1979) also found a good correlation between the amount of hyphae produced in soil and the production of water-stable aggregates. However, Lynch and Bragg (1985) have expressed reservations about the direct role of fungi in soil stabilization. They felt that the role of hyphae may be indirect, serving as a substrate for other polysaccharide-producing micro-organisms. Polysaccharides have been observed on the surface of hyphae of vesicular-arbuscular mycorrhizal fungi (Tisdall and Oates, 1979). It is possible that the external hyphae, along with plant roots, produce a framework for aggregation, while bacterial polysaccharides cement the soil particles together.

### C. External hyphae as a source of inoculum

Warner and Mosse (1980, 1983) and Warner (1984) conducted a series of experiments using a fabric membrane system that restricted growth of infected roots, but allowed growth of external hyphae, into the surrounding soil. From these studies they concluded that hyphae alone could initiate infections. However, hyphae detached from roots lost viability rapidly, though colonization of organic fragments extended the infectivity of hyphae. Nonetheless, hyphae in the soil may be an important source of inoculum in systems where hyphal networks develop (Brundrett *et al.*, 1985; Birch, 1986).

### III.  Distribution of external hyphae

#### A.  Morphology and development of external hyphae

The dimorphic nature of the external hyphae of vesicular-arbuscular mycorrhizal fungi has been known for many years. Mosse (1959) and Nicolson (1959) described coarse, thick-walled hyphae having diameters $> 20 \mu m$ and fine, thin-walled hyphae with diameters in the range $2-10 \mu m$. Coarse hyphae are characteristic of vesicular-arbuscular mycorrhizal fungi; they are usually aseptate and have distinct angular projections. Fine hyphae appear to be short-lived, are often septate, and cannot be distinguished readily from other hyphae in the soil.

Sutton (1973) described three phases in the development of vesicular-arbuscular mycorrhizal fungal infections in crop plants (a lag phase, followed by a phase of exponential development, and finally a constancy or plateau phase). More recently, the development of vesicular-arbuscular mycorrhizal fungal infections in individual roots has been described (Brundrett, 1985). However, only a few researchers have quantified the development of the external hyphae in relation to the internal phases of development, and these data are not always in agreement (see Sylvia, 1990). Many more experiments, conducted over time, are needed to clarify the relationship between the development of internal and external hyphae.

#### B.  Hyphal production in soil

Fungal hyphae are abundant in soil (Kjøller and Struwe 1982); however, the proportion of total hyphae in native soils that belong to mycorrhizal fungi is not known. Estimates of the amount of vesicular-arbuscular mycorrhizal fungal hyphae in soil vary greatly. Besides the inherent differences among experimental systems, a major problem in summarizing data on external hyphae is that different components of the external phase have been measured. Some authors measured only hyphae attached to roots (rhizoplane hyphae), others measured hyphae extracted from soil, and still others reported both these values, combined or separately.

Data on external hyphae are often reported on a root-length basis. The range of maximum values reported for rhizoplane hyphae are 3 to $6 \, \mathrm{cm \, cm^{-1}}$ of root (Diem *et al.*, 1986; Sylvia, 1988) while the range of maximum values for soil hyphae are $< 1$ to $592 \, \mathrm{m \, cm^{-1}}$ of root (Sanders and Tinker, 1973; Tisdall and Oades, 1979; Abbott *et al.*, 1984; Abbott and Robson, 1985b; Sylvia, 1986). The unusully high value reported by Sylvia (1986) was from a native sand dune where rooting density was very low. On a soil-mass basis, the range of maximum values reported for soil hyphae are $< 1$ to $26 \, \mathrm{mg}^{-1}$ of soil (Tisdall and Oades, 1979; Abbott et al., 1984; Abbott and Robson, 1985a,b; Sylvia, 1986; Schubert et al., 1987; Sylvia, 1988). In rhizosphere soil, Allen and Allen (1986) recorded values up to $54 \, \mathrm{mg}^{-1}$.

The biomass of external hyphae has also been estimated by several workers. On a soil basis, Nicolson and Johnston (1979) recorded up to $2.5 \, \mathrm{g \, litre}^{-1}$ of dune sand, but their sample included an unknown amount of debris. Bethlenfalvay and Ames (1987) found $40 \, \mu\mathrm{g \, g}^{-1}$ of soil using a chitin assay in a pot-culture study. On a root basis, Bethlenfalvay et al. (1982) reported $47 \, \mathrm{mg \, g}^{-1}$ of root, but this estimate included spores. Sanders et al. (1977) found $3.6 \, \mu\mathrm{g \, cm}^{-1}$ of infected root length. Kucey and Paul (1982) estimated that attached hyphae represented 2% of the total root biomass, but they failed to report actual values. Miller et al. (1987) estimates ranged from 1.1 to $5.2 \, \mathrm{mg \, plant}^{-1}$.

## C.  Factors affecting distribution of external hyphae

Little is known about the effects of various biotic and abiotic factors on the distribution of external hyphae of vesicular-arbuscular mycorrhizal fungi. Many of the chemical, physical, and biotic properties of the soil that influence plant response to vesicular-arbuscular mycorrhizal fungal colonization act directly on the external phase of the symbiosis. For example, phosphorus concentration is known to affect the function of mycorrhiza. Same et al. (1983) found that small additions of phosphorus to phosphorus-deficient plants stimulated the internal phase of root colonization, whereas higher phosphorus applications reduced root colonization. Abbott et al. (1984) reported that external hyphae had a similar response to phosphorus; small additions stimulated, whereas large additions suppressed, hyphal growth. Schwab et al. (1983) and Miller et al. (1987) also found that high phosphorus application reduced the proliferation of external hyphae. Other biological, chemical and physical properties of soil that have been shown to affect external hyphae are pH, copper concentration, organic matter content, and water (see Sylvia, 1990). Furthermore, vesicular-arbuscular mycorrhizal fungi vary considerably in their innate capacity to produce external hyphae (Sanders et al., 1977; Graham et al., 1982; Abbott and Robson, 1985b). There is now a critical need to clarify the relationships between root colonization, production of external hyphae and plant-growth response.

## IV.  Methods of quantification

### A.  Indirect methods for total hyphae

*1.  Colonization of "receiver" plants*

Several researchers have used colonization of a "receiver" plant, separated from an inoculated "donor" plant by root-free soil, to estimate the rate of hyphal growth in soil (Warner and Mosse, 1983; Schüepp *et al.*, 1987; Miller *et al.*, 1989). The essence of this method is that a volume of soil is maintained free of roots by utilizing non-biodegradable fabric with a pore size (50–100 μm) that excludes roots, but which allows free passage of fungal hyphae. A tripartate system was constructed where one compartment contained the inoculated donor plant, the middle compartment contained root-free soil, and the third compartment contained the receiver plant (Fig. 1). Using such a system one can estimate the distance that hyphae grow through the soil; however, it will underestimate the rate of hyphal growth since colonization is not instantaneous upon contact of hypha and root. In a system with abundant soil hyphae, Brundett *et al.* (1985) found a 1-day lag between

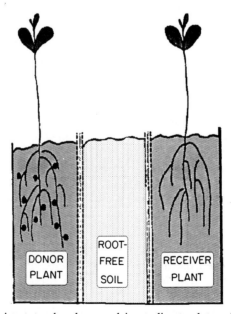

**Fig. 1.**  Typical tripartate chamber used in studies to determine hyphae growth rates of vesicular-arbuscular mycorrhizal fungi. Adapted from Schüepp *et al.* (1987) with permission.

hyphal contact and penetration and a 2- to 3-day lag for arbuscule development.

## 2.  Soil aggregation

Graham et al. (1982) estimated relative hyphal development by weighing plant roots with adhering soil. They reasoned that the weight of the root ball would be proportional to the amount of hyphae that bound the unit together. The soil mass was air-dried and then roots were shaken vigorously to remove air-dry soil that was not firmly attached to the root system. The soil that remained attached was washed from the root system into a beaker of water, and after it settled, the water was decanted and the soil residue was dried and weighed. The roots were blotted dry and their fresh masses determined. The amount of soil adhering to roots was expressed as the amount of dry soil attached in milligrams per gram fresh mass of root. Using this method, Graham et al. (1982) obtained a correlation between estimates of hyphal development and plant-growth response. However, it is not possible to compare hyphal development among different soils or plant species using this method since many factors affect the quantity of soil in the root ball (Kough and Linderman, 1986).

## 3.  Chitin determination

Chitin determinations have been used to estimate hyphal biomass in soil under controlled experimental conditions (Pacovsky and Bethlenfalvay, 1982; Ahmadsad, 1984; Bethlenfalvay and Ames, 1987). However, its utility in native soil is limited because chitin is ubiquitous in nature, where it is found in the cell walls of many fungi and the exoskeletons of insects, and certain soils contain physical and chemical properties which prevent the technique from working properly (Jarstfer and Miller, 1985).

For this method, dried soil samples were mixed with concentrated KOH (120 g KOH per 100 g $H_2O$) and autoclaved for 1 h to degrade chitin to chitosan. Subsamples of the soil–KOH suspension were transferred to centrifuge tubes and ice-cold ethanol (75%) and celite (diatomaceous silica, suspended in 75% ethanol and layered over the first ethanol wash) were added. If hyphae of vesicular-arbuscular mycorrhizal fungi are abundant in the soil it may be necessary to dilute the supernatant with KOH before adding ethanol. The sample was centrifuged at c. 5000 g at 4 °C for 10 min. The pellet was then resuspended and washed sequentially in ethanol (40%), HCl (0.01 N, pH 2), and distilled water. The pellet was suspended in 1.5 ml of water and assayed

colorimetrically by the procedure of Ride and Drysdale (1971). Equal
volumes (1.5 ml) of the chitosan suspension, 5% (w/v) $NaNO_2$, and 5%
$KHSO_4$ were transferred to a centrifuge tube, shaken for 15 min, and
centrifuged at 1500 g for 2 min at 2 °C. Two 1.5-ml subsamples were
removed from the tubes and 0.5 ml of 12.5% $NH_4SO_3NH_2$ was added to
each, shaken for 5 min, and then 0.5 ml of 0.5% 3-methyl-2-benzothia-
zolone hydrochloride (MBTH, prepared daily) was added. This was
heated in a boiling water bath for 3 min, cooled, and 0.5 ml of 0.5%
$FeCl_3$ (100 ml contained 0.83 g $FeCl_3.6H_2O$, stored at 4 °C and
discarded after 3 days) was added. After 30 min, the optical density at
650 nm was read.

A standard curve was constructed by utilizing purified chitin (Sigma
Chemical Company, St Louis, MO, USA) as an internal standard.
Various dilutions of chitin (e.g. 1, 0.5, and 0.25 mg) were mixed with a
suspension of 50 ml of moist soil in 40 ml of concentrated KOH and
processed as above. An additional suspension without chitin was included to determine the background level of chitin in the soil. By
subtracting the chitin content of controls from soils containing vesicular-
arbuscular mycorrhizal fungi, the biomass of vesicular-arbuscular mycorrhizal fungi was calculated. Bethlenfalvay et al. (1981) estimated that
there was 85 $\mu$g of chitin $mg^{-1}$ dry weight of fungus.

## B.   Direct methods for total hyphae

### 1.   Filtration-gridline method

Attempts have been made to estimate hyphal development directly by
extracting hyphae from soil and quantifying lengths by a modified
gridline-intersect method (Abbott et al., 1984; Sylvia, 1986; Miller et
al., 1987). Soil cores of known volume were collected at random,
thoroughly mixed, and subsamples (e.g. 10 g) were removed, suspended
in water (e.g. 500 ml), and passed through a sieve with 250-$\mu$m pores.
The filtrate was blended for 15 s and a portion (e.g. 25 ml) was passed
through a cellulose membrane with 0.45-$\mu$m pores. The exact quantity
of soil, water, and suspension should be determined empirically so that
a thin layer of soil is deposited on the filter membrane without excessive
clogging. The membrane was flooded with a trypan blue solution (0.5 g
trypan blue, 500 ml de-ionized water, 170 ml lactic acid, and 330 ml
glycerin) and rinsed with de-ionized water. The membrane was then cut
to fit on a microscope slide and observed at 300 × through an eyepiece
Whipple disc that had 10 × 10 lined grid. The total length of hyphae was
estimated by the gridline-intersect method (Newman, 1966). The length

of lines on the grid was measured with a stage micrometer. By knowing the mass of soil in the original core, the mass of soil suspended in water, the volume of the suspension transferred to the filter, and the proportion of the total membrane observed microscopically, it is possible to estimate the length of hyphae per gram of the original soil sample. Due to the spatial variability of hyphae on the membranes, it is advisable to quantify 20 to 60 fields per membrane.

O'Keefe and Sylvia (submitted) modified this technique to allow observation of spatially fixed roots and hyphae. Soil cores were collected from pot cultures using 22-mm diameter by 18-cm long, bevelled-end PVC tubes. Cores were impregnataed with a 50 g litre$^{-1}$ gelatin solution that contained 25 ml litre$^{-1}$ of the trypan blue solution as follows. Cores were packed at both ends with polyester fibre, rubber stoppers with glass tubes were placed in each end of the core, and the cores were placed upright in an incubator at 34 °C to keep the gelatin from solidifying. The gelatin solution was placed above the incubator on a hot plate and allowed to run down through a tube connected to the bottom of the core. The solution flowed upwards through the core until the colour of the solution coming out through the top of the core was the same as the colour of the solution going in. The cores were then clamped at the bottom, removed from the incubator, and placed at 4 °C to solidify. Impregnated soil was pushed out of the PVC tubes in 5-mm increments using a tight-fitting wooden dowel and sectioned with a razor blade. Sections were inspected microscopically and hyphae adjacent to roots were quantified using an eyepiece Whipple disc by the gridline-intersect method.

The major problem with the filtration-gridline method is that hyphae of non-mycorrhizal fungi often are not distinguishable from those of vesicular-arbuscular mycorrhizal fungi. Several researchers have used hyphal diameter to distinguish vesicular-arbuscular mycorrhizal fungi from other fungi in the soil. For example, Graham et al. (1986) and Bethlenfalvay and Ames (1987) counted only hyphae with diameters $> 5 \mu$m as vesicular-arbuscular mycorrhizal fungal hyphae. However, Abbott and Robson (1985b) found that most hyphae in pot cultures inoculated with vesicular-arbuscular mycorrhizal fungi had diameters between 1 and 5 $\mu$m. They concluded that it was not possible to distinguish hyphae of vesicular-arbuscular mycorrhizal fungi from hyphae of non-mycorrhizal fungi by morphological or staining criteria. One approach has been to subtract the length of hyphae in control treatments from vesicular-arbuscular mycorrhizal fungi-inoculated treatments; however, this assumes that no interactions occur among the hyphae in the soil.

## 2. Immunofluorescence assay

Serological techniques have the potential for specific detection of mycorrhizal hyphae in soil. Kough and Linderman (1986) used immuno-fluorescence assay (IFA) to detect hyphae of vesicular-arbuscular mycorrhizal fungi in soil. Antigens were prepared by washing spores with sterile, de-ionized water and incubating them for 2 weeks at 6 °C in an antibiotic solution (streptomycin, 200 $\mu$g ml$^{-1}$ and gentamycin, 100 $\mu$g ml$^{-1}$ in sterile, buffered saline). They were then washed, dried for 3 days, and checked for sterility on nutrient media. Clean spores were mixed 1:1 (w/v) with buffered saline and ground in a tissue macerator. Particulate and soluble fractions were separated by centrifugation (13 000 $g$ for 3 min). The particulate fraction was washed four times and suspended in buffered saline and stored at $-20$ °C. One-half millilitre of the particulate solution was injected intravenously and another intramuscularly into rabbits several times over a period of 96 days after which time blood samples were taken and antibody titres determined. The fungal hyphae to be examined were placed on a membrane filter and washed with buffered saline. Serum was incubated with hyphae at 40 °C for 1 h. The samples were then rinsed with buffered saline. A commercially prepared fluorescein-labelled goat anti-rabbit serum was incubated with the samples at 40 °C for 1 h and washed with buffered saline. Samples were observed microscopically with ultraviolet illumination (450–490 nm wavelength).

## C. Methods for detection of active hyphae

A common shortcoming of the above methods is that they fail to distinguish the active or functioning portion of the hyphal mass from inactive hyphae. However, since phosphorus uptake and translocation by fungi is an active process (Beever and Burns, 1980), quantification of active hyphae is important. Recently, stains that differentiate actively metabolizing cells from inactive cells have been used to estimate the "viability" of vesicular-arbuscular mycorrhizal fungal hyphae. Sylvia (1988) found that a solution containing iodonitratetrazolium (INT) and NADH could be used to locate active hyphae of two *Glomus* spp.; INT is reduced by the electron transport system of living cells and results in formation of a red colour. Schubert *et al.* (1987) used fluorescein diacetate (FDA) in their studies to identify functioning hyphae; FDA is hydrolysed within living cells, releasing fluorescein which can be detected with ultraviolet illumination.

For the INT method, a membrane with extracted hyphae (see filtration-gridline technique above) was flooded with a solution containing INT and incubated for 8 h at 28 °C. The INT solution consisted of equal parts of INT (1 mg ml$^{-1}$), NADH (3 mg ml$^{-1}$), and 0.2 M Tris buffer (pH 7.4). The membrane was counter-stained with 0.5% trypan blue, rinsed with de-ionized water, and cut to fit on a microscope slide. The proportions of active (hyphal contents stained red) and inactive hyphae were determined by the gridline-intersect method. For the FDA method, a stock solution was prepared by dissolving 5 mg FDA in 1 ml of acetone, then adding 0.1 M phosphate buffer of pH 7.4 to give a final concentration of 50 $\mu$m ml$^{-1}$. Hyphae were separated from soil and placed on a microscope slide. These were stained with the FDA solution for 5 min at 21 °C and observed immediately with ultraviolet illumination.

## V. Limitations of current methods and strategies for their improvement

There is no completely satisfactory method to quantify external hyphae of vesicular-arbuscular mycorrhizal fungi in soil. Three major problems have yet to be overcome: (1) there is no reliable method to distinguish vesicular-arbuscular mycorrhizal fungi from the myriad of other fungi in the soil; (2) assessment of the viability and activity of hyphae is problematic; and (3) meaningful quantification is very time-consuming. Improved methods are essential to further our understanding of the ecophysiology of the hyphal system of vesicular-arbuscular mycorrhiza. Development of techniques such as sectioning of intact soil cores (Skinner and Bowen, 1974) and autoradiography (Owusu-Bennoah and Wild, 1979) should greatly increase our understanding of the function of hyphae in natural systems. Use of polyclonal (Aldwell and Hall, 1986; Kough and Linderman, 1986) and monoclonal (Morton et al., 1987) antibodies offers promise for monitoring specific fungi in native soil, while computerized image-analysis systems should speed measurement and analysis of data. The development of these techniques should allow significant advances to be made in our knowledge of the ecology of external hyphae in the future.

## References

Abbott, L. K. and Robson, A. D. (1985a). Austral. J. Soil Res. 23, 253–261.
Abbott, L. K. and Robson, A. D. (1985b). New Phytol. 99, 245–255.

Abbott, L. K., Robson, A. D. and De Boer G. (1984). *New Phytol.* **97**, 437–446.

Ahmadsad, I. (1984). *Angew Bot.* **58**, 359–364.

Aldwell, F. E. B. and Hall, I. R. (1986). *Trans. Br. Mycol. Soc.* **87**, 131–134.

Allen, E. B. and Allen, M. F. (1986). *New Phytol.* **104**, 559–571.

Beever, R. E. and Burns, D. J. W. (1980). *Adv. Bot. Res.* **8**, 127–219.

Bethlenfalvay, G. J. and Ames, R. N. (1987). *Soil Sci. Soc. Am. J.* **51**, 834–837.

Bethlenfalvay, G. J., Pacovsky, R. S. and Brown, M. S. (1981). *Soil Sci. Soc. Am. J.* **45**, 871–874.

Bethlenfalvay, G. J., Brown M. S. and Pacovsky, R. S. (1982). *New Phytol.* **90**, 537–543.

Birch, C. P. D. (1986). In *Physiological and Genetical Aspects of Mycorrhizae, Proceedings of the 1st European Symposium on Mycorrhizae* (V. Gianinazzi-Pearson and S. Gianinazzi, eds), pp. 233–237. INRA, Paris.

Bonfante-Fasolo, P. (1984). In *VA Mycorrhiza* (C. Ll. Powell and D. J. Bagyaraj, eds), pp. 5–33. CRC Press, Boca Raton, FL.

Brundrett, M. C., Piche, Y. and Peterson, R. L. (1985). *Can. J. Bot.* **63**, 184–194.

Chiariello, N., Hickman, J. C. and Mooney, H. A. (1982). *Science* **217**, 941–943.

Cooper, K. M. (1984). In *VA Mycorrhiza* (C. Ll. Powell and D. J. Bagyaraj, eds), pp. 155–186. CRC Press, Boca Raton, FL.

Diem, H. G., Gueye, M. and Dommergues, Y. R. (1986). In *Physiological and Genetical Aspects of Mycorrhizae, Proceedings of the 1st European Symposium on Mycorrhizae* (V. Gianinazzi-Pearson and S. Gianinazzi, eds), pp. 227–232. INRA, Paris.

Eissenstat, D. M. (1990). *Oecologia* **82**, 342–347.

Fitter, A. H. (1985). *New Phytol.* **99**, 257–265.

Francis, R. and Read, D. J. (1984). *Nature* **307**, 53–56.

Francis, R., Finlay, R. D. and Read, D. J. (1986). *New Phytol.* **102**, 103–111.

Graham, J. H. and Syvertsen, J. P. (1984). *New Phytol.* **97**, 277–284.

Graham, J. H., Linderman, R. G. and Menge, J. A. (1982). *New Phytol.* **91**, 183–189.

Graham, J. H., Timmer, L. W. and Fardelmann, D. (1986). *Phytopathology* **76**, 66–70.

Hardie, K. and Leyton, L. (1981). *New Phytol.* **89**, 599–608.

Heap, A. J. and Newman, E. I. (1980). *New Phytol.* **85**, 169–171.

Hirrel, M. C. and Gerdemann, J. W. (1979). *New Phytol.* **83**, 731–738.

Jarstfer, A. G. and Miller, R. M. (1985). In *Proceedings of the 6th North American Conference on Mycorrhizae* (R. Molina, ed.), p. 410. Forest Research Laboratory, Oregon State University, Corvallis, OR.

Kjøller, A. and Struwe, S. (1982). *Oikos* **39**, 391–422.

Koske, R. E., Sutton, J. C. and Sheppard, B. R. (1975). *Can. J. Bot.* **53**, 87–93.

Kough, J. L. and Linderman, R. G. (1986). *Soil Biol. Biochem.* **18**, 309–313.

Kucey, R. M. N. and Paul, E. A. (1982). *Soil Biol. Biochem.* **14**, 413–414.

Lynch, J. M. and Bragg, E. (1985). In *Advances in Soil Science*, pp. 135–171. Springer-Verlag, New York.

Miller, D. D., Bodmer, M. and Schüepp, H. (1989). *New Phytol.* **111**, 51–59.

Miller, R. M., Jarstfer, A. G. and Pillai, J. K. (1987). *Am. J. Bot.* **74**, 114–122.

Morton, J. B., Wright, S. F. and Sworobuk, J. E. (1987). In *Mycorrhizae in the Next Decade: Practical Applications and Research Priorities, Proceedings of the 7th North American Conference on Mycorrhizae* (D. M. Sylvia, L. L. Hung and J. H. Graham, eds), p. 317. Institute of Food and Agricultural Sciences, Gainsville, FL.

Mosse, B. (1959). *Trans. Br. Mycol. Soc.* **42**, 439–448.

Nelsen, C. E. (1987). In *Ecophysiology of VA Mycorrhizal Plants* (G. Safir, ed.), pp. 71–91. CRC Press, Boca Raton, FL.

Newman, E. I. (1966). *J. Appl. Ecol.* **3**, 139–145.

Newman, E. I. (1988). *Adv. Ecol. Res.* **18**, 243–270.

Nicolson, T. H. (1959). *Trans. Br. Mycol. Soc.* **42**, 421–438.

Nicolson, T. H. and Johnston, C. (1979). *Trans. Br. Mycol. Soc.* **72**, 261–268.

Owusu-Bennoah, E. and Wild, A. (1979). *New Phytol.* **82**, 133–140.

Pacovsky, R. S. and Bethlenfalvay, G. J. (1982). *Plant Soil* **68**, 143–147.

Rhodes, L. H. and Gerdemann, J. W. (1975). *New Phytol.* **75**, 555–561.

Ride, J. P. and Drysdale, R. B. (1971). *Physiol. Plant Pathol.* **1**, 409–420.

Same, B. I., Robson, A. D. and Abbott, L. K. (1983). *Soil Biol. Biochem.* **15**, 593–597.

Sanders, F. E. and Sheikh, N. A. (1983). *Plant Soil* **71**, 223–246.

Sanders, F. E. and Tinker, P. B. (1973). *Pestic. Sci.* **4**, 385–395.

Sanders, F. E., Tinker, P. B., Black, R. L. B. and Palmerley, S. M. (1977). *New Phytol.* **78**, 257–268.

Schubert, A., Marzachi, C. Mazzitelli, M., Cravero, M. C. and Bonfante-Fasolo, P. (1987). *New Phytol.* **107**, 183–190.

Schüepp, H., Miller, D. D. and Bodmer, M. (1987). *Trans. Br. Mycol. Soc.* **89**, 429–435.

Schwab, S. M., Menge, J. A. and Leonard, R. T. (1983). *Am. J. Bot.* **70**, 1225–1232.

Skinner, M. F. and Bowen, G. D. (1974). *Soil Biol. Biochem.* **6**, 57–61.

Smith, S. E. and Walker, N. A. (1981). *New Phytol.* **89**, 225–240.

Sutton, J. C. (1973). *Can. J. Bot.* **51**, 2487–2493.

Sutton, J. C. and Sheppard, B. R. (1976). *Can. J. Bot.* **54**, 326–333.

Sylvia, D. M. (1986). *Mycologia* **78**, 728–734.

Sylvia, D. M. (1988). *Soil Biol. Biochem.* **20**, 39–43.

Sylvia, D. M. (1990). In *Rhizosphere Dynamics* (J. E. Box and L. H. Hammond, eds), pp. 144–167. Westview Press, Boulder, CO.

Tinker, P. B. and Gildon, A. (1983). In *Metals and Micronutrients, Uptake and Utilization by Plants* (D. A. Robb and W. S. Pierpoint, eds), pp. 21–32. Academic Press, New York.

Tisdall, J. M. and Oades, J. M. (1979). *Austral. J. Soil Res.* **17**, 429–441.

van Kessel, C., Singleton, P. W. and Hoben, H. J. (1985). *Plant Physiol.* **79**, 562–563.

Viets, F. G. (1972). In *Water Deficits and Plant Growth* (T. T. Kozlowskii, ed.), Vol. III, pp. 217–239. Academic Press, New York.

Warner, A. (1984). *Trans. Br. Mycol. Soc.* **82**, 352.

Warner, A. and Mosse, B. (1980). *Trans. Br. Mycol. Soc.* **74**, 407–410.

Warner, A. and Mosse, B. (1983). *Trans. Br. Mycol. Soc.* **80**, 353–354.

Whittingham, J. and Read, D. J. (1982). *New Phytol.* **90**, 277–284.

# 4

# A Cuvette System Designed to Enable Monitoring of Growth and Spread of Hyphae of Vesicular-arbuscular Mycorrhizal Fungi External to Plant Roots

HANNES SCHÜEPP and MAJA BODMER

*Swiss Federal Research Station, CH-8820 Wädenswil, Switzerland*

DIANE DOUD MILLER

*Department of Horticulture, Ohio State University, 2001 Fyffe Court, Columbus, OH 43210, USA*

METHODS IN MICROBIOLOGY
VOLUME 24   ISBN 0-12-521524-X

Copyright © 1992 by Academic Press Limited
All rights of reproduction in any form reserved

# I.  Introduction

Vesicular-arbuscular mycorrhizal fungi live in close contact with host plants by forming abundant fungal structures (arbuscules, vesicles, hyphae) among and within the root cells. At the same time the mycelium of the symbiotic fungi spreads externally from the roots to colonize the soil. Warner and Mosse (1980) demonstrated that fungal mycelium could spread from confined inoculated areas outside the root into host plant roots. Hepper and Warner (1983) concluded that organic matter was important for saprophytic growth of hyphae of vesicular-arbuscular mycorrhizal fungi.

A technique which enables indirect monitoring of the hyphal spread of vesicular-arbuscular mycorrhizal fungi was developed by Schüepp and co-workers (1987b), based on the concept of Warner and Mosse (1983). A cuvette system was used in which the colonization of "trap" or "receiver" plants separated from inoculated "donor" plants by root-free soil compartments, was periodically examined to measure indirectly the spread of external hyphae through soil. The cuvette system has also been used to monitor hyphal spread through field-cut soil monoliths, fitted into the cuvettes (Schüepp and Bodmer, 1991).

# II.  Methods

## A.  Components, construction and operation of the cuvette system

The system designed to monitor the hyphal growth of vesicular-arbuscular mycorrhizal fungi through root-free soil zones (Fig. 1) consists of cuvette sections which are 2 cm thick, 15 cm wide and 15 cm deep (Schüepp *et al.*, 1987b). Adjacent cuvette sections are confined on both sides by a sandwich of polyamide nets, composed of a net of 80 $\mu$m mesh size placed between two pieces of polyamide net of 1 mm mesh size. The fine mesh restricts root growth between cuvette sections but allows unrestricted passage of fungal hyphae. The coarse nets prevent damage to the fine net while filling the cuvette compartments with soil or substrates, and also help to maintain accurate section widths. The cuvette sections and the solid plate dividers, which are used to separate different treatments, are of stainless steel. The cuvette holder, the base rods and the set screws are of galvanized steel.

The first cuvette in the sequence is planted with a "doner", a suitable mycorrhizal host plant which has been pre-inoculated with a mycorrhizal fungus (Fig. 2). Adjacent to the "donor" cuvette is the root-free soil

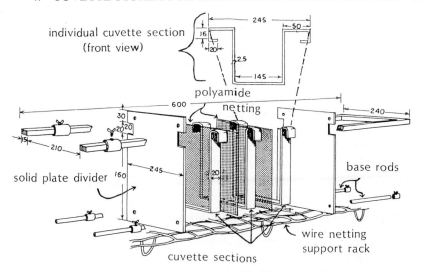

**Fig. 1.**   Expanded view of the cuvette system with dimensions (in mm) (Schüepp *et al.*, 1987b).

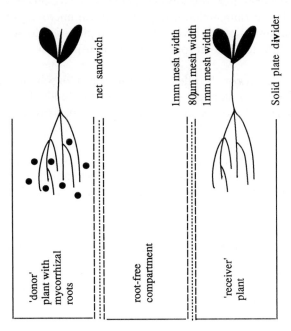

**Fig. 2.**   Cuvette system for monitoring the hyphal spread of vesicular-arbuscular mycorrhizal fungi through root-free compartments (Schüepp *et al.*, 1987b).

compartment through which hyphae must pass if colonization is to occur in the "receiver" cuvette. In the "receiver" cuvette a non-inoculated seedling of the test plant is planted.

It is necessary to disassemble the cuvettes to harvest the plants satisfactorily and assess the vesicular-arbuscular mycorrhizal coloniz-ation. It is therefore most efficient to provide sufficient additional replicates to check periodically the progress of the mycorrhiza formation in the receiver plants. To utilize space and cuvette sections more economically, it is possible to have receiver plants on both sides of "donor" plants. Additionally, cuvettes containing non-inoculated con-trols are added for monitoring of possible contamination.

## B.  Preparation of soil samples and substrates

When the experimental protocol dictates eliminating indigenous mycor-rhizal propagules, the test soil is sterilized. Either chemical sterilization with basamid (active ingredient dazomet, 98%, BASF Company) at a rate of 1 g basamid litre$^{-1}$ soil or $\gamma$-irradiation at a dose of 25 kGy with the soil sealed in plastic bags can be used. $\gamma$-Irradiation has less effect on the physicochemical properties of the soil. Re-inoculation of the sterilized soil is recommended using selected soil micro-organisms, either from pure cultures isolated from the soil sample prior to sterilization or by adding a filtrate (2–5 $\mu$m pore size) of a suspension of non-sterilized test soil.

Each cuvette section is filled with the selected soil or substrate by inserting a sheath of stainless steel, open at both ends, between the net sandwiches. The section is then filled using a specially designed 15 cm wide funnel and the sheath surrounding the section is removed. This produces very uniform cuvette sections.

The growth of hyphae through the root-free cuvette from "donor" to "receiver" varies greatly in extent and rate depending on the substrate or soil amendment added to the intermediate cuvette (Schüepp *et al.*, 1987b).

## C.  Use of soil monoliths cut from the field in the cuvette system

The cuvette system may be used to monitor the ability of mycorrhizal fungi to spread through natural, non-sterilized soil sections cut from the field. Soil monoliths of "undisturbed" soil 2 cm thick are cut in the field, transferred into the cuvette system, and confined on both sides by the hyphae-penetrable net sandwich as previously described.

Soil monoliths are cut from the field using the following procedure.

First, using a spade, a 20 cm deep hole is cut in the soil. Then parallel to a vertical wall of this hole, 3–4 cm thick square soil sections with a side length of about 20 cm are cut. Each soil section is then laid on a board for undisturbed transport to the laboratory. In the laboratory, the monoliths are shaped, using a knife, to fit tightly into the cuvettes. Final trimming takes place when each monolith is placed into its cuvette, care being taken to ensure that the monolith fills the cuvette completely to prevent there being any holes through which the mycorrhizal hyphae may bypass the soil monolith.

These field-cut soil monoliths have been utilized in several experimental arrangements in the cuvette system (Fig. 3). In the first arrangement, non-mycorrhizal test plants were planted, in γ-irradiated soil, in "receiver" cuvettes located on both sides of a field-cut soil monolith. Within a few weeks mycorrhiza were present in the receiver plants. Vesicular-arbuscular mycorrhizal hyphae, originating either from indigenous spores or from root pieces in the soil monolith, were able to grow through the net sandwich to colonize "receiver" plant roots in the adjacent cuvettes.

In the second experimental arrangement, the non-mycorrhizal test plants were separated from the cuvette containing the monolith of field-cut soil by a 2 cm wide cuvette filled with γ-irradiated soil. The receiver plant roots were not colonized even after several weeks. It appears that the vesicular-arbuscular mycorrhizal hyphae do not have a sufficient energy supply from spores or root piece fragments to traverse the 2 cm wide root-free cuvette.

The third experimental arrangement was identical to the second except that "donor" plants, pre-inoculated with a known mycorrhizal fungus from a pot culture, were placed in the first of a sequence of five cuvettes. Within a few weeks the mycorrhiza present in the "donor" plants traversed the adjacent root-free cuvette, the monolith and the second root-free cuvette, to colonize receiver plant roots heavily in the fifth cuvette.

## III.   Applications of the cuvette system

### A.   Monitoring behaviour of vesicular-arbuscular mycorrhizal hyphae in root-free soil

Little is known about the capability of hyphae of vesicular-arbuscular mycorrhizal fungi to grow saprophytically in the soil, but we believe the cuvette system provides a tool to facilitate the study of their potential to

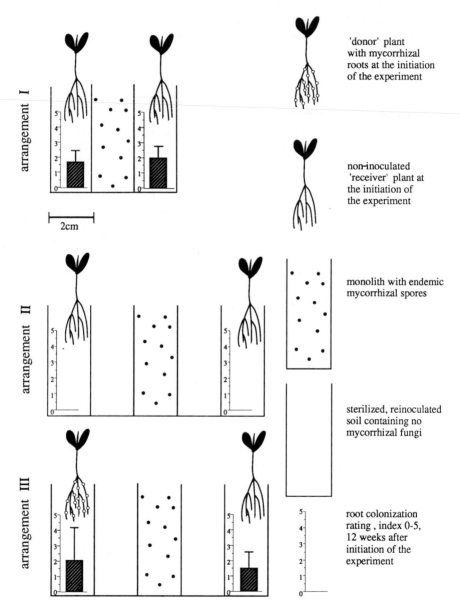

**Fig. 3.** Different experimental designs for evaluation of mycorrhizal mycelial growth through a root-free monolith cut in the field (Schüepp and Bodmer, 1991).

do so. Research is needed to elucidate the dynamics of competitive saprophytic growth of external hyphae of vesicular-arbuscular mycorrhizal fungi in relation to nutrition flux from the roots. The cuvette system provides the potential for this and for many different types of studies of hyphal extension of vesicular-arbuscular mycorrhizal fungi.

### B.  Investigation of vesicular-arbuscular mycorrhizal hyphae in natural, undisturbed soil

Studies in which soil monoliths are inserted in the cuvette system suggest that hyphae of vesicular-arbuscular mycorrhizal fungi can spread rapidly through natural soil. This is interesting because it is known that in axenic conditions most bacteria and fungi are able to penetrate and colonize sterilized soil rapidly whereas in natural soil, sustainable colonization with a micro-organism is hardly ever successful. Consequently, many attempts to use antagonists for biological control have failed. Although the ecosystem of the soil monoliths is confined by the dimension of the cuvettes, microbe presence and activity will, at least initially, be similar to that in natural conditions. Experiments, using field-cut soil monoliths in the cuvettes, can be designed to study the impact of crop rotation, farming practices, and soil conditions on spread of mycorrhiza.

### C.  Elucidating the role of vesicular-arbuscular mycorrhizal fungi in turnover of minerals and organic material in the soil ecosystem

The significance of vesicular-arbuscular mycorrhiza for plant nutrition and plant health is increasingly recognized. The complex interactions of the hyphae external to the plant roots with the soil ecosystem are, however, far from being understood. Root-free soil cuvettes are a useful tool to investigate accumulation and immobilization of macro- and microelements in the hyphae as well as the transport of nutrients from the soil to the roots. The root-free cuvettes can be amended with selected compounds in various concentrations or in isotope form to study translocation of external hyphae. The method can also be modified to determine whether external hyphae can cause significant nutrient fluxes among plants.

### D.  Monitoring side effects of xenobiotic substances

Changes in amount of vesicular-arbuscular mycorrhizal infection may arise as a side effect of xenobiotic substances. The response of mycorrhizal fungi to plant protection measures is very complex and, ideally, it should be monitored in long-term field studies. In addition to field studies, however, effects of pesticides on specific processes of growth and development of vesicular-arbuscular mycorrhiza can be elucidated under controlled conditions. Examples include effects of pesticides on spore germination, on infection and colonization of the root system, and on spread of hyphae through the soil. By carefully adjusting the experimental conditions, and by timing the evaluations according to stage and development of the vesicular-arbuscular mycorrhizal fungus, specific reactions and responses to the pesticides can be monitored using the cuvette system.

The cuvette system can be used to enable monitoring of growth of external hyphae through soil amended with selected pesticides, or through a field-cut soil monolith from pesticide field trials. Recent results (Schüepp and Bodmer, 1991) indicate that fungicides can stimulate or inhibit the spread of hyphae of vesicular-arbuscular mycorrhizal fungi through root-free soil compartments by altering the competition in the soil ecosystem.

A more comprehensive understanding of the complex interactions of vesicular-arbuscular mycorrhiza in the soil ecosystem is needed in order to evaluate potential impacts of pesticides on the symbiosis. The methods described in this chapter are valuable in studying effects of pesticide residues on vesicular-arbuscular mycorrhiza.

### E.   Understanding uptake of heavy metals by plants

The significance of vesicular-arbuscular mycorrhiza in soil contaminated with heavy metals is a research topic with environmental importance. Vesicular-arbuscular mycorrhiza may increase the uptake of strongly adsorbed compounds or elements and, in contrast, decrease uptake of heavy metals into the plant parts above ground by adsorbing them in mycorrhizal roots (Schüepp *et al.*, 1987a). Much research is needed to understand movements of heavy metals in the soil ecosystem. The cuvette technique may be useful to investigate the transfer of heavy metals by external hyphae.

### F.   Determining interactions among rhizospheres of different plant species

Almost all species of plants are able to form mycorrhiza. Actual mycorrhiza formation, however, varies considerably from species to species. The cuvette system may be used to investigate the effects of the rhizospheres of different plants. For example, cuvettes containing "donor" and "receiver" plants of the same species could be separated by a cuvette containing a different plant species.

External hyphae not only extend the range over which nutrient uptake can occur, but also may connect roots, allowing for the transfer of nutrients between plants. The cuvette system could, therefore, be used to study rhizospheres of different plant species, side by side, connected by hyphae of vesicular-arbuscular mycorrhizal fungi.

## G. Comparing competitive ability of different vesicular-arbuscular mycorrhizal fungi and isolates

Not all vesicular-arbuscular mycorrhizal fungi are equal in host plant root infectivity, or in effectivity in stimulating host plant growth. The cuvette system can be used to monitor these differences among species and isolates, and could be used as a screening technique to select superior symbiotic fungi. These would be valuable in any inoculation programme and would be especially desirable with micropropagated plants. Currently we have little knowledge with which to prescribe a plant–vesicular-arbuscular mycorrhizal fungus combination for a particular soil.

## IV. Conclusion

The cuvette system described in this chapter facilitates study of the growth and spread of external vesicular-arbuscular mycorrhizal hyphae. The system is composed of individual cuvette units which may be assembled in many experimental arrangements. The net sandwich separating cuvettes provides sufficient integrity to prevent contamination among units. With careful watering contamination by splashing can be prevented. The cuvettes are in close contact and contain a large volume of soil which reduces variation in moisture and temperature among treatments.

The monitoring of the dynamics of spread of external mycorrhizal hyphae using the cuvette technique should ideally be supplemented by biomass studies in which total length of mycelium is measured.

## References

Hepper, C. and Warner, A. (1983). *Trans. Br. Mycol. Soc.* **81** 155–156.

Schüepp, H. and Bodmer, M. (1991). *Toxicol. Environ. Chem.* **30**, 193–199.
Schüepp, H., Dehn, B. und Sticher, H. (1987a). *Angew. Bot.* **61**, 85–96.
Schüepp, H., Miller, D. D. and Bodmer, M. (1987b). *Trans. Br. Mycol. Soc.* **89**, 429–435.
Warner, A. and Mosse, B. (1980). *Trans. Br. Mycol. Soc.* **74**, 407–410.
Warner, A. and Mosse, B. (1983). *Trans. Br. Mycol. Soc.* **80**, 353–354.

# 5

# Ergosterol Analysis as a Means of Quantifying Mycorrhizal Biomass

JAN-ERIK NYLUND and HÅKAN WALLANDER

*Department of Forest Mycology and Pathology, Swedish University of Agricultural Sciences, PO Box 7026, S-750 07 Uppsala, Sweden*

## I. Introduction

The quantification of ectomycorrhiza has always been a problem. A traditional procedure, used by Melin and his colleagues, is the counting of mycorrhizal root tips. The results are usually presented as percentages of the total number of root tips, but this approach does not take into account the influence of mycorrhiza infection on the root system—a mycorrhizal plant usually has far more root tips than an uninfected one. Another approach, used where soil cores have been extracted, has been

METHODS IN MICROBIOLOGY
VOLUME 24   ISBN 0-12-521524-X

Copyright © 1992 by Academic Press Limited
All rights of reproduction in any form reserved

to count the number of root tips per unit volume of soil, but this method suffers from the same basic shortcoming. For physiological studies, root tip counting has proved to be even more unsatisfactory. We have observed cases where counts indicating 100% mycorrhizal infection have corresponded to a wide range of fungal biomass values as measured using other types of assay (Salmanowicz and Nylund, 1988).

The alternative method has long been the chitin assay (in its two basic variants, cf. Vignon *et al.*, 1986), which is based on a quantitative determination of a compound present in the mycobiont but not in the host, and which is assumed to be proportional to the total mycobiont biomass. Although well developed, this assay has been used as a principal tool in only a few studies. A fundamental problem concerns the concept of fungal biomass: while the chitin content may be assumed to be roughly proportional to the total amount of cell wall, the amount of cell wall is certainly not proportional to the amount of cytoplasm, which is normally concentrated at the tips, leaving the bulk of the hyphae highly vacuolated (cf. Schnürer and Paustian, 1986). This may be less of a problem in a mycorrhizal root, where the degree of vacuolation, judging from numerous electron micrographs (e.g. Nylund and Unestam, 1982), is relatively low and more or less uniform.

Another fungus-specific compound, ergosterol, is a principal component of membranes (Weete, 1974), and should therefore provide a better correlation with the "metabolically active" biomass of a fungus. There appear, however, to be no reports as yet of correlations between ergosterol and cytoplasm mass. This may be because some membrane components seem to adhere to the cell wall even after the cytoplasm has withdrawn.

Assays of ergosterol were first employed by Seitz and co-workers (Seitz *et al.*, 1977, 1979) who were investigating potential mycotoxin-producing fungal infections in stored grain. After a few years this method was used in other types of research, for example, for estimating the fungal fraction of detritus (Lee *et al.*, 1980), wheat leaf resistance to a fungal pathogen (Griffiths *et al.*, 1985), soil fungal biomass (Matcham *et al.*, 1985; Grant and West, 1986; West *et al.*, 1987), forest soil mycoflora (Zelles *et al.*, 1986a and b), the biomass of salt marsh fungi in decaying plant material (Newell *et al.*, 1987, 1988). It was also tested by Osswald and co-workers (Osswald *et al.*, 1986, Osswald and Elstner, 1987) in an attempt to quantify needle and shoot fungi in trees supposedly damaged by atmospheric pollution.

## II. Technique

### A. Development

The basic procedures involve sample extraction, saponification, partitioning, purification of the sample and ergosterol determination with high-performance liquid chromatography (HPLC). In the following account, the development of the method will be traced, and the procedures currently used in our laboratory will be described in detail. The discussion is not intended to review all procedures but simply to outline the technical developments which have led to our present technique.

Seitz *et al.* (1977) extracted their samples with methanol, saponified under reflux boiling for 30 min, partitioned the saponified sample using a petroleum ether extraction, and purified it on a specially prepared column of anhydrous sodium sulphate and silica gel which was eluted using methylene chloride–isopropanol (99:1). The eluate was analysed by HPLC using a Porasil column and methylene chloride–isopropanol (99:1) as the liquid phase. The omission of the saponification step reduced total yields by 38%, indicating that about one-third of the ergosterol in their material was bond/esterified. Their total recovery (based on assays of pure ergosterol added to fungus-free grain) was $93 \pm 5\%$. They also compared the HPLC technique with ultraviolet spectroscopy of sterol fractions obtained by thin-layer chromatography (TLC) and found the former procedure to be more sensitive. In a later publication Seitz *et al.* (1979) simplified the column purification and changed to a Bondapak $C_{18}$ column with water–methanol (95:5) as the liquid phase.

Zelles *et al.* (1987) completely eliminated the column purification but otherwise used the procedures of Seitz *et al.* (1979) for analysis of their soil samples.

Newell *et al.* (1987) also slightly modified the procedure, partitioning the saponified sample with pentane, and microfiltering it while dissolved in methylene chloride–isopropanol. In a subsequent study (Newell *et al.*, 1988) various aspects of the procedure were examined. Pre-storage oven drying at 50 °C reduced the yield by 75% and normal deep freezing followed by thawing reduced it by 14% compared with lyophilization. In contrast, immediate immersion and continued storage of the sample in methanol increased the amount of extracted ergosterol. Reflux extraction did not improve yield in comparison with homogenization for 2 min in methanol. Combined reflux extraction and saponification of the unhomogenized material reduced yields by one-third. The use of nitrogen instead of air when drying the pentane fraction did not affect the recovery. When the dried pentane fractions were redissolved in several different solvent mixtures, pure methanol proved to be the most suitable. For the preparation of standards, analyses of purified ergosterol dissolved in methylene chloride–methanol (1:1) showed that no decay occurred during storage in the dark for up to two weeks at room temperature. Exposure to fluorescent light for a few hours had no effect, while some hours of sunlight brought about a conversion to ergocalciferol. It can be concluded that exposure of samples to normal laboratory lighting during preparation is not detrimental.

Recently, Johnson and McGill (1991) presented a method of combined extraction and saponification, followed by petroleum ether partitioning, drying and redissolving in ethyl acetate; they used testosterone as an internal standard to monitor processing losses. No data are given regarding the yield versus separate extraction with homogenization. In another recent paper, Martin *et al*. (1991) scaled down and speeded up the procedure: very small samples (~ 25 mg) were extracted in cold absolute ethanol using mortar and centrifugation. The supernatant was used for HPLC without further processing.

## B.  Current procedure

The first application of the ergosterol assay in ectomycorrhiza research was reported by Salmanowicz and Nylund (1988). In the method used currently in our laboratory, methanol is used as extractant rather than ethanol. Boiling is carried out in closed test tubes without reflux boiling. Samples are not pre-separated prior to injection. The modifications are described below. Volumes and sample sizes have to be adapted to suit the material and expected biomasses for every experiment, taking the sensitivity of the equipment into account.

### 1.  Sample preparation

*(a) Roots.* Selected roots are frozen in liquid nitrogen, freeze-dried, weighed, and ground in a ball mill. 10–50 mg samples are extracted in 10 ml of 95% ethanol in a test tube, shaken at room temperature for 2 min, centrifuged, the supernatant collected and extracted again for 2 min. (For a recent modification, see below).

*Comments.* The drying serves to preserve the sample, but also provides a dry weight reading upon which calculations can be based.

Freeze-drying gives no loss of ergosterol (Newell, 1988), in contrast to other drying procedures.

Martin *et al.* (1991) extracted single roots directly with alcohol without further pre-treatment; this is suitable if absolute quantities in individual roots are to be determined, but problems such as the appearance of contamination peaks can be encountered with this method.

*(b) Mycelium in substrate.* Roots are separated before analysis. Samples are frozen but not freeze-dried, then shaken with 95% ethanol for 2 h at room temperature, filtered and rinsed once in ethanol. The total extract volume is determined and a 10 ml sample (depending on concentration) is used for further analysis.

*Comments.* No checks have been carried out for losses which could occur due to the use of less efficient freezing procedures, see Newell *et al.* (1988), discussed above.

When brick pellets (Leca) are used as a potting medium, the pellets are covered with ethanol. When sand is used, the ethanol volume must be large enough for the substrate to disperse while shaking.

When peat is used as the substrate, contamination occurs leading to additional peaks which obscure the ergosterol peak during HPLC. A purification procedure has not been developed at this stage and similar problems are encountered with humus-rich soil.

Using brick pellets, further extractions or rinses do not increase recovery.

An alternative procedure used for analysis of large samples of fungus-infested grain involves extraction by reflux boiling for 2 h in methanol (J. Schnürer, pers. commun.).

## 2. Saponification

An aliquot of 2 ml 60% KOH (v/w) is added to the 10 ml sample in a test tube (approx. 30 ml volume, with screw cap) and a small crystal of pyrogallic acid added. The tube is firmly capped because pressure builds up. Heat to 100 °C for 15–20 min (cf. Johnson and McGill, 1991). Cool, add 2 ml water. Neutralize with HCl (roughly 2.3 ml 8 M HCl).

*Comments.* Saponification serves to convert sterol esters to free sterol. Basically, the analysis can be done using the originally free ergosterol alone (cf. Martin, in press). However, this step also helps in purifying the sample. Without the added water, the two phases in the next step do not separate.

There are no obvious consequences of omitting the neutralization step. Analyses of grain samples can be carried out without addition of acid (J. Schnürer, pers. commun.). The use of $H_2SO_4$ instead of HCl results in heavy $K_2SO_4$ precipitation which cannot subsequently be dissolved in ethanol, and which interferes with the procedure. Some precipitate is occasionally also found using HCl but the quantities are small enough not to cause interference.

## 3. Hexane extraction

An aliquot of 5 ml hexane (a technical grade, mixture of isomers is adequate) is added to the sample which is shaken for 20 s. The hexane is collected and the procedure is repeated. The hexane fractions are combined and evaporated to dryness below 40 °C in a conical bottle.

*Comments*. It is essential to mix thoroughly, cf. Newell *et al*. (1988). Centrifugation will speed up separation and facilitate complete recovery. The second extract contains only about 5% of the sample. Pasteur pipettes can be used to collect the fraction from the test tube with sufficient precision.

## 4. Recent modification of steps 1–3

The powder is saponified directly in 10 ml of ethanol, shaken in hexane, centrifuged and the hexane fraction collected. The hexane extraction is repeated.

## 5. Final purification

The extract is dissolved in a known quantity of 99.5% ethanol (0.5 ml as routine; there should be at least about 10 $\mu$g ergosterol ml$^{-1}$) and filtered through a 0.45 $\mu$m membrane. Special HPLC filters with low rest volume are required.

*Comments*. Previously the samples were purified with Sep-Pak $C_{18}$ pre-columns, but no improvement in resolution or column life was observed. This step, which always implies a risk with regard to recovery rate, and is time-consuming, was therefore eliminated.

## 6. HPLC separation

Using a reverse-phase column (e.g. Waters Nova-Pak $C_{18}$) and pure methanol as the mobile phase, 20 $\mu l$ samples (or other volume depending on the ergosterol concentration), were injected and run at 1.6 ml min$^{-1}$. Peaks are detected with a UV detector at 280 nm; the peak should appear (depending on the column) after 3–4 min.

*Comments.* A methanol–water gradient gives better separation but doubles the time required. If this procedure is used, particular care must be taken to de-aerate the carriers.

Other sterols of fungal, plant or bacterial origin are retained in the purified sample but only ergosterol is detectable at 280 nm because of its particular structure (a conjugated pair of double bonds).

## 7. Standard curve

To prepare pure ergosterol from commercial preparations (we used Sigma), dissolve as much as possible in 50 ml boiling 99.5% ethanol. Cool slowly to room temperature to precipitate crystals, and recover them by filtering through a glass sinter filter (the crystals are large enough to be collected with a tweezer from the filter). Repeat the procedure. Freeze-dry or dry in vacuum (evaporator), weigh and redissolve to prepare standard. Use the standard solutions without further steps for injection. The resulting standard curves are linear within the set sensitivity range, while different ranges on our instrument required separate curves for high accuracy.

*Comments.* Much ergosterol will be lost during recrystallization.

The accuracy of the instrument should be tested from time to time; we have found the readings to be very stable.

### III. Evaluation of the method

#### A. Sensitivity and replicability

Depending on the HPLC column and instrumentation, approximately 1 $\mu g$ ergosterol per ml injection fluid is the lowest absolute amount detectable using the techniques described (cf. Newell *et al.*, 1988); a calibrated range of 0.1–0.5 $\mu g$ is suitable for most instruments (cf.

Martin *et al.*, 1991). Using the technique of Martin *et al.* (1991) samples as small as a few mg (dry wt) can be used.

When analysing for a standard added to non-mycorrhizal root samples we achieved a recovery varying from 85 to 98%. Grant and West (1986) worked with microbes in soil, using an advanced technique, incorporating $^{14}$C through photosynthesis into their litter substrate and analysing for ergosterol losses; their recovery figures ranged from 88 to 96%, depending on the soil tested. With the elimination of the pre-column treatment, we now regularly obtain about 95% recovery. Sample degradation during processing because of light or other factors is negligible (cf. Newell *et al.*, 1988); replicability is therefore very good.

Since plant material contains no ergosterol (cf. analyses by Salmanowicz and Nylund, 1988; Johnson and McGill, 1991), there is no background disturbance of the sensitivity, and large sample sizes with small fungal components can be readily analysed.

## B.  Variation in ergosterol levels within the same species

Examination of pure cultures of mycorrhizal fungi of different age revealed little variation in ergosterol concentration unless the cultures were many months old (Salmanowicz and Nylund, 1988). This agrees well with the findings of Seitz *et al.* (1979), where two *Aspergillus* and one *Alternaria* spp. reached a relatively constant level after two days of culture, up to the end of the 3-week period of study. Newell *et al.* (1987), working with ascomycetes growing on litter, found larger variations which depended on species and substrate, with peaks where the phase of logarithmic growth was supposed to culminate. They also noted that the composition of the medium had a considerable influence on the content of some species but not on others. They ascribed this to possible recovery problems. However, another explanation may be that a fungus grows in more concentrated form on some media than on others: ergosterol would be expected to reflect the amount of metabolizing cytoplasm, not of inactive mass (see above). Newell *et al.* remark that conditions inducing partial autolysis are likely to have a strong effect on the ergosterol level.

In view of the likelihood that intramatrical fungal tissue in mycorrhiza is in a steady state, it is reasonable to assume that the ergosterol content of a given species is itself largely stable in that material. More variation may be encountered in extramatrical mycelium growing from the mycorrhizal roots, particularly in axenic culture where dying hyphae are not decomposed or devoured by the microfauna.

## C.  Variation in ergosterol levels between species

Newell *et al*. (1987) reviewed published data on fungal ergosterol content, and found considerable variation, reporting concentrations ranging from 1 to about 50 $\mu$g mg$^{-1}$ fungal tissue. Even their own decomposer ascomycetes varied widely between each other, ergosterol values ranging from 4.2 to 9.9 $\mu$g mg$^{-1}$ in liquid culture, with one species deviating by as much as 16.4 $\mu$g mg$^{-1}$. Olsen (1973), studying several pathogens and saprophytes, found values ranging from 1 to 7 $\mu$g mg$^{-1}$. A. Hütterman (pers. commun.), having tested a number of mycorrhizal species, also found the variation to be too large to enable generalized conclusions to be made about biomass of unknown fungi. In fruiting bodies (references in Salmanowicz and Nylund, 1988), a wide range of values has been found; some mycorrhizal fungi tested by ourselves yielded from 2.62 to 6.69 $\mu$g mg$^{-1}$ of ergosterol.

Seitz *et al*. (1979), however, found much less variation among grain-colonizing imperfect fungi (*Alternaria* and *Aspergillus*). Our own observations (Salmanowicz and Nylund, 1988) demonstrated a total range of $\pm12\%$ around the mean among four ectomycorrhiza fungi grown at periods from 2 weeks to 4 months, which must be considered fully satisfactory in relation to other sources of variation in the assay.

## D.  Comparison with the chitin assay

Comparative analysis of ergosterol and chitin contents of living and dead mycorrhiza formed on plants grown *in vitro* in peat or pellets showed that ergosterol disappeared entirely from dead roots, while the chitin levels remained largely unchanged. Boiling of living mycelium for 15 min reduced the ergosterol content by 50% while the chitin again remained unchanged. Newell *et al*. (1988) propose that these differences might form the basis of a technique for comparing the ratio between living and dead fungal biomass in litter or soil.

Working *in vitro*, Seitz *et al*. (1979) compared the chitin and ergosterol contents in growing cultures of two *Aspergillus* and one *Alternaria* spp. on liquid and moistened grain substrates. They found a close correlation between the two components over a 3-week period in two of the isolates, while chitin content did not adequately relate to growth in *Aspergillus amstelodami*. The chitin assay was also less sensitive, and did not enable detection of growth before 48 h of culture, while the ergosterol did so after 24 h. The threshold detection levels in their assays were 0.04–0.09 $\mu$g ergosterol and 0.02–0.04 $\mu$g glucosamine. The ergosterol assay was also considered to be faster.

Matcham *et al.* (1985), again using *in vitro* methods, this time with *Agaricus bisporus*, obtained a reasonably close ($r = 0.97$) correlation between chitin assay (alkaline digestion) data and weight increase after 28 days in liquid culture. After this stage, however, the chitin readings fell. The ergosterol figures were linearly correlated to growth for the 56 days of the experiment, giving a very high correlation rate ($r = 0.997$). In grain culture, the surface extension of the culture was compared with the assay data on ergosterol content. The correlation was still high; yet no regression equations were provided and the graphs show that a non-linear formula might be more suitable. Using liquid culture contents to estimate the fungal biomass in grain cultures, the chitin technique gave a 70% higher figure than the ergosterol technique. The authors confirmed the observation of Seitz *et al.* (1979) that the ergosterol assay was more sensitive and also faster than that involving chitin.

Studying microbial biomass in soil, Grant and West (1986) compared ergosterol, chitin (acid digestion) and diaminopimelic acid assay data. They concluded that the two latter "are unsatisfactory indicators of absolute levels of microbial biomass", and that ergosterol, in contrast to the other two compounds, reflected cellular content, and not soil material. In a subsequent study, West *et al.* (1987) found 0.30–0.59% ergosterol in the fungal mass estimated microscopically and by other techniques, a range which corresponded well to those which we have observed. Changes in glucosamine levels were not consistent with changes in fungal biomass or between soils. The linear correlation coefficients between glucosamine and ergosterol were only 0.674, between fungal surface (microscopy) and ergosterol 0.864, and between surface and glucosamine, 0.716. Such results again indicate that a large portion of the measured chitin/glucosamine was derived from the non-living organic matter.

In another soil study, Zelles *et al.* (1987b) obtained contradictory results using the two methods. However, no other method for determination of fungal biomass was used to assess which of the two techniques gave better results.

In the only comparison of the two techniques available at the time of preparing the chapter, Johnson and McGill (1991) found that estimates of intramatrical mycelium using chitin (alkaline hydrolysis) and ergosterol assays were not correlated ($r = 0.15$); the latter figures in this case were much larger. The experimental error of the assay was also larger for chitin than for ergosterol. The latter procedure was half as labour-intensive as the former, and in their study was found to be more sensitive in detecting biomass changes after application of phosphorus fertilizer to the seedlings.

## E.  Applications in mycorrhiza research

Apart from the quoted study by Johnson and McGill (1991), other published reports appear to be lacking. Preliminary data on nitrogen effects on mycorrhiza development in Scots pine in a semi-hydroponic system have been published (Wallander and Nylund, 1989). Biomass estimates using the ergosterol technique were very well correlated with nitrogen levels in the shoot.

## F.  General conclusions

The basic shortcomings of the method are those of variation in the ergosterol content depending on growing conditions, and interspecies variation. As long as axenic cultures or mycorrhiza formed with a single species of fungus are studied, as is the case in our ongoing work, these problems are negligible. As a quantitative tool in studies of natural or multi-species mycorrhiza, the results are inevitably less precise, but in our opinion are still far more dependable than any root-tip counting approach. When compared with chitin assay, the ergosterol elimination is relatively free of fundamental problems. To achieve a more complete evaluation, it will be necessary to have more studies of pure cultured fungi as well as of mycorrhiza where the two techniques considered, together with counting and other methods (stereology, respiration measurement and FDA analysis; cf. Schnürer and Rosswall, 1982), are used simultaneously.

## Acknowledgement

We wish to thank Dr J. Schnürer for valuable discussions and criticism of the manuscript.

## References

Grant, W. D. and West, A. W. (1986). *J. Microbiol. Meth.* **6**, 47–53.
Griffiths, H. M., Gareth-Jones, D. and Akers, A. (1985). *Ann. Appl. Biol.* **107**, 293–300.
Johnson, B. N. and McGill, W. B. (1991). *Can. J. For. Res.* (in press).
Lee, C., Howarth, R. W. and Howes, B. L. (1980). *Limnol. Oceanogr.* **25**, 290–303.

Martin, F., Delaruelle, Ch., Hilbert, J.-L. and Costa, G. (1991). *New Phytol.* (in press).

Matcham, S. E., Jordan, B. R. and Wood, D. A. (1985). *Appl. Microbiol. Biotechnol.* **21**, 108–112.

Newell, S. Y., Miller, J. D. and Fallon, R. D. (1987). *Mycologia* **79**, 688–695.

Newell, S. Y., Arsuffi, T. L. and Fallon, R. D. (1988). *Appl. Environ. Microbiol.* **54**, 1876–1879.

Nylund, J.-E. and Unestam, T. (1982). *New Phytol.* **91**, 63–79.

Olsen, R. (1973). *Physiol. Plant.* **28**, 507–515.

Osswald, W. E. and Elstner, E. F. (1987). *Allge. Forstzeit.* **27–29**, 693–694.

Osswald, W. E., Höll, W. and Elstner, E. F. (1986). *Z. Naturforschung* **41c**, 542–546.

Salmanowicz, B. and Nylund, J.-E. (1988). *Eur. J. For. Pathol.* **18**, 291–298.

Schnürer, J. and Paustian, K. (1986). In *Perspectives in Microbial Ecology, Proceedings of the 4th International Symposium on Microbial Ecology* (F. Megusar and M. Gantar, eds). Slovene Society for Microbiology, Ljubljana.

Schnürer, J. and Rosswall, T. (1982). *Appl. Environ., Microbiol.* **43**, 1256–1261.

Seitz, L. M., Mohr, H. E., Burroughs, R. and Sauer, D. B. (1977). *Cereal Chem.* **54**, 1207–1217.

Seitz, L. M., Sauer, D. B., Burroughs. R., Mohr, H. E. and Hubbard, J. D. (1979). *Phytopathology* **69**, 1202–1203.

Vignon, C., Plassard, D., Mousain, D. and Salsac, L. (1986). *Physiol. Veg.* **24**, 201–207.

Wallander, H. and Nylund, J. E. (1989). *Agric. Ecosystems. Environ.* **28**, 547–552.

Weete, J. D. (1974). *Fungal Lipid Biochemistry*, pp. 151–209. New York, Plenum Press.

West, A. W., Grant, W. D. and Sparling, G. P. (1987). *Soil Biol. Biochem.* **19**, 607–612.

Zelles, L., Hund, K. and Stepper, K. (1986a). *Z. Pflanzener. Bodenkunde* **150**, 249–252.

Zelles, L., Hund, K. and Stepper, K. (1986b). *Z. Pflanzener. Bodenkunde* **150**, 253–257.

# 6

# Establishment of Vesicular-arbuscular Mycorrhiza in Root Organ Culture: Review and Proposed Methodology

GUILLAUME BÉCARD

*United States Department of Agriculture, ARS, ERRC, 600 East Mermaid Lane, Philadelphia, PA 19118, USA*

YVES PICHÉ

*Département des Sciences Forestières, Centre de Recherche en Biologie Forestière, Faculté de Foresterie et de Géomatique, Université Laval, Ste-Foy, Québec G1K 7P4, Canada*

METHODS IN MICROBIOLOGY
VOLUME 24   ISBN 0-12-521524-X

Copyright © 1992 by Academic Press Limited
All rights of reproduction in any form reserved

## I. Introduction

Vesicular-arbuscular mycorrhizal fungi are obligate biotrophs which have so far resisted all attempts to be cultivated axenically (in pure culture). This lack of independent growth has not prevented vesicular-arbuscular mycorrhizal fungi from becoming distributed world-wide as a symbiotic partner of most vascular plants, under a wide variety of pedologic and climatic conditions.

Numerous studies have demonstrated the beneficial role of vesicular-arbuscular mycorrhizal fungi in improving plant growth by increasing resistance to drought and disease, as well as by enhancing nutrient absorption efficiency (Stribley, 1987; Nelsen, 1987). The relief of stress due to fungal contributions to plant growth has been emphasized in the literature, but plant involvement in vesicular-arbuscular mycorrhizal fungal growth remains poorly understood.

Cultivation of vesicular-arbuscular mycorrhizal fungi under axenic conditions continues to be a preoccupation and represents one of the most challenging goals of modern plant biology. Using roots growing *in vitro* as plant partners, Mosse and Hepper (1975) and Miller-Wideman and Watrud (1984) provided new scientific insights on this question when they succeeded in establishing vesicular-arbuscular mycorrhizal symbiosis *in vitro*.

In recent years, culture of isolated roots has been given a new impetus with the use of roots genetically transformed by the Ri plasmid of *Agrobacterium rhizogenes* (Tepfer, 1984). Rapid growth of axenic "hairy roots" led Mugnier and Mosse (1987) to inoculate such roots with vesicular-arbuscular mycorrhizal fungi. They observed some vesicular-arbuscular mycorrhizal colonization on transformed roots of *Convolvulus sepium* inoculated with germinated spores of *Glomus mosseae* (Mugnier and Mosse, 1987). Some evidence of independent growth of *G. intraradices* has also been observed (Mosse, 1988). These authors thus demonstrated that root organ cultures have potential for growing vesicular-arbuscular mycorrhizal fungi *in vitro*. The method used, however, did not result in the reproducible generation of the entire life-cycle of the fungus including the production of viable spores.

Bécard (1989) presented an in depth evaluation of the root organ culture method and improved the procedures so that typical vesicular-arbuscular mycorrhiza can now be obtained on transformed as well as non-transformed roots, leading to complete control of the life-cycle of a few species of vesicular-arbuscular mycorrhizal fungi.

The purpose of this chapter is to provide a detailed description of the procedures of the method and to suggest new avenues, in order to stimulate research on the biology of these symbiotic fungi.

## II.  Root organ culture

### A.  Classical method

Isolated roots can be cultivated continuously *in vitro* using methods thoroughly described by White (1943, 1963). Clonal root cultures of some 15 plant species have been established successfully (Butcher, 1980). Initiation of root cultures requires seed germination under axenic conditions. Emerging radicles are then excised and cultivated on White's medium, in either solid or liquid form.

One of the standard seed surface sterilization procedures is to soak them in sodium hypochlorite solution (1–3% available chlorine) for 30 min at room temperature with thorough stirring. Many other disinfecting solutions can be used, each with different immersion times (Butcher, 1980). Surface sterilization can be improved by pre-treatment with 70% ethanol (1–2 min) and/or with a surfactant (Tween 20 or 80). Ultimately, the goal is to achieve the best sterilization without altering germination.

Seeds can be germinated in the dark in sterile Petri dishes containing either moistened filter papers placed on water agar or, if possible, a rich microbiological culture medium in order to detect contaminants.

White's medium was first developed to grow tomato roots and has since been viewed as the starting point for the development of media for other species. It is a completely defined medium that contains macro- and micro-elements, one main source of carbon (sucrose) and vitamins. It has been modified by Boll and Street (1951) by the addition of copper and molybdenum and by Sheat *et al.* (1959) using iron chelated by sodium ethylenediamine tetraacetate. This Fe-EDTA preparation allows for the use of a wider pH range (up to 7.5).

Despite the aforementioned success in continuous culture of roots of some plant species, it must be emphasized that this still remains impossible with the majority of dicotyledons, especially woody species, and also with most monocotyledons and gymnosperms. Existing media probably fail to meet qualitative and quantitative nutritional requirements of roots from many species, as evidenced by the improvement of root growth following supplementation with additional substances (Butcher and Street, 1964; Torrey, 1986). The sugar alcohol, myo-inositol, for example, improved the root growth of *Comptonia peregrina* (Goforth and Torrey, 1977). Sometimes the growth of roots can be enhanced by adding to the culture media growth-regulating substances such as natural auxin, or synthetic substances which have a similar mode of action (Zeadan and MacLeod, 1984; Robbins and Hervey, 1978; Lazzeri and Dunwell, 1984). Growth can even be dependent upon these auxins. In fact, the level of natural auxin established in the cells of isolated roots may frequently be far less than that required for most active growth and development. This deficiency seems to be especially pronounced when the supply of sugar is reduced (Butcher and Street, 1964).

Great advances have been made in recent years with *in vitro* cultivation of isolated roots, especially the culture of genetically transformed roots following infection of plant tissues by *Agrobacterium rhizogenes*.

## B. Transformed roots

The bacterium *A. rhizogenes* is a soil-borne pathogen causing "hairy root" disease in dicotyledonous plants. When wounded tissues are infected with the bacteria, they form large numbers of roots (hairy roots) in 2–3 weeks (Fig. 1a,b). In a system analogous to crown gall tumours induced by the much studied *A. tumefaciens*, hairy roots synthesize opines (Tepfer and Tempé, 1981). Their constitutive cells have integrated copies of T-DNA (Transfer DNA) which occurs in a large plasmid of *A. rhizogenes*, the Ri (Root-inducing) plasmid (Chilton et al., 1982). The different regions of T-DNA encoding for opines, auxin and other loci involved in the production of roots are the subjects of active investigations (Tepfer, 1989).

After emerging from the infected tissue, the transformed roots can be subcultured as excised roots. On solid culture media, the fast-growing root apices can be freed of the original bacterial inoculum by the use of antibiotics such as carbenicillin or ampicillin. After several transfers without detectable contamination, a clonal culture, derived from a single root, is established. An example of production of such a clonally transformed root culture, established from carrot tissues, was described by Bécard and Fortin (1988) (Fig. 2). Slices (5 mm thick) of peeled and sterilized (95% ethanol for 10 s, then 1% NaOCl for 15 min) fresh carrot roots are surface inoculated with either $A_4$ or 15834 (ATCC) bacterial cultures. The inoculum can be taken directly by the loopful from slants or from a bacterial suspension (approx. $10^9$ bacterial cells ml$^{-1}$ of sterile water) derived from a culture in exponential phase on Difco Nutrient Agar. Bacteria have a higher virulence on the distal face of the slices (corresponding to the apical part of the carrot) because of a higher endogenous auxin level (Ryder et al., 1985). The inoculated slices are incubated in the dark at 28 °C on water agar in Petri dishes. Different protocols can be used with other plant organs (stem, leaf, cotyledon) of different dicotyledonous plant species. A list of about 100

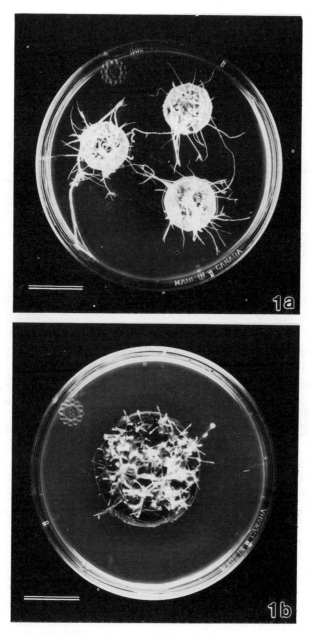

**Fig. 1.** Hairy roots growing on wounded plant tissues. The surface sterilized slices of carrot (a) and sugar beet (b) were inoculated with *Agrobacterium rhizogenes* strain 15834 (ATTC) 3 weeks earlier. Bar, 2 cm.

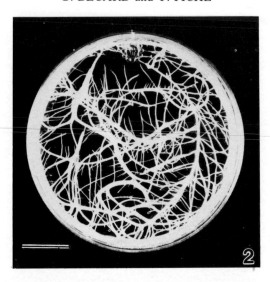

**Fig. 2.**   Culture of a single Ri T-DNA transformed root of carrot on M medium
(Bécard and Fortin, 1988) after 10 days of incubation at 26 °C. Bar, 2 cm.

plant species for which stable transformed root cultures have been
obtained has been presented by Tepfer (1989). The list is rapidly
lengthening, especially since many laboratories are developing "hairy
root" technology for the production of plant secondary metabolites
(Hamill *et al.*, 1987).

One desirable characteristic of these transformed roots is their ability
to quickly form numerous lateral roots. Tepfer (1984) concluded that
they are better adapted to growth in culture than normal roots, and that
they also survive longer periods without subculture. Possibly the new
genome of transformed roots affects its capacity for auxin production.
Another characteristic sometimes observed in transformed roots is the
inversion of their geotropic mode of growth. Nevertheless, it appears
that transformed roots have the same synthetic capacity as roots of the
plant from which they were obtained; the chromosome number in the
transformed cells is the same as in the cells of the parent plant, in
contrast to those of disorganized callus cultures (Hamill *et al.*, 1987).

Successful transformation can be confirmed by the detection of opines
in root tissues (Tepfer and Tempé, 1981). These metabolites are
indicative of the presence of Ri T-DNA in the root cells (Tepfer, 1989).
A better confirmation of transformation can be obtained by molecular
hybridization (Chilton *et al.*, 1982).

Various culture media have recently been used for growing such Ri T-DNA transformed roots: Murashige and Skoog (MS) medium (Hamill *et al.*, 1987; Mugnier, 1988); Gamborg B5 medium (Hamill *et al.*, 1987; Parr *et al.*, 1988); culture media according to Torrey (1954) and Monnier (1976); and finally White's medium (Bécard and Fortin, 1988). The last medium was preferred to MS, even when diluted, because it allowed significantly better growth of the roots. Specifically, the presence of ammonium in MS caused a rapid (less than two weeks) drop in the pH of the culture medium which was detrimental to the root growth. On White's medium, nitrogen exclusively in the form of nitrate is assimilated, which counteracts the acidification of the culture medium following root growth. In this way, the culture medium is buffered and maintains pH at 6 for several months (G. Bécard and Y. Piché, unpubl. res.).

## III. Fungal inocula

### A. Nature of inocula

Different forms of inocula have been used to form vesicular-arbuscular mycorrhiza with root organ cultures: sporocarpic chlamydospores of *Glomus mosseae* (Mosse and Hepper, 1975; Mugnier and Mosse, 1987), segments of mycorrhizal roots infected with different *Glomus* spp. (Strullu and Romand, 1986), internal vesicles extracted from these latter infected roots (Strullu and Romand, 1987), non-sporocarpic azygospores of *Gigaspora margarita* and chlamydospores of *Glomus intraradices* (Miller-Wideman and Watrud, 1984; Bécard and Fortin, 1988; Bécard and Piché, 1989a, b, 1990; Chabot, 1990). Virtually any kind of vesicular-arbuscular mycorrhizal fungus may provide propagules amenable to *in vitro* propagation. One should, however, select those species for which spore production, sterilization and germination can be easily obtained. These criteria are more often encountered with non-sporocarpic spores because they are isolated directly and purified after wet sieving the soil of the mycorrhizal rhizosphere. Thus, they are available in large numbers and are easier to select and surface sterilize than sporocarpic spores, which must be extracted by dissection from the sporocarps before sterilization. There is good evidence that mycorrhizal roots which have to be selected from an entire root system give unpredictable responses as a source of fungal inocula. The reason for this unpredictability stems from their content of internal vesicles. There is also no available method to date for reproducibly sterilizing these mycorrhizal infected roots with 100% success.

## B. Sterilization and conservation of inocula

The surface sterilization of propagules used as starting inocula is a crucial prerequisite for successful *in vitro* mycorrhiza formation. It is this step which causes many problems for researchers working with vesicular-arbuscular mycorrhizal fungi under controlled conditions.

Many procedures have been used for surface sterilization (Mosse, 1962; Mertz *et al.*, 1979; Tommerup and Kidby, 1980; Macdonald, 1981; Strullu and Romand, 1986). Most scientists prefer chloramine T (2%) solutions with traces of surfactant and antibiotic(s) such as streptomycin and gentamycin (Mosse, 1962). The success in utilizing these disinfectant solutions is highly dependent on how they are used and on which material.

The fungal material to be sterilized must be purified as much as possible since most contaminants come from old spores or debris. Therefore the fungal material, following extraction from soil by wet sieving (Gerdeman and Nicolson, 1963), should be subjected to density gradient centrifugation, in order to remove dead spores and other debris. This method isolates healthy spores that are generally easier to surface sterilize. The centrifugation method of Furlan *et al.* (1980) has been effectively used for several years for various genera of vesicular-arbuscular mycorrhizal fungi. This method utilizes Renografin-60 (E. R. Squibb and Sons Ltd, Montréal, Canada), a contrast agent used for medical diagnostic procedures, which is non-toxic to fungal propagules. Basically, the surface sterilization procedure involves 20 min treatment in chloramine T, and its efficiency has been demonstrated by Mertz *et al.* (1979).

## 1. First step

In a blood collection tube (vacutainer, Becton Dickinson, Rutherford, NJ, USA) the spores are first washed in 0.05% Tween 20 for 1 min, then treated by agitation at 4 °C in 2% chloramine T for two periods of 10 min. A vacuum is created in the tube by using a syringe (20 cm$^3$) needle through the rubber plug. This removes the dissolved gasses from the different materials in suspension, thereby improving effectiveness of chloramine T. A sterile mixture of 200 mg litre$^{-1}$ streptomycin and 100 mg litre$^{-1}$ gentamycin is used for the four subsequent rinses. By this method, the spores have undergone several thorough washings, since the different solutions are removed in a laminar flow hood by centrifuging (30 s) in a clinical centrifuge.

## 2. Second step

In the upper portion of a sterile 0.22 μm filter holder apparatus (15 ml) the spores are treated as previously, but without the Tween 20 washing. A sterile Pasteur pipette is used to stir the spore solutions frequently. The different liquid phases are removed automatically by aspiration using a vacuum pump, thus minimizing the risk of accidental contamination. Then, propagules (less than 100 per Petri dish) are spread out on water agar and the Petri dishes are stored at 4 °C in inverted positions.

Thousands of individual sterilized propagules are then available for up to several months for experiments. Mertz et al. (1979) suggested at least one week of storage at 4 °C between the two steps of sterilization to maximize the action of the antibiotic. Spores retain their viability for much longer (several months) storage times. The two steps, however, may be carried out without the intervening storage period if the last rinsing of propagules is made with antibiotics.


## IV. Dual culture

### A. Establishment of dual culture

Dual cultures of root and fungus are generally established on solid media. This is the simplest method of co-cultivating the two organisms. Mosse and Hepper (1975) did not find any advantage in using a divided plate, where the distal parts of the roots were grown in a mineral salts medium lacking sugar and vitamins while their proximal ends were grown in a complete medium. Root inoculations have been established in liquid culture but with some difficulty (Mosse and Hepper, 1975). Mugnier and Mosse (1987) used a divided culture system to grow the proximal ends of the roots in a complete medium, while growing the root tips in a separate compartment containing water agar supplemented with peat. Only the root tips were then inoculated with spores. Such elaborate systems facilitate study of nutrient transfer between the two compartments.

Pre-germinated spores of pieces of mycorrhizal roots showing hyphal regrowth have been used to inoculate root systems by placing them close to emerging lateral roots. This is a preferred inoculation method when germination requires special conditions or when sterilization is not sufficiently controlled. We currently use a simplified method, wherein a single ungerminated spore is used to inoculate a single root system. The germ tube growth of G. margarita is negatively geotropic and as such it is possible to choose in advance which root tissue it will first contact by placing a pre-selected zone of the root over the growing germ tube. This is done in horizontally incubated Petri dishes containing germinated spores in agar, over which the corresponding root is placed (Bécard and Fortin, 1988). Initial contact is followed by a predictable root colonization (Bécard and Fortin, 1988). This standardized inoculation allows, for the first time, quantitative testing of effects of different media compositions on mycorrhiza formation and calculation of the percentage of contacts resulting in root colonization. In addition, it allows relationships between the degree of root colonization and the growth of extraradical hyphae to be determined. The preferred system is to use vertically incubated Petri dishes with one germinated spore in the middle and one root placed perpendicularly to the growing germ tube. The elongation of extraradical hyphae is then observed in two dimensions. Elongation is measured (mm) non-destructively, using a calibrated grid (Bécard and Piché, 1989a,b, 1990).

## B. Culture media

Adjusting the appropriate medium for co-cultivating the two organisms, roots and fungus, is the most important factor for successful vesicular-arbuscular mycorrhiza formation in root organ culture. A compromise must be found between the requirements of actively growing roots, which need a rich complex medium, and those of the extraradical phase of the vesicular-arbuscular mycorrhizal fungus, which normally grow in a rhizosphere (i.e. in a relatively nutrient-poor medium). Different examples of culture media for such dual cultures are given in Table I. For comparison, the composition of White's medium and Murashige and Skoog's medium (MS), which are currently the most frequently used for plant tissue culture, have also been given. It must be mentioned that these different media used for dual culture are not totally comparable. Miller-Wideman and Watrud (1984) used a 1/10 diluted MS medium for inducing initiation of roots from tomato seedling explants (the shoot–radicle transition zone). Vesicular-arbuscular mycorrhiza were obtained using *G. margarita* on these roots but the nutritional influence of the culture medium is difficult to evaluate because of its interaction with the initial explants. The medium used by Mugnier and Mosse (1987) for dual culture of transformed roots of *Convolvulus sepium* and *G. mosseae* was an adaptation of MS medium and was only present in one compartment, in order to "feed" the proximal end of each root. These authors obtained vesicular-arbuscular mycorrhiza with roots growing on peat-supplemented water agar. Strullu and Romand (1986) used a

**TABLE I**

Composition (in mM) of media used for plant tissue culture and for dual culture of isolated roots and vesicular-arbuscular mycorrhizal fungi

| | MS[a] | White[b] | Mosse and Hepper (1975) | Miller-Wideman and Watrud (1984) | Strullu and Romand (1986) | Mugnier and Mosse (1987)[c] | Bécard and Fortin (1988) | Mosse (1988) |
|---|---|---|---|---|---|---|---|---|
| $N(NO_3^-)$ | 40 | 3.2 | 3.2 | 4 | 20.2 | 1.12 | 3.2 | 2.9 |
| $(NH_4^+)$ | 20 | – | – | 2 | – | 0.625 | – | – |
| P | 1.25 | 0.14 | 0.07 | 0.125 | 0.3 | 0.25 | 0.035 | 0.15 |
| K | 20 | 1.7 | 1.7/1.77 | 2 | 3.3 | 0.75 | 1.735 | 1.80 |
| Ca | 3.0 | 1.2 | 1.2/1.26 | 0.3 | 8.6 | 0.6 | 1.2 | 1.00 |
| Mg | 1.5 | 3 | 3 | 0.15 | 1.5 | 0.3 | 3 | 0.17 |
| S | 1.6 | 4.4 | 3 | 0.16 | 2.5 | 0.32 | 3 | 0.17 |
| Cl | 6 | 0.9 | 0.9 | 0.6 | – | 1.2 | 0.9 | 0.85 |
| Na | 0.2 | 3 | 0.012 | 0.02 | 0.2 | 0.04 | 0.02 | 0.04 |
| Fe | 0.1 | 0.02 | 0.012 | 0.01 | 0.1 | 0.02 | 0.02 | 0.0012 |
| Mn | + | + | + | + | + | + | + | + |
| Zn | + | + | + | + | + | + | + | + |
| B | + | + | + | + | + | + | + | + |
| I | + | + | + | + | – | + | + | – |
| Mo | + | + | + | + | + | + | + | + |
| Cu | + | + | + | + | + | + | + | + |
| Co | + | – | – | – | – | + | – | – |
| Ni | – | – | – | – | – | – | – | – |
| Al | – | – | – | – | – | – | – | – |

| | MS[a] | White[b] | Mosse and Hepper (1975) | Miller-Wideman and Watrud (1984) | Strullu and Romand (1986) | Mugnier and Mosse (1987)[c] | Bécard and Fortin (1988) | Mosse (1988) |
|---|---|---|---|---|---|---|---|---|
| Myo-inositol[d] | 100 | – | – | 10 | – | 1 | 50 | 100 |
| Nicotinic acid[d] | 0.5 | 0.5 | 0.5 | 0.05 | 1 | 0.2 | 0.5 | 0.5 |
| Pyridoxine–HCl[d] | 0.5 | 0.1 | 0.1 | 0.05 | 1 | 0.2 | 0.1 | 0.5 |
| Thiamine–HCl[d] | 0.1 | 0.1 | 0.1 | 0.01 | 1 | 0.2 | 0.1 | 0.1 |
| Glycine[d] | 2 | 3 | 3 | 0.2 | – | – | 3 | – |
| Ca pantothenate acid[d] | – | – | – | – | 1 | 0.2 | – | – |
| Biotin[d] | – | – | – | – | 0.001 | 0.002 | – | 0.2 |
| Cyanocobalamine[d] | – | – | – | – | 0.4 | – | – | – |
| Sucrose (g litre$^{-1}$) | 30 | 20 | 20 | 3? | 15 | 12 | 10 | 30 |
| pH | 5.7 | 4.8 | 7 | 6.8 | 7 | 6.9[e] | 5.5 | 6.2 |

[a] Murashige and Skoog (1962).
[b] Modified White's medium from Butcher (1980)
[c] Only for proximal ends of the roots. Vesicular-arbuscular mycorrhiza inoculation was made on water aga (supplemented with peat).
[d] Values given in mg litre$^{-1}$
[e] After sterilization.

medium that was relatively rich, especially in nitrogen, for isolating different vesicular-arbuscular mycorrhizal fungi from mycorrhizal strawberry roots and for reassociating them with tomato plants and excised roots of *Allium cepa* and *Solanum lycopersicum*. It is not clear whether the latter roots were continuously grown on the culture medium. Mosse and Hepper (1975) and Bécard and Fortin (1988) used a similarly modified White's medium to form vesicular-arbuscular mycorrhiza, with *G. mosseae* in clonal clover root cultures and with *G. margarita* in a clonal culture of transformed carrot roots, respectively. It appears more logical to use White's medium as a basic medium since it has been developed for root organ culture, rather than searching for a convenient adaptation of MS medium. The latter is considerably richer than White's medium (see Table I) and must be diluted. Moreover, the presence of $NH_4^+$ ions can have deleterious effects on the roots (see above). Two major differences distinguish the composition of the two modified White's media: phosphorus and sucrose are at half-strength concentration in the medium of Bécard and Fortin (1988). These authors showed that reducing levels of these two components was a determining factor in the achievement of successful colonization. The fact that a high phosphorus level is detrimental to vesicular-arbuscular mycorrhizal establishment is an indication that the root organ culture system is behaving normally, despite its artificial basis. The intolerance of the fungus to a high concentration of sugar in the medium is a further indication of its normal behaviour. In good agreement with these results, the medium of Mosse (1988) (Table I), rich in phosphorus and sucrose, provided nutritive conditions to stimulate "independent" growth of *G. intraradices*, but simultaneously suppressed mycorrhizal infection of transformed carrot roots.

The negative effect of sodium on the development of mycorrhizal colonization was first observed by Mosse and Phillips (1971) for *Endogone mosseae* in the root of *Trifolium parviflorum* and has been confirmed by Bécard and Fortin (1988). Both modifications of White's medium (Mosse and Hepper (1975); Bécard and Fortin (1988)), therefore, are almost completely devoid of this element. We have used this medium, described in detail by Bécard and Fortin (1988), for three years with success (Bécard and Piché, 1989a,b, 1990). It has also been convenient for dual culture of *G. intraradices* with transformed roots of carrot or normal roots of tomato (Chabot, 1990). The non-transformed tomato roots, however, needed to be regularly subcultured on complete White's medium. We have recently adopted a new agar substitute (Gel-Gro, ICN Biochemicals, Cleveland, OH 44128, USA) which is a naturally derived, highly purified polysaccharide that contains fewer impurities and may be used at one-quarter to one-third the concentration compared with other agars. Gel-Gro allows better growth of the roots compared with the agar (Difco Bacto agar) previously used, and is transparent, allowing better quality microscopic observations of the growing fungus.

## V.    Results obtained by this methodology

### A.    Observational data

All the aforementioned systems produced typical mycorrhizal infections with formation of appressoria, intraradical hyphae, vesicles (except for *G. margarita*) and arbuscules. In fact, all the stages occurring in the development of vesicular-arbuscular infections in whole plants have been found. It has been observed, however, that the infection process could be arrested at any stage (Mosse and Hepper, 1975; Mugnier and Mosse, 1987). The status of non-host among some non-mycorrhizal plant species seems to be maintained in root organ culture (Bécard and Piché, 1990). The infection rate, as a proportion of infected root length, has been calculated to be 17% at 9 weeks after inoculation with *G. margarita* (Miller-Wideman and Watrud, 1984), 50% in the most successful cultures with *G. mosseae* (Mosse and Hepper, 1975) and 40% in old cultures with *G. margarita* (G. Bécard and Y. Piché, unpubl. res.). The latter percentage has been calculated by direct observation of the culture plates using transmitted light, which precludes the need for staining the roots. With this method, the infected roots appear darker (Fig. 3).

Recently developed lateral roots appear to be preferred infection sites (Mosse and Hepper, 1975; Miller-Wideman and Watrud, 1984) and most infections start 0.5–1 cm behind the root tips (Mosse and Hepper, 1975). Bécard and Fortin (1988) observed that the elongation zone of the main root, where lateral root primordia were forming, was a preferential site for primary infection by the germ tubes of germinating spores of *G. margarita*.

One of the major advantages of the root organ culture system is that many observations on extraradical fungal development can be made non-destructively using a binocular lens with incident or transmitted light or using an inverted microscope. The formation, prior to infection, of fan-like hyphae on the root surface, along with small vegetative spores or auxiliary cells, branching hyphae reminiscent of arbuscules, and spores of *G. margarita*, can be seen directly. Direct observations have also established that hyphae in agar were not "attracted" by roots

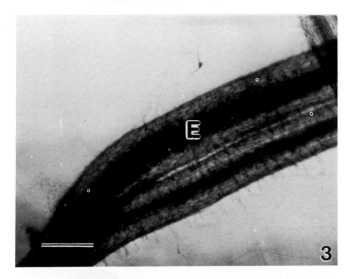

**Fig. 3.**  Dual culture of *G. margarita* and Ri T-DNA transformed roots of carrot observed non-destructively. One root segment is colonized by the endophyte (E) and looks darker when observed under a microscope with transmitted light. Bar, 300 μm.

(Mosse and Hepper, 1975; Bécard and Fortin, 1988) and that formation of *G. margarita* spores occurs within 5 days (Fig. 4a–d).

Researchers are just beginning to discover many interesting features of vesicular-arbuscular mycorrhizal fungi, including the function and ontogeny of their different structures. It is essential to culture several other species of vesicular-arbuscular mycorrhizal fungi in order to compare their fundamental characteristics.

## B.  Experimental data

The dual culture of vesicular-arbuscular mycorrhizal fungi with isolated roots is a very simplified model of a rhizosphere, where most of the chemical, physical and biological parameters are controlled. This is a useful experimental tool for physiological studies involving the movement of substances from the plant into the symbiotic fungus. These types of studies have been neglected in the past, but are obviously highly relevant to the establishment of vesicular-arbuscular mycorrhizal fungi in pure culture.

In preliminary experiments, Mugnier and Mosse (1987) demonstrated the usefulness of their system by studying translocation of radioisotopes

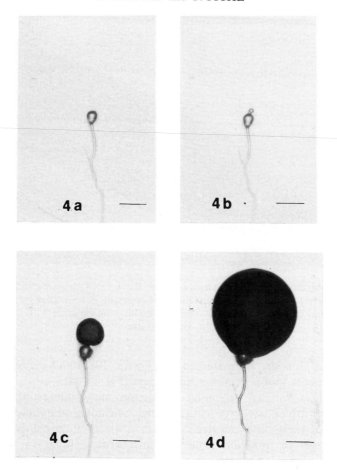

**Fig. 4.** *In vitro* formation of an azygospore of *G. margarita* in dual culture with Ri T-DNA transformed roots of carrot observed the first (a), second (b), third (c) and fifth (d) days. All the photographs are shown at the same magnification. Bar, 100 μm.

from roots to the extraradical part of the fungus. They found movement of carbon-, sulphur- and phosphorus-containing compounds.

Through successive analytical steps, Bécard and Piché (1989a, b) have found two mechanisms contributed by roots which influence growth of the mycosymbiont during the acquisition of its biotrophic status. The first mechanism involves the synergistic fungal growth promotion by $CO_2$ and root exudates. In fact, it is now possible to produce significant monoxenic fungal growth from germinating spores of *G. margarita* by replacing roots with these two abiotic factors. The use of $CO_2$ incubators (containing 2% $CO_2$), for example, allows for efficient control of spore germination and hyphal growth of several vesicular-arbuscular mycorrhizal fungi (G. Bécard and Y. Piché, unpubl. res.). It remains to be learnt how to replace factors involved in the second mechanism according to Bécard and Piché (1989a) for obtaining continuous pure culture of the endophytes.

Using a root organ culture assay, Gemma and Koske (1988) have shown that aerial germ tubes of *Gigaspora gigantea* are attracted by prospective host roots (tomato and corn) and less by non-host roots (beet and kohlrabi). This is another example of the type of vesicular-arbuscular mycorrhiza-related physiological phenomena which can be studied using the root organ culture.

## C. Continuous culture

The continuous culture of a vesicular-arbuscular mycorrhizal fungus in root organ culture means that the endophyte is maintained *in vitro* indefinitely. It must be subcultured in order to maintain an increase in its biomass.

*In vitro* subcultures of mycorrhizal roots have been obtained only by Strullu and Romand (1986), in a system of successive isolations of some *Glomus* spp. from mycorrhizal roots and their reassociation with different root systems. In this way, it is possible to create a fungal library and keep vesicular-arbuscular mycorrhizal fungal strains under artificial conditions. Many questions remain concerning the amount of biomass produced by the endophytes under these conditions and, whether long periods of time can be allowed to elapse between the subcultures without altering fungal viability.

We maintain *G. margarita in vitro* by another method. Under our conditions, the fungus produces about 500 spores per Petri dish (90 × 15 mm) in one year from three mother spores used as inoculum. Moreover, the daughter spores can be maintained for more than two years in the initial Petri dishes at 28 °C without loss of viability. During this time, if desiccation of the dishes is prevented, thousands of cream-white coloured spores of this fungus in pure culture are available for different purposes, including re-initiating dual cultures. This method has the advantage of maintaining the mycosymbiont without much manipulation and potentially constitutes a system for producing high quality vesicular-arbuscular mycorrizal inoculum.

## VI.  Conclusion

The use of root organ cultures in the study of vesicular-arbuscular mycorrhiza appears to be very promising. In this chapter, the use of transformed carrot roots inoculated with a single spore of *G. margarita* is proposed as a simple experimental system which allows reproducible observations of all stages of vesicular-arbuscular mycorrhizal development, including the extraradical phases (Bécard and Fortin, 1988). The control of colonization of selected root parts by *G. margarita* germ tubes is the key to making quantitative and consistent measurements of the initiation and development of a vesicular-arbuscular mycorrhizal symbiosis. The system has been useful for determining symbiotic factors provided by the root which govern fungal growth ($CO_2$, root exudates, nutrients from host cells to arbuscules. Bécard and Piché, 1989a, b). Additionally, the use of a single system allows the comparison and integration of data from one experiment to another.

Structural studies of early colonization by vesicular-arbuscular mycorrhizal fungi as well as cytological analysis of hyphal development and spore development should also be feasible in the near future. The *in vitro* root organ culture system can also be used as a simplified rhizosphere model for ecological studies. Used as a bioassay, it enables the investigation of the effects of numerous environmental stresses on the establishment of vesicular-arbuscular mycorrhiza. These include heavy metal toxicity, extremes of temperature and pH, simulated acid rain, pesticides, plant pathogens, competition with other species and more.

Molecular biology applied to characterize vesicular-arbuscular mycorrhizal fungal species is now practical since it is possible to amplify a given DNA probe from a minimum quantity of fungal DNA using the polymerase chain reaction (PCR). The accuracy of this approach, however, strongly depends upon the purity of the sample. Spores produced under axenic conditions are an ideal starting material to ensure this condition. In the near future, sophisticated physical methods (nuclear magnetic resonance) are likely to be used for analysing critical transfer of compound(s) from host cells to the vesicular-arbuscular mycorrhizal endophyte during hyphal growth stimulation. These advanced techniques are now possible because dual axenic cultures have minimized technical difficulties in manipulating vesicular-arbuscular mycorrhizal fungi *in vitro* and also sufficient quantities of clean vesicular-arbuscular mycorrhizal material can be produced. Essentially, transformed root systems colonized by vesicular-arbuscular mycorrhizal spores could be, for mycorrhizologists, the "white mouse" for analytical studies involving vesicular-arbuscular mycorrhizal fungi and their responses to different host and environmental stimuli. It is very tempting at this point to speculate that the pure culture of vesicular-arbuscular mycorrhizal fungi will be achieved in the 1990s, as researchers creatively apply modern technologies to a specific model system of vesicular-arbuscular mycorrhiza.

## Acknowledgements

We thank J. A. Fortin of Université de Montréal, B. Balaji of University of Madras, L. Hutchison of Université Laval, R. L. Peterson and M. Chapple of the University of Guelph, and L. W. Doner of ERRC, USDA, Philadelphia for their helpful critical comments of the manuscript. We also thank J. Lei for excellent technical assistance and M. Pelchat for her computer skills.

## References

Bécard, G. (1989). Thèse PhD, Université Laval. 92 pp.
Bécard, G. and Fortin, J. A. (1988). *New Phytol.* **108**, 211–218.
Bécard, G. and Piché, Y. (1989a). *New Phytol.* **112**, 77–83.
Bécard, G. and Piché, Y. (1989b). *Appl. Environ. Microbiol.* **55**, 2320–2325.
Bécard, G. and Piché, Y. (1990). *Can. J. Bot.* **68**, 1260–1264.
Boll, W. G. and Street, H. E. (1951). *New Phytol.* **50**, 52–75.
Butcher, D. N. (1980). In *Tissue Culture Methods for Plant Pathologists* (D. S. Ingram and J. P. Helgelson, eds), pp. 13–17. Blackwell Scientific, Oxford.
Butcher, D. N. and Street, H. E. (1964). *Bot. Rev.* **30**, 513–586.
Chabot, S. (1990). Mémoire MSc, Université Laval.
Chilton, M., Tepper, D. A., Petit, A., David, C., Casse-Delbart, F. and Tempé, J. (1982). *Nature* **295**, 432–434.
Furlan, V., Bartschi, H. and Fortin, J. A. (1980). *Trans Br. Mycol. Soc.* **75**, 336–338.
Gemma, J. N. and Koske, R. E. (1988). *Trans. Br. Mycol. Soc.* **91**, 123–132.
Gerdemann, J. W. and Nicolson, T. H. (1963). *Trans. Br. Mycol. Soc.* **46**, 235–244.
Goforth, P. L. and Torrey, J. G. (1977). *Am. J. Bot.* **64**, 476–482.
Hamill, J. D., Parr, A. J., Rhodes, M. J. C., Robins, R. J. and Walton, N. J. (1987). *Biotechnology* **5**, 800–804.
Lazzeri, P. A. and Dunwell, J. M. (1984). *Ann. Bot.* **54**, 351–361.
Macdonald, R. M. (1981). *New Phytol.* **89**, 87–93.
Mertz, S. M., Heithaus III, J. J. and Bush, R. L. (1979). *Trans. Br. Mycol. Soc.* **72**, 167–169.
Miller-Wideman, M. A. and Watrud, L. S. (1984). *Can. J. Microbiol.* **30**, 642–646.

Monnier, M. (1976). *Rev. Cytol. Biol. Vég.* **39**, 1–120.
Mosse, B. (1962). *J. Gen. Microbiol.* **27**, 509–520.
Mosse, B. (1988). *Can. J. Bot.* **66**, 2533–2540.
Mosse, B. and Hepper, C. M. (1975). *Physiol. Plant Pathol.* **5**, 215–223.
Mosse, B. and Phillips, J. M. (1971). *J. Gen. Microbiol.* **69**, 157–166.
Mugnier, J. (1988). *Plant Cell Rep.* **7**, 9–12.
Mugnier, J. and Mosse, B. (1987). *Phytopathology* **77**, 1045–1050.
Murashige, T. and Skoog, F. (1962). *Physiol. Planta.* **15**, 473–497.
Nelsen, C. E. (1987). In *Ecophysiology of Mycorrhizal Plants* (G. R. Safir, ed.), pp. 71–91. CRC Press, Boca Raton, FL.
Parr, A. J., Peerless, A. C. J., Hamill, J. D., Walton, N. J., Robins, R. J. and Rhodes, M. J. C. (1988). *Plant Cell Rep.* **7**, 309–312.
Robbins, W. J. and Hervey, A. (1978). *Am. J. Bot.* **65**, 1132–1134.
Ryder, M. H., Tate, M. E. and Kerr, A. (1985). *Plant Physiol.* **77**, 215–221.
Sheat, D. E. G., Fletcher, B. H. and Street, H. E. (1959). *New Phytol.* **58**, 124–141.
Stribley, D. P. (1987). In *Ecophysiology of Mycorrhizal Plants* (G. R. Safir, ed.), pp. 59–70. CRC Press, Boca Raton, FL.
Strullu, D. G. and Romand, C. (1986). *C. R. Acad. Sci. Paris III* **303**, 245–250.
Strullu, D. G. and Romand, C. (1987). *C. R. Acad. Sci. Paris III* **305**, 15–19.
Tepfer, D. (1984). *Cell* **37**, 959–967.
Tepfer, D. (1989). In *Plant Molecular Biology* (D. von Wettstein and N.-H. Chua, eds), pp. 294–342. Plenum Press, New York.
Tepfer, D. and Tempé, J. (1981). *C. R. Acad. Sci. Paris III* **292**, 153–156.
Tommerup, I. C. and Kidby, D. K. (1980). *Appl. Environ. Microbiol.* **39**, 1111–1119.
Torrey, J. G. (1954). *Plant Physiol.* **29**, 279–287.
Torrey, J. G. (1986). In *New Root Formation in Plants and Cuttings* (M. B. Jackson, ed.), pp. 31–66. Kluwer Academic Publishers, Dordrecht, The Netherlands.
White, P. R. (1943). *A Handbook of Plant Tissue Culture.* J. Cattel, Lancaster, PA.
White, P. R. (1963). *The Cultivation of Animal and Plant Cells*, 2nd edn. Ronald Press, New York.
Zeadan, S. M. and Macleod, R. D. (1984). *Ann Bot.* **54**, 77–85.

# 7

# Cytology, Histochemistry and Immunocytochemistry as Tools for Studying Structure and Function in Endomycorrhiza

S. GIANINAZZI and V. GIANINAZZI-PEARSON

*Laboratoire de Phytoparasitologie, INRA-CNRS, Station de Génétique et d'Amélioration des Plantes, INRA, B.V. 1540, 21034 Dijon Cédex, France*

## I. Introduction

In mycorrhiza, as in other plant–microbe associations, the interactions between host and mycosymbiont are determined by a permanent dia-

METHODS IN MICROBIOLOGY
VOLUME 24   ISBN 0-12-521524-X

Copyright © 1992 by Academic Press Limited
All rights of reproduction in any form reserved

logue between the genomes of the two associates (Gianinazzi-Pearson and Gianinazzi, 1989a). This dialogue leads to modifications in both fungal and root tissues which culminate in morphological integration and functional compatibility, both of which are fundamental prerequisites for a successful outcome of the symbiosis (Gianinazzi-Pearson and Gianinazzi, 1988). Morphological integration reaches its extreme in endomycorrhiza. Here, the fungus penetrates the plant cell wall and develops in intimate contact with the plant protoplast, while the host, in response to this intrusion, modifies its behaviour to both control and exploit fungal activity to its own advantage (Gianinazzi and Gianinazzi-Pearson, 1990).

In this chapter we describe some microscopical approaches for analysing the structure of endomycorrhizal infections, identifying modifications in protein expression (enzyme activities) and localizing specific molecules at both the tissue and cellular level. Techniques have been chosen to illustrate how they can not only increase our understanding of how endomycorrhizal symbionts interact with each other, but also furnish potential tools for diagnosing the functional state of the symbiosis.

## II.    Basic methods for observing endomycorrhizal associations using the light microscope

### A.    Infection distribution and structure

Non-destructive observations can be made of vesicular-arbuscular endomycorrhizal infections in root pieces using ultraviolet light (Ames et al., 1982). Although this can be useful for selecting materials for biochemical analysis or electron microscope preparation, the technique is limited to young root tissues in which only living arbuscules can usually be detected. Destructive techniques using stains such as trypan blue or chlorazole black E, applied after preliminary treatment of root tissues in potassium hydroxide (Phillips and Hayman, 1970; Brundrett et al., 1983), reveal all fungal structures: intracellular coils in the case of ericoid and orchid endomycorrhiza, or vesicles and arbuscules as in vesicular-arbuscular mycorrhiza (Figs 1A–C). These preparations are semi-permanent and the extent of an endomycorrhizal infection within a root system, as well as the impact of factors upon it, can be monitored using methods such as those described by Phillips and Hayman (1970), Giovannetti and Mosse (1980), Trouvelot et al. (1986) or McGonigle et al. (1990).

These simple procedures also give information about the influence of the host plant upon fungal morphology. For example, in most herbaceous plants, vesicular-arbuscular mycorrhizal fungi form arbuscules by repeated branching of a single hypha which arises from a much developed intercellular hyphal network (Fig. 1C). In some species (e.g. *Gentiana*, and some woody plants), there is no intercellular mycelium and arbuscules are more difficult to distinguish because they develop directly as fine hyphal branches from a system of large coiled intracellular hyphae which spread from one cortical cell to another (Fig. 1D).

Cross-sections of fresh or resin-embedded material, stained with trypan or toluidine blue, illustrate how the fungal infection is controlled by the host tissues. In fact, proliferation of vesicular-arbuscular mycorrhizal fungi, with the formation of arbuscules and vesicles, only occurs in the inner root cortex and not in epidermal, hypodermal or exodermal tissues where hyphae are large and undivided, sometimes forming coils (Fig. 1E).

## B. Physiological activity

To understand what the endomycorrhizal infection means in terms of function, it is necessary to use staining techniques which localize a given type of compound/molecule (non-vital staining) or an enzyme activity (vital staining), and which reflect a physiological function in the infection.

## 1. Fungal tissue

Whilst destructive staining techniques yield information about the total amount of root tissue colonized by an endomycorrhizal fungus, they give no idea how much of the infection is physiologically active during plant development or under different environmental conditions.

Of the various non-vital stains that can be used at the light microscope level to detect compounds in endomycorrhizal mycelium both inside and outside roots (Jensen, 1962; Nemec, 1981), those staining lipids or polyphosphate, for example, give useful information concerning the physiology of the fungal symbiont. The presence of lipids indicates that the fungus is actively accumulating carbon obtained from the plant and that of polyphosphate, which is the form in which the fungus accumulates phosphate before releasing it to the plant, suggests that the fungal mycelium is transferring phosphate to the root tissues (Callow *et al.*, 1978).

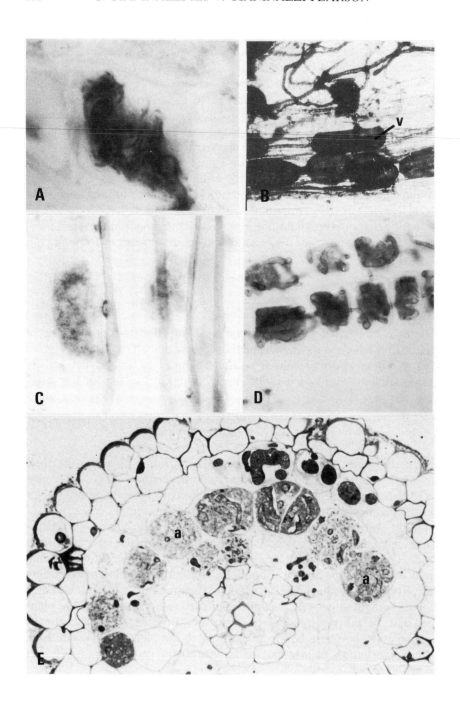

Vital stains can be used to determine the localization of enzyme activities (Pearse, 1968). Succinate dehydrogenase (SDH) is a mitochondrial enzyme involved in the Kreb's cycle (MacDonald and Lewis, 1978) and staining for its activity provides a method for estimating the amount of living fungal mycelium within roots. In cross-sections of roots infected by vesicular-arbuscular mycorrhiza, live arbuscules can be easily distinguished from senescent ones (Fig. 2A) and in root pieces, estimations of SDH-stained fungal mycelium, as compared to that stained by trypan blue, gives the proportion of the infection that is metabolically active. Using this technique, it has been possible to show rapid effects of pesticides on fungal activity and different studies have shown that, apart from the very early stages of mycorrhizal development (up to 6–8 weeks), trypan blue staining in fact overestimates active infection as revealed by SDH activity (Ocampo *et al.*, 1982; Kough *et al.*, 1987; Smith *et al.*, 1990; Smith and Gianinazzi-Pearson, 1990; Abdel-Fattah, 1991). Vital staining can also be used to determine the distribution of activity of alkaline phosphatase (Fig. 2B), an enzyme considered to be involved in fungal phosphate metabolism (Gianinazzi-Pearson and Gianinazzi, 1978, 1983, 1988). Light microscope studies of this enzyme activity suggest that it is inducible; it is not often detectable in hyphae growing out from spores, where its activity is limited to a subapical zone (Fig. 2C). Following infection of a host root it only appears in mycelium after the initial stages of endomycorrhiza formation (Tisserant, 1990).

These and other fungal enzymes (for example, NADH disphorase or esterases) are also useful markers for analysing the proportion of metabolically active external mycelium developing at the root surface or in soil (see Sylvia, Chapter 3, this volume).

## 2. *Plant tissue*

Distinct morphological changes in plant tissue, following endomycorrhizal infection, are not evident at the light microscope level. However, by vital staining of root sections, it is possible to show that peroxidase activity associated with epidermal and hypodermal cells is enhanced in

---

**Fig. 1.** Light microscope observations of endomycorrhizal roots after trypan (A–D) or toluidine blue (E) staining. (A) Hyphal coils of *Hymenoscyphus ericae* in *Vaccinium myrtillus* ($\times$ 530). (B) Vesicles (v) of *Glomus fasciculatum* in wheat ($\times$ 220). (C) Arbuscule of *G. mosseae* in onion ($\times$ 310). (D) Hyphal coils of *G. mosseae* in gentiana ($\times$ 340). (E) Cross-section of resin-embedded gentiana root infected by *G. mosseae* with arbuscules (a) in parenchymal cortex ($\times$ 860).

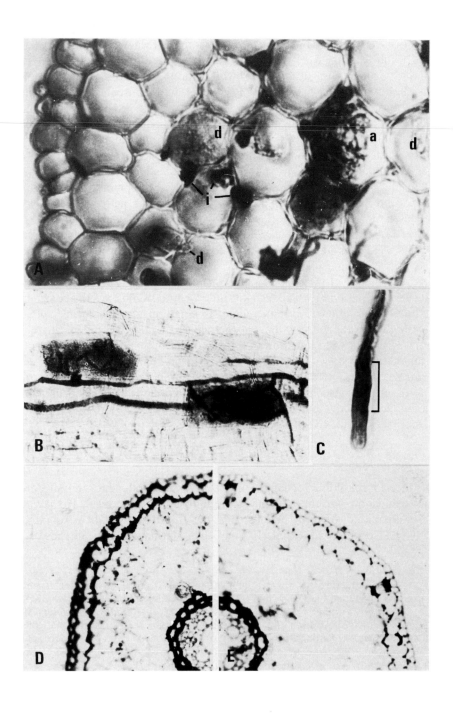

endomycorrhizal roots (Fig. 2D). Since peroxidase is considered to be associated with defence reactions in plant tissues (Kuc, 1982), this may be indicative of modifications contributing to a greater resistance of endomycorrhizal roots to certain pathogens (Bgyaraj, 1984). The lack of peroxidase activity in the infected parenchymal cells indicates that the enzyme is not induced here by the symbiotic fungus (see also Spanu and Bonfante-Fasolo, 1988). This confirms other observations suggesting that endomycorrhizal infection either fails to elicit, or elicits only weakly, the defence reactions of infected host cells (Gianinazzi and Gianinazzi-Pearson, 1990; see below and Bonfante-Fasolo and Spanu, Chapter 8, this volume).

### III. Ultracytochemical and ultracytoenzymological approaches

The morphological modifications in the endomycorrhizal symbionts are accompanied by changes in their cellular and macromolecular organization. These are observed using electron microscope techniques which involve fixing cell structures, embedding tissues in a support resin and, for simple structural observations, the staining of thin sections with contrasting agents. More specific information concerning the macromolecular nature of and physiological activities associated with cell components can be obtained using relatively simple techniques, for example, to detect structural polysaccharides or proteins, or to reveal enzyme activities (see Hall, 1978; Aldrich and Todd, 1986).

### A. Structural components

The most obvious influence of the plant on the ultrastructural organization of the fungal symbiont is related to modifications in the fungal wall (see Bonfante-Fasolo and Spanu, Chapter 8, this volume). These are particularly evident in ericoid endomycorrhizal fungi as they grow

---

**Fig. 2.** Light microscope observations of histochemically localized enzyme activities in vesicular-arbuscular mycorrhizal roots. (A) Succinate dehydrogenase staining associated with living intercellular (i) and arbuscular (a) hyphae but not with dead (d) hyphae (*G. mosseae* × *Allium cepa*) (× 630) (Smith and Gianinazzi-Pearson, unpublished data). (B) Alkaline phosphatase activity throughout intercellular and arbuscular hyphae (*G. fasciculatum/Platanus acerifolia*) (× 375), and (C) limited to the tip of a germ tube (*Gigaspora margarita*) (× 550) (courtesy of B. Tisserant, INRA, Dijon). (D) Enhanced peroxidase activity in outer cells of a vesicular-arbuscular mycorrhizal root (*G. mosseae/A. cepa*) as compared to (E), a non-mycorrhizal root (× 120).

around and within host roots. The latter induce extraradical hyphae to produce an abundant fibrillar sheath which ultracytochemical tests (periodic acid-thiocarbohydrozide silver proteinate (PATAg), Swift and phosphotungstic acid (PTA) reactions) show to consist of polysaccharides, proteins and glycoproteins (Fig. 3A,B). This fibril production by ericoid fungi appears to be essential for successful infection, probably by ensuring adhesion of the external hyphae to the surface of the host root (Gianinazzi-Pearson et al., 1986). However, when infection has occurred the host plant exerts a controlling influence over fibril synthesis by the intracellular hyphae so that the fibrillar sheath is no longer produced within living host cells (Bonfante-Fasolo et al., 1984).

Modifications in fungal wall metabolism are less dramatic in vesicular-arbuscular mycorrhiza; they are characterized by a gradual thinning out of hyphal walls which is accompanied by a simplification in their macromolecular structure (Gianinazzi-Pearson et al., 1981; Bonfante-Fasolo and Grippiolo, 1982; Jacquelinet-Jeanmougin et al., 1987). Extraradical mycelium and hyphae in epidermal or hypodermal tissues develop a multi-layered wall with a fibrillar texture. Polysaccharide distribution (PATAg test) varies from one wall layer to another, whilst proteins (Swift reaction) are more evenly distributed throughout the fungal wall. PATAg and Swift reactions also indicate that once the vesicular-arbuscular mycorrhizal fungus infects the cortical cells of the host, the walls formed by the arbuscular hyphae remain monolayered, (Figs 3C–F) and are amorphous in nature. This simplified wall structure, which is similar to that found in hyphal tips, may lead to an increased plasticity of the vesicular-arbuscular mycorrhizal fungal wall, which in turn contributes to the change in fungal morphogenesis (arbuscule formation) in the parenchymal host cells.

Numerous structural modifications occur in the host cells with infection (increases in cytoplasmic volume, numbers of organelles, extent of membrane systems) and the sequence of cytological events involved is remarkably similar in different types of endomycorrhiza. These have already been described in several reviews (see for example Scannerini and Bonfante-Fasolo, 1983; Gianinazzi-Pearson, 1984) and for the purpose of the present chapter, only those occurring at the symbiotic intracellular host–fungus interface will be discussed. As infection occurs, host membrane extending from the plasmalemma to surround the

---

**Fig. 3.** Ultracytochemical localization of PATAg-positive polysaccharides (A,C,D) and proteins (Swift reaction) (B,E,F). (A,B) Fibrillar sheath (fs) of *H. ericae* (A, × 30000; B, × 60000). (C) PATAg-positive polysaccharidic wall material (wm) surrounding infecting hypha in vesicular-arbuscular mycorrhiza (*G.*

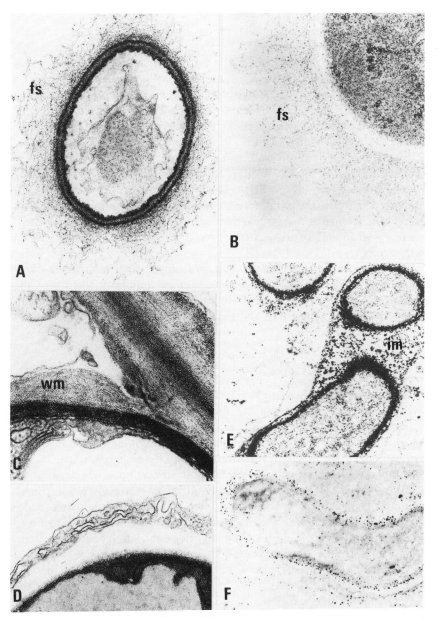

*fasciculatum/O. umbellatum*) ($\times$ 60 000). (D) $H_2O_2$ control of (C). (E) Scattered proteinaceous fibrillar material in the interfacial matrix (im) around fine arbuscule branches in vesicular-arbuscular mycorrhiza (*G. mosseae/A. cepa*) ($\times$ 20 000). (F) Protease control of (E).

invading fungus deposits wall-like matrix material against the fungal hypha (Dexheimer et al., 1979; Scannerini and Bonfante-Fasolo, 1979; Bonfante-Fasolo and Gianinazzi-Pearson, 1982; Serrigny, 1985). This material is continuous with and is structurally organized in the same way as primary cell wall (Fig. 3C). The similarity between the two is confirmed by chemical (dimethylsulphoxide (DMSO), ethylenediamino-tetroacetic acid (EDTA)) or enzymic (cellulase, pectinase, protease) extractions associated with reactions for polysaccharide or protein localization (PATAg, Swift) (Table I). As the intracellular hyphae grow, however, the amount of wall material deposited in the host–fungus interface decreases so that it often contains only scattered PATAg- and Swift-positive fibrils (Fig. 3E). This interference with the cell-wall building activity of the symbiotic host membrane is not associated with changes in its structure, cytochemical nature (PTA, PATAg and Swift reactivity—Table I) or associated enzyme activities (see below) (Gianinazzi-Pearson et al., 1984).

The reciprocal interference of host and fungus with each other's wall metabolism in endomycorrhiza is no doubt an important feature of the symbiosis. It is likely to contribute to compatibility between the symbionts by facilitating nutrient and signal exchange between them.


## B.  Morphofunctional aspects

Molecules having a functional significance in the physiology of endomycorrhizal associations can be localized at the cellular level using cytochemical or cytoenzymological procedures. The localization of an enzyme activity by electron microscopy requires immobilization of the enzyme and conservation of its activity during tissue preparation.

In the case of phosphatases, for example, detection of their activity is based on the *in situ* capture and precipitation by heavy metal ions (lead, cerium) of phosphate ions cleaved from substrates. In ericoid endomycorrhiza, such an approach has been used to show that intense acid phosphatase activity (inhibited by fluoride and molybdate) becomes associated with the fibrillar sheath of hyphae as they grow close to a root (Fig. 4A). Possible explanations for this are that enzyme activity or synthesis is stimulated either by the host tissues themselves or by the low phosphate levels existing close to the root surface. When the ericoid fungus develops within living host cells, however, very little or no acid phosphatase activity can be detected on the fungal surface (Fig. 4B) (Gianinazzi-Pearson et al., 1986; Lemoine et al., 1990). As in the case of fibril synthesis by hyphae, the host plant exerts a specific control over

**TABLE I**

Cytochemistry of the host–fungus interface in vesicular-arbuscular mycorrhiza (Dexheimer *et al.*, 1979; Scannerini and Bonfante-Fasolo, 1979; Gianinazzi-Pearson *et al.*, 1981; Bonfante-Fasolo *et al.*, 1981).

| Cytochemical test | Polysaccharides | | | | | Proteins | | Phosphotungstic acid |
|---|---|---|---|---|---|---|---|---|
| | PATAg | DMSO | EDTA | Cellulase | Pectinase | Swift | Protease | |
| *Interface* | | | | | | | | |
| Fungal wall | ++[a] | – | – | – | – | ++ | + | + |
| Matrix material | | | | | | | | |
| Living fungus | + | ++ | +/– | ++ | ++ | ++ | ++ | – |
| Dead fungus | + | ++ | + | ++ | ++ | ++ | ++ | – |
| Host membrane | + | – | – | – | – | + | + | ++ |
| *Cell wall* | | | | | | | | |
| Primary wall matrix | + | ++ | +/– | ++ | ++ | ++ | ++ | – |
| Middle lamella | + | +/– | ++ | + | + | +/– | +/– | – |

[a] – unaffected; +/– weakly affected; + positive reaction; ++ strong positive reaction.

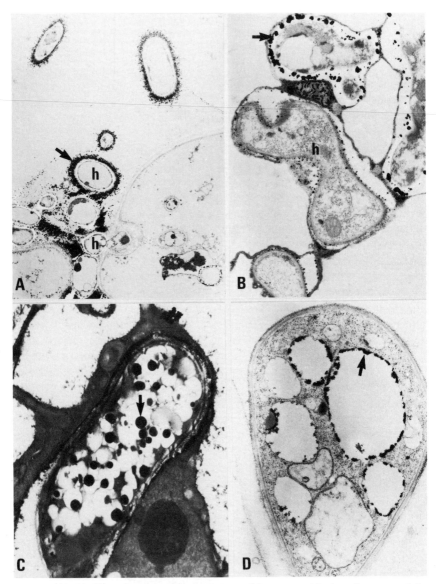

**Fig. 4.** (A,B) Ultracytochemical detection of intense acid phosphatase activity (arrow) associated with the hyphal surface of *H. ericae* (h) growing over a root of *E. tetralix* (A) and less or no activity within host cells (B), (A, × 4000; B, × 15000). (C) Electron-dense polyphosphate granules (arrow) within vacuoles of an intracellular hypha in vesicular-arbuscular mycorrhiza (*G. fasciculatum*/

the physiological activity of the fungal symbiont. Techniques to determine whether such host control affects enzyme protein expression or, more directly, synthesis of the fungal molecule itself, are discussed below.

Using ultracytochemical and ultracytoenzymological techniques, it has also been possible to demonstrate that in vesicular-arbuscular mycorrhiza, polyphosphate accumulation and alkaline phosphatase activity, observed using the light microscope, both occur in the vacuoles of symbiotic fungal hyphae. This vacuolar localization of polyphosphate (seen as osmiophilic granules in electron microscope preparations — Figure 4C) has led to the suggestion that the fungal vacuoles play an essential role in the active transport of phosphate along hyphae (Tinker, 1978; Gianinazzi-Pearson and Gianinazzi, 1983, 1988). The coincidental presence of an alkaline phosphatase activity in these vacuoles and its frequent association with the fungal tonoplast (Fig. 4D) may reflect an eventual involvement of this enzyme also in the phosphate transport mechanism.

Changes can be detected in the physiological activity of root cortical cells as the intracellular hyphae of an endomycorrhizal fungus comes into close contact with the host protoplast. Two types of membrane-linked enzymes that have been investigated in endomycorrhiza are: neutral phosphatases, which are considered to be associated with sites of wall-precursor production, and energy-generating ATPases (Marx et al. 1982; Jeanmaire et al., 1985; Gianinazzi-Pearson et al., 1984, 1991). In vesicular-arbuscular mycorrhiza-forming plant species, where these enzymes have been more closely studied, neither system is particularly active along the peripheral plasmalemma in differentiated parenchymal cortical cells. This is to be expected since wall extension processes in these cells have terminated and they are not directly concerned with energy-requiring absorption processes. In contrast, the activity of both types of enzyme is intense along the perisymbiotic host membrane surrounding the intracellular hyphae of an endomycorrhizal fungus developing in these root cells.

Considerable ATPase activity can be detected cytochemically along the fungal plasmalemma and the host membrane, and in the interfacial matrix associated with living hyphae in different types of endomycorrhiza (Marx et al., 1982; Gianinazzi-Pearson et al., 1984; Serrigny and

---

R. idaeus) (× 9000) (courtesy of D. Morandi, INRA, Dijon). (D) Electron microscope visualization of alkaline phosphatase activity (arrow) along the tonoplast of an intercellular hypha in vesicular-arbuscular mycorrhiza (G. fasciculatum/P. acerifolia) (× 20 000) (courtesy of B. Tisserant, INRA, Dijon).

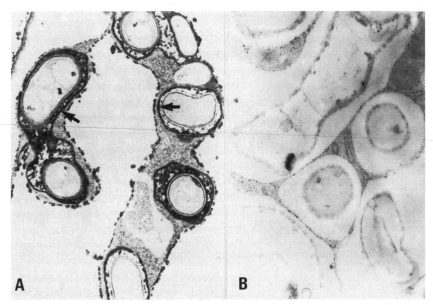

**Fig. 5.** (A) Ultracytochemical localization of host membrane-bound ATPase activity (arrow) around arbuscule branches in a vesicular-arbuscular mycorrhizal infection (*G. mosseae/A. cepa*) (× 15000). (B) Vanadate inhibition of activity (× 17000).

Dexheimer, 1985) (Fig. 5). This enzyme activity disappears with senescence of the intracellular hyphae. In a critical study comparing the effect of ATPase and non-specific phosphatase inhibitors in vesicular-arbuscular mycorrhiza, Gianinazzi-Pearson *et al.* (1991) concluded that plasmalemma $H^+$-ATPase is most probably a component of the ATPases activated on the host and fungal membranes. The simplified interface formed by endomycorrhizal fungi and host cells is considered to be the site of a two-way traffic of nutrients in which photosynthetic compounds pass from plant to fungus whilst nutrients, such as phosphate, pass from the fungus to the host cells. Active absorption (of either C or P) by the fungal or plant membranes at the interface requires energy and the presence of $H^+$-ATPase indicates that an energy-generating enzyme system is present.

Neutral phosphatase activity (TPPase, IDPase) also occurs around the living fungus where little or no host wall material is deposited in the interfacial matrix (see above) (Gianinazzi-Pearson *et al.*, 1984; Jeanmaire *et al.*, 1985; Serrigny and Dexheimer, 1985). This suggests that although the host membrane appears to have lost its wall-building activity, it in fact maintains its production of cell wall precursors in reaction to fungal development. It is their assembly into organized wall material that is somehow impeded. This interpretation is strengthened by the immunocytochemical localization of probable cell wall precursors in the host–fungus interface (see below).

The formation of a living interface, where fungal and plant wall structures are reduced to a minimum and in which opposed symbiont membranes possess an energy-generating enzyme system ($H^+$-ATPase) for active transmembrane transport, must greatly facilitate the bidirectional movement of solutes between the symbionts. This situation contrasts sharply with that observed in biotrophic pathogen interactions where there is inhibition of host membrane ATPase activity in the presence of a fungus and only one-way nutrient transport towards the pathogen (Gay, 1984; Woods and Gay, 1987).

## IV. Immunocytochemical characterization of structural and functional cell components

A number of commercially available probes such as fluorescent, gold- or enzyme-conjugated lectins and enzymes can be used to determine the presence or absence of non-specific cell components in the plant and fungus during endomycorrhizal interactions (Jabaji-Hare et al., 1990; see Bonfante-Fasolo and Spanu, Chapter 8, this volume). However, if the production and localization of more specific molecules is to be studied, it is necessary to use an immunological approach. Monoclonal or polyclonal antibodies are raised either directly against a purified molecule or against a population of cell components, the individual active antigens of which can be subsequently identified (Wang, 1986). Once antibodies have been characterized, corresponding antigens can be localized immunocytochemically at the tissue or cellular level by light or electron microscopy. Different techniques exist for tissue preparation and cellular localization of antigens (see Wang, 1986; Hayat, 1989; and Mayer et al., Chapter 14, Methods in Microbiology Vol. 23); each has its advantages and limitations which need to be considered when being applied to different material. Those outlined here have proved most reliable for preserving both cell structure in endomycorrhizal tissues and antigenicity to various homologous or heterologous antibody probes. The examples which follow illustrate how, using indirect immunogold labelling with light and electron microscopy, it is possible not only to identify modifications in the molecular configuration of endomycorrhizal symbionts, but also to define relationships between protein synthesis and protein expression (enzyme activity) during cellular interactions.

## A. Detection of fungal components (homologous antibodies)

Serological detection with light or electron microscopy represents a potential tool for differentiating between individual species or strains of endomycorrhizal fungi associated with roots, the mycelia of which are otherwise virtually indistinguishable. It has only been applied in a few studies of ericoid and vesicular-arbuscular mycorrhizal fungi, using antibodies raised against antigens from either hyphae or spores (Kough *et al.*, 1983; Mueller *et al.*, 1986; Ravolanirina, 1990). Although such antibody preparations sometimes cross-react with heterologous antigens in hyphae of different endomycorrhizal species (probably due to non-proteinaceous immunogens in original extracts—unpubl. data), they can give some insight into fungal behaviour during development of the symbiotic state. For example, the distribution of hyphae of a vesicular-arbuscular mycorrhizal fungus within infected roots can be clearly shown by light microscope immunolocalization using an homologous polyclonal antibody preparation raised against the soluble fraction of spore extracts of this same fungus (Fig. 6A). Corresponding antigens are only present within living hyphae where they are mainly associated with the cyto-plasmic compartment (Fig. 6E), so that positive reactions observed in the light microscope reflect the presence of living symbiotic mycelium. Furthermore, by increasing the dilution of the antibodies a differential intensity in the immunolabelling of hyphae can be obtained: whilst labelling remains intense over external hyphae and coils in outer cells, it becomes extremely weak over the intercellular hyphae and arbuscules in parenchymal cortical cells (Fig. 6C). This reduction in detectable anti-genicity as the mycelium develops within the root tissues suggests that molecular changes are occurring in the cell contents of the fungus during its symbiotic association with the host plant.

There are likewise very few examples where immunocytochemistry has been applied to studies of the synthesis and localization of specific

---

**Fig. 6.** Light (A–D) (silver enhancement technique) and electron (E,F) micro-scope immunolocalization of antigens reacting with polyclonal antibodies raised against soluble spore extracts of *G. margarita* in vesicular-arbuscular mycorrhizal roots (*G. margarita/V. vinifera*). (A) External (e), intercellular (i) and arbuscu-lar (a) hyphae are immunolabelled with antibodies diluted to 1:100 ($\times$ 135). (B) Pre-immune serum control of (A). (C) Differential immunolabelling of external hyphae (e), hyphal coils (hc) and arbuscules (a) with a 1:10000 antibody dilution ($\times$ 500). (D) Pre-immune serum control of (C). (E,F) Immunogold labelling of antigens (1:10000 antibody dilution) associated with the cytoplasm (c) of living (E) but not dead (F) hyphae ($\times$ 30000) (V. Gianinazzi-Pearson, F. Ravolanirina and S. Gianinazzi, unpubl. data).

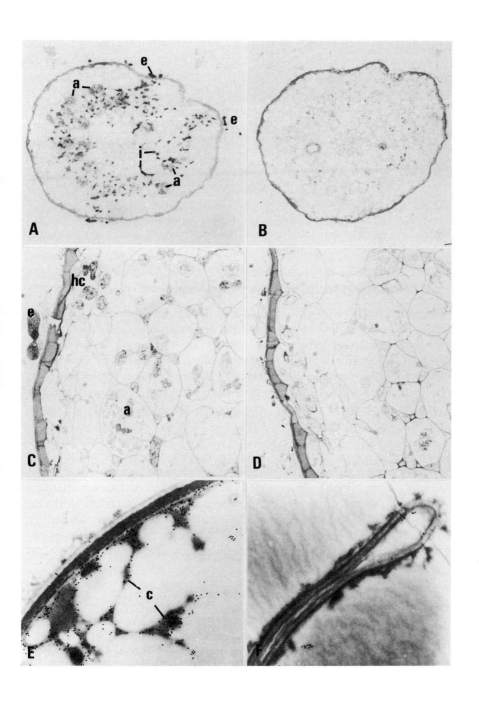

fungal molecules in endomycorrhizal associations. The usefulness of this type of approach is illustrated by investigations of the molecular basis of variations in wall-bound acid phosphatase activity observed in ericoid fungi growing with host roots (see above). After immunolabelling with polyclonal antibodies raised against the purified enzyme (a mannose glycoprotein, Straker et al., 1989) light and electron microscope examination clearly demonstrates that the association of the phosphatase molecule with the wall and fibrillar sheath of mycelium growing over the host root (Gianinazzi-Pearson and Gianinazzi, 1989b; Lemoine et al., 1990) (Fig. 7A), coincides with the localization of intense phosphatase activity on the hyphal surface. Phosphatase antigens can, however, also be detected in the walls of hyphae developing within living host cells (Fig. 7B), where there is little or no enzyme activity. These observations suggest that the host influence on fungal metabolism is through a direct inhibition of the wall acid phosphatase activity rather than through a repression of enzyme synthesis itself.

## B.    Detection of plant components (heterologous antibodies)

Immunocytochemistry also provides a means of probing endomycorrhizal tissues for plant molecules which are characteristically produced in other plant/micro-organism interactions, but which may be synthesized in such low amounts in endomycorrhiza that they cannot be detected in extracts by biochemical immunoassays. Corresponding antibodies must be well characterized in order to avoid any misinterpretation, and the following examples illustrate the interest of using such heterologous probes.

Several monoclonal antibodies have been obtained by immunization with peribacteroid membrane fractions of pea nodules, some of which react to antigen epitopes in the plant plasmalemma (Brewin et al., 1985; Bradley et al., 1988). The use of such antibodies to localize common antigens in the plant-derived membrane surrounding an endomycorrhizal fungus can give information about its nature and indicate any similarities with the symbiotic membrane around bacteroids in nodules. For example, an oligosaccharide antigen recognized by the antibody MAC 206

**Fig. 7.**    Electron microscope immunogold localization (arrows). (A,B) Wall- and fibril-associated acid phosphatase (1/50 polyclonal antibody dilution) in external (A) and internal (B) hyphae of an ericoid endomycorrhiza (*H. ericae/V. corymbosum*) (A, × 30 000; B, × 22 000) (courtesy of M. C. Lemoine, INRA, Dijon). (C) $b_1$ protein antigen around a living vesicular-arbuscular mycorrhizal

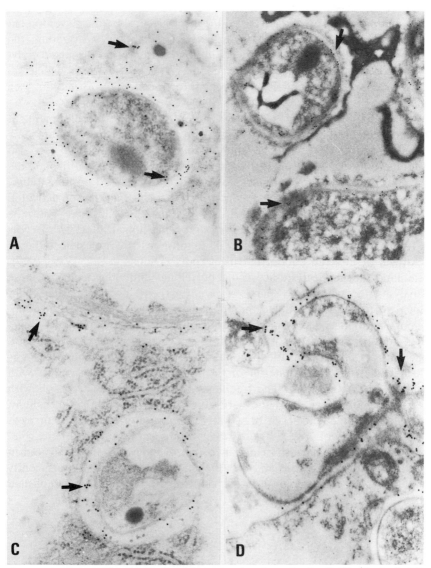

fungus in host cells detected using monoclonal antibody (1:50 dilution) (*G. mosseae/N. tabacum*) ($\times$ 35 000). (D) An oligosaccharide epitope reacting with a monoclonal antibody (AFRC MAC 206) (1:100) on the host-derived membrane and matrix material in the arbuscule interface in vesicular-arbuscular mycorrhiza (*G. intraradices/P. sativum*) ($\times$ 40 000).

can be detected along the host membrane and in the interfacial matrix around fine arbuscule branches, where it is more abundant than on the peripheral plasmalemma (Gianinazzi-Pearson et al., 1991) (Fig. 6D). This antigen may be part of the cell coat or glycocalyx, the development of which increases along the plant-derived membrane surrounding the endomycorrhizal fungus, and it could contribute to the dispersed polysaccharidic material characteristic of the interfacial matrix, which accumulates due to fungal interference with cell wall formation (see above).

Investigations into host processes controlling the development of endomycorrhizal fungi indicate that plant defence mechanisms are only weakly activated (Gianinazzi and Gianinazzi-Pearson, 1990; see also Bonfante-Fasolo and Spanu, Chapter 8, this volume). In particular, the synthesis of plant molecules such as pathogenesis-related proteins (chitinase, glucanase, $b_1$), which are considered to be markers of activated defence mechanisms (Gianinazzi, 1984; Carr and Klessing, 1989), is stimulated during host control over pathogenic fungal infection of roots (Tahiri-Alaoui et al., 1990), yet these proteins are not detectable in extracts of infected endomycorrhizal roots (Dumas et al., 1989, 1990). Immunocytochemical studies using polyclonal antibodies confirm these analyses for chitinase (Spanu et al., 1989). In contrast, when light microscopy is used to detect immunolabelling of $b_1$-protein with monoclonal antibodies, corresponding antigen in host cells containing living vesicular-arbuscular mycorrhizal fungus is clearly revealed (Gianinazzi-Pearson et al., 1988). Electron microscopy shows that the protein is localized within host material of the interfacial matrix in contact with living hyphae (Fig. 6C). The frequency of antigenic sites decreases with arbuscule senescence and they are absent from intercellular mycelium, adjacent uninfected cells or fungal mycelium in pure culture (Gianinazzi, 1991; and unpubl. res.). These observations suggest that defence mechanisms to vesicular-arbuscular mycorrhizal fungi are only activated in individually infected cells and that their level of expression is sufficiently weak to be compatible with the biotrophic relationship established between the symbionts.

## V.  Conclusion

Cytological, histochemical and immunocytochemical techniques provide powerful tools for studying the structure and function of symbionts and their interactions in endomycorrhiza. Knowledge of the physical and chemical nature of cellular and subcellular structures permitting functional compatability in endomycorrhizal associations has been considerably improved and subcellular localization of enzyme activities has given some insight into their physiological significance for the symbiotic condition. Furthermore, by combining immunolocalization with ultracytoenzymology it is possible to understand how the synthesis and the expression of the molecules involved are regulated.

Little is known about endomycorrhiza-specific gene products and no probes are presently available for the plant molecules involved. Heterologous antibodies from other plant/micro-organism systems are proving useful for analysing molecular modifications at the cellular level. The application of *in situ* hybridization using RNA or DNA probes to endomycorrhizal research could provide more direct information about genes and gene expression in the symbionts. However, although some genes related to the endomycorrhiza symbiosis have been identified (Duc *et al.*, 1989), none have as yet been cloned. Until they are available, heterologous probes obtained from other plant–microbe interactions may be applied to analyse the endomycorrhizal symbiosis.

## VI.  Protocols

### A.  Cytochemical and histochemical techniques for light microscopy

*1.  General staining* (Phillips and Hayman, 1970; Brundrett *et al.*, 1983)

For fresh sectioned (roots) or unsectioned (hyphae) material, or whole roots.

1. For whole roots, digest first in 10% KOH, 1 h, 90 °C, rinse in $H_2O$ followed by 5 min in 5% lactic acid.
2. Stain 15 min, 90 °C in 0.1% trypan blue or 0.4% chlorazole black E in lactophenol or lactoglycerol.
3. Rinse and mount on slides in glycerol.

Highly pigmented or suberized roots can be cleared by either adding a few drops of $H_2O_2$ to the KOH after digestion or placing stained roots in diluted bleach (1% active chloride).

*2.  Structural histochemistry* (Jensen, 1962; Pearse, 1968; Nemec, 1981)

Many stains exist for carbohydrates, proteins, lipids, etc. Examples are:

*(a) Neutral lipids.* For fresh material (hyphae) or cleared (10% KOH, 90 °C, 1 h) roots; e.g. Sudan III stains red.

1. Stain in 0.1% trypan blue plus 0.1% Sudan III or IV in lactophenol or lactoglycerol.
2. Rinse and mount on slides in glycerol.

*(b) Polyphosphate.* For fresh sectioned (roots) or unsectioned (hyphae) material, or resin embedded sections; polyphosphate stains pink.

1. Stain in 0.1% toluidine blue, pH 1–2, at room temperature.
2. Rinse and mount on slides in glycerol.

*3. Enzyme histochemistry — vital stains* (Pearse, 1968)

For fresh material.

*(a) Succinate dehydrogenase* (Kough *et al.*, 1987; Smith and Gianinazzi-Pearson, 1990). For root pieces or sections, hyphae; enzyme activity indicated by a dark purple stain.

1. Incubate 4–5 h, or overnight, at room temperature in 50 mM Tris-HCl, pH 7.4, plus 0.5 mM $MgCl_2$, 1 mg $ml^{-1}$ nitro blue tetrazolium (NBT), 0.25 M $Na_2$ succinate. Rinse with $H_2O$.
2. For root pieces, clear 10–15 min in boiling 20% chloral (fume cupboard).
3. Mount on slides in glycerol.

*(b) Alkaline and acid phosphatase* (Gianinazzi-Pearson and Gianinazzi, 1976; Tisserant, 1990). For root pieces, sections or hyphae; enzyme activity indicated by a violet-black stain.

1. Root pieces may require prior digestion in a cellulase–pectinase solution for 1–4 h, depending on plant species.
2. Incubate 4–5 h, or overnight, at room temperature in either 0.05 M Tris-citric acid, pH 8.5–9.2, plus 1 mg $ml^{-1}$ Na $\alpha$-naphthyl acid phosphate, 1 mg $ml^{-1}$ Fast Blue, 0.05% $MgCl_2$, 0.05% $MnCl_2$, or 0.1 M acetate, pH 4–5 plus 1 mg $ml^{-1}$ Na $\alpha$-naphthyl acid phosphate, 1 mg $ml^{-1}$ Fast Garnet, 0.05% $MgCl_2$. Rinse with $H_2O$.
3. For pigmented roots, clear in diluted bleach (1% active chloride).
4. Mount on slides in glycerol.

*(c) Peroxidase.* Root sections; enzyme activity indicated by a blue stain which becomes black on storing.

1. Incubate sections at room temperature in a solution of 1 part $NH_4Cl$, 1 part 5% EDTA, 6 parts saturated benzidine, 1 part 3% $H_2O_2$, prepared just before use. Rinse with $H_2O$.
2. Mount on slides in (lacto)glycerol.

## B.  Ultrastructural, cytochemical and cytoenzymological techniques for electron microscopy

*1.  General preparation of samples and staining of thin sections*

1. Pre-fixation: cut root pieces 3–4 mm long, or for fungal colonies on agar $3 \times 1$ mm thick blocks, and immerse in 2–2.5% glutaraldehyde in 0.1 M cacodylate or PIPES buffer, pH 6.8–7.2, 3 h at room temperature or overnight at $+4\,°C$ (at first, place briefly under vacuum to eliminate air and ensure infiltration). Rinse with the same buffer ($4 \times 10$ min).
2. Post-fixation: 1% osmium tetroxide ($OsO_4$) in 0.1 M cacodylate buffer, pH 6.8–7.2, 1 h at room temperature. Rinse with distilled $H_2O$.
3. Dehydration at room temperature in ethanol or acetone: 30% $\times$ 10 min, 50% $\times$ 20 min, 70% $\times$ 20 min (can be stored at 4 °C at this stage), 95% $\times$ 20 min, 100% $\times$ 30 min (twice) and polypropylene oxide $\times$ 30 min (twice).
4. Infiltration with resin (e.g. Epon 812) at room temperature: 75% propylene oxide–25% resin (3 h), 50% propylene oxide–50% resin (overnight), 25% propylene oxide–75% resin (3 h), pure resin (overnight and 1 h, 37 °C).
5. Transfer specimens to embedding moulds and orientate them. Cure 48 h at 55–60 °C.
6. Cut semi-thin sections (0.5 $\mu$m), dry on a slide and stain with 1% toluidine blue, pH 11 (1 g toluidine blue, 1 g sodium borate, 100 ml $H_2O$): cover sections with a drop of stain, heat gently 0.5–2 min without boiling, rinse with $H_2O$. Check material in light microscope.
7. Collect ultra-thin sections (80–90 nm) on Cu grids.
8. Post-staining of ultra-thin sections:
   (a) Float grids, section-side down, 15–30 min at room temperature on a drop of 3% uranyl acetate in 50% ethanol (Valentines, 1961); shade from direct sunlight. Rinse under a jet of distilled $H_2O$.

(b) Float grids on a drop of lead citrate (Reynolds, 1963) for 10–15 min at room temperature; prevent the formation of lead carbonate as precipitate with a trap for $CO_2$ (NaOH) in the container. Rinse with distilled $H_2O$ and dry on filter paper.

2.  *Cytochemistry* (Dexheimer *et al.*, 1979; Scannerini and Bonfante-Fasolo, 1979)

*(a) Polysaccharides.* Periodic acid–thiocarbohydrazide (TCH)–silver proteinate method: PATAg or Thiéry reaction (Thiéry, 1967) for detection of vicinal glycol groups (1–4 polysaccharides).

1. Mount sections on gold or nickel grids and perform staining at room temperature.
2. Treat sections with 1% aqueous periodic acid (30 min), rinse with distilled $H_2O$.
3. Float the sections on 0.2% TCH in 20% aqueous acetic acid (5 h).
4. Rinse with 20%, 10%, 5%, 1% aqueous acetic acid (5 min × 2) then with distilled $H_2O$.
5. Treat sections with 1% aqueous silver proteinate (30 min) in the dark. Rinse with distilled $H_2O$.

*Controls.* Omit either Step 2 or Step 3, or replace step 2 by 10% $H_2O_2$ (30 min) or include an incubation with 0.01% aqueous sodium borohydride (30 min) between Steps 3 and 4.

*Dimethylsulphoxide (DMSO) extraction of xylose-rich components.* Incubate root samples at room temperature 7–16 h in pure DMSO after pre-fixation, treat to embedding and do PATAg reaction.

*Ethylenediaminotetraacetic acid (EDTA) extraction of galacturonic acid components.* Incubate root samples 7–16 h in 1% aqueous EDTA after pre-fixation, treat to embedding and do PATAg reaction.

*(b) Proteins.* Procedure for detection of proteins rich in sulphydryl groups (Swift, 1968)

1. Mount ultra-thin sections on gold or nickel grids and perform staining at room temperature.
2. Float sections on a silver methenamine solution (90 min) 45 °C in the dark in a covered container. The silver methenamine solution is made up as follows: Solution A, 5 ml 5% silver nitrate + 100 ml 3% hexamethylene tetramine; Solution B, 10 ml 1.44% boric

acid + 100 ml 1.9% borax. Immediately before use, mix 25 ml Solution A, 5 ml Solution B and 25 ml distilled $H_2O$. Rinse with distilled $H_2O$.

*Controls.* Before Step 2 digest with pronase, pH 7.2, 58 units, 4 h, 37 °C, after a preliminary oxidation (30 min) with 10% $H_2O_2$ (or 1% periodic acid).

*(c) Glycoproteins* (Huet *et al.*, 1974). Stain ultrathin sections (5 min) with 10% aqueous PTA at room temperature no further washing, dry sections rapidly on filter paper.

*Control.* Bleach sections with 10% $H_2O_2$ (or 1% periodic acid) (30 min) followed by digestion with pronase, pH 7.2, 58 units, 4 h, 37 °C and perform the PTA staining.

*(d) Plasmalemma* (Roland *et al.*, 1972). Stain sections 2 min with 1% phosphotungstic acid in 10% chromic acid. This is a specific stain for the plasmalemma.

3. *Cytoenzymology* (Gianinazzi *et al.*, 1979: Marx *et al.*, 1982; Jeanmaire *et al.*, 1985)

1. Pre-fixation: cut root pieces 3–4 mm long, of for fungal colonies on agar 3 × 1 mm thick blocks, and immerse 1 h in ice-cold 1.5–2% glutaraldehyde, or 0.1% glutaraldehyde–4% formaldehyde (poorer fixation), in 0.1 M cacodylate buffer, pH 7.2; place *briefly* under vacuum to eliminate air.
2. Rinse in ice-cold 5% saccharose–cacodylate buffer (3 × 30 min), 5% saccharose–distilled $H_2O$ (2 × 30 min) then ice-cold 5% saccharose buffer used for the enzyme reaction (see below).
3. Select material using a binocular microscope (maintain cold in 5% saccharose buffer) and cut into 0.5 mm sections, discarding end portions.
4. Incubate sections overnight at 4 °C, then 30 min–1 h at 37 °C in the following solutions:
   (a) Acid phosphatase: 0.1 M acetate buffer, pH 4, 1.8 mM $PbNO_3$, 5% saccharose, 10 mM $\beta$-glycerophosphate or 4 mM $\alpha$-naphthyl acid phosphate.
   *Controls.* Omit enzyme substrate, or add 20 mM NaF or 150 $\mu$M Na molybdate to the reaction medium.

(b) Alkaline phosphatase: 0.1 M Tris-maleate buffer, pH 8.5, 1.8 mM $PbNO_3$, 5% saccharose, 10 mM $\beta$-glycerophosphate or 4 mM $\alpha$-naphthyl acid phosphate.
*Controls.* Omit enzyme substrate or add 4 mM KCN to the reaction medium.

(c) Neutral phosphatase: 0.1 M Tris-maleate buffer, pH 7.2, 1.8 mM $PbNO_3$, 5% saccharose, 3 mM $MnCl_2$, 1 mg ml$^{-1}$, thiamine pyrophosphate (TPP) or inosine diphosphate (IDP).
*Controls.* Omit enzyme substrate or add 100 $\mu$M molybdate to reaction medium (there are no specific inhibitors for neutral phosphatases).

(d) ATPase: 0.1 M Tris-maleate buffer, pH 7.2, 1.8 mM $PbNO_3$, 5% saccharose, 5.9 mM $MgCl_2$, 1 mg ml$^{-1}$ ATP (Na salt).
*Controls.* Omit enzyme substrate or add 100 $\mu$M of either vanadate or diethylestilbestrol (DES) to the reaction medium.

5. Rinse in ice-cold 5% saccharose–distilled $H_2O$.
6. Post-fix, dehydrate, infiltrate with resin, cut and post-stain sections in 3% uranyl acetate as described in general preparation of samples (Section VI.B.1).

Enzyme activity is indicated by an electron-dense lead phosphate precipitate.

## C.  Immunocytochemistry

*1.  General preparation of samples* (Wells, 1985; Straker *et al.*, 1989; Gianinazzi-Pearson *et al.*, 1990)

1. Fix 3–4 mm root pieces, or $3 \times 1$ mm thick agar blocks, in 2% glutaraldehyde, or 0.1% glutaraldehyde–4% paraformaldehyde (poorer fixation), in 0.1 M cacodylate or PIPES buffer, pH 6.8–7.2, 4 h, +4 °C. Rinse with the same buffer ($4 \times 5$ min).
2. Treat samples with 50 mM $NH_4Cl$ in 10 mM phosphate-buffered saline, 1 h, +4 °C.
3. Dehydrate in ethanol: at 4 °C, 30% $\times$ 30 min, then at −15 °C, 50% $\times$ 60 min,   70% $\times$ 60 min,   90% $\times$ 60 min,   95% $\times$ 60 min, 100% $\times$ 30 min (twice).
4. Infiltrate with acrylic resin at −15 °C (e.g. LR White Medium grade resin, which is less toxic than Lowicryl K4M, + 0.5% benzoin methyl ether): 1 volume resin–1 volume 100% ethanol $\times$ 1 h, 2 volumes resin–1 volume 100% ethanol $\times$ 1–3 h, 3

volumes resin–1 volume 100% ethanol × overnight, pure resin × 8 h (twice) and overnight.
5. Place samples in resin-filled Beem capsules and polymerize in an
aluminium foil-lined box under direct or indirect UV light.
6. For light microscopy, collect semi-thin sections (0.5 $\mu$m) on 1%
gelatin-coated slides; for electron microscopy, collect ultra-thin
sections (85–95 nm) on collodion-carbon coated gold or nickel
grids.


## 2. Indirect immunolabelling

*(a) For light microscopy (Silver enhancement technique).* This technique
has the advantage that, unlike immunofluorescence, it does not require
special microscope equipment.

1. Cover sections on slides (30 min, room temperature) with Tris-buffered saline (TBS) (10 mM Tris, 154 mM NaCl, pH 7.4) plus 1%
BSA and 0.05% Tween 20. Non-specific background labelling may
be reduced by adding 1–5% heat-inactivated normal goat serum
(56 °C, 30 min), or by preceding this step with an incubation in
50 mM glycine/TBS.
2. Incubate with primary antibody (e.g. rabbit IgG) at the appropriate dilution in TBS–BSA–Tween 20, overnight, +4 °C, in a humid
chamber. Wash with TBS–Tween 20.
3. Incubate with 5 nm gold-conjugated goat anti-rabbit IgG (Janssen
Pharmaceutica or Biocell) diluted 1:20 in TBS–BSA–Tween 20,
1 h, room temperature. Wash with TBS then $H_2O$.
4. Intensification of gold signal by the silver enhancement reaction
(kits from Janssen or Biocell) (Danscher and Noorgaard, 1983;
Holgate *et al.*, 1983), 4–8 min, room temperature; control the
reaction under light microscope.
5. Counterstain with 2% Basic Fuchsin, let dry, mount with immersion oil and cover-slip.

Antibody–antigen labelled sites are indicated by a black deposit of
silver grains.


*(b) For electron microscopy.*

1. Place grids in 20 $\mu$l droplets of reagents and perform Steps 1–5 as
for light microscopy except use 10 or 15 nm gold-conjugated goat
anti-rabbit IgG.

2. Post-stain with 2% aqueous uranyl acetate, 10 min, room temperature.

*Controls.* Either omit Step 2 or replace the primary antibody in Step 2 by immunodepleted antiserum or pre-immune serum.
Gold particles localize antibody–antigen labelled sites.

## Acknowledgements

The authors thank J. Lherminier for useful discussions, J. Beurteaux for assistance with the photography, and M. C. Lemoine, B. Tisserant and D. Morandi for access to unpublished results.

## References

Abdel-Fattah, G. M. (1991). *Some ecological and physiological studies on vesicular-arbuscular (VA) mycorrhizal fungi.* Ph.D. Thesis, Mansoura University, Egypt. 202 pp

Aldrich, H. C. and Todd, W. J. (eds) (1986). *Ultrastructure Techniques for Microorganisms.* Plenum Press, New York and London.

Ames, R. N., Ingham, E. R. and Reid, C. P. P. (1982). *Can. J. Microbiol.* **28**, 351–355.

Bagyaraj, J. B. (1984). In *VA Mycorrhiza* (C. L. Powell and J. B. Bagyaraj, eds), pp. 131–154. CRC Press, Boca Raton, FL.

Bonfante-Fasolo, P. and Gianinazzi-Pearson, V. (1982). *New Phytol.* **91**, 691–704.

Bonfante-Fasolo, P. and Grippiolo, R. (1982). *Can. J. Bot.* **60**, 2302–2312.

Bonfante-Fasolo, P., Dexheimer, J., Gianinazzi, S., Gianinazzi-Pearson, V. and Scannerini, S. (1981). *Plant Sci. Lett.* **22**, 13–21.

Bonfante-Fasolo, P., Gianinazzi-Pearson, V. and Martinengo, L. (1984). *New Phytol.* **98**, 329–333.

Bradley, D. J., Wood, E. A., Larkins, A. P., Galfre, G., Butcher, G. W. and Brewin N. J. (1988). *Planta* **173**, 149–160.

Brewin, N. J., Robertson, J. G., Wood, E. A., Wells, B., Larkins, A. P., Galfre, G. and Butcher, G. W. (1985). *EMBO J.* **4**, 605–611.

Brundrett, M. C., Piché, Y. and Peterson R. L. (1983). *Can. J. Bot.* **62**, 2128–2134.

Callow, J. A., Capaccio, L. C. M., Parish, G. and Tinker, P. B. (1978). *New Phytol.* **80**, 125–134.

Carr, J. P. and Klessing, D. F. (1989). In *Genetic Engineering* (J. K. Setlow, ed.), pp. 65–109. Plenum Press, New York and London.

Danscher, G. and Noorgaard, R. J. O. (1983). *J. Histochem. Cytochem.* **31**, 1394–1398.

Dexheimer, J., Gianinazzi, S. and Gianinazzi-Pearson, V. (1979). *Z. Pflanzenphysiol.* **92**, 191–206.

Duc, G., Trouvelot, A., Gianinazzi-Pearson, V. and Gianinazzi, S. (1989). *Plant Sci.* **60**, 215–222.

Dumas, E., Gianinazzi-Pearson, V. and Gianinazzi, S. (1989). *Agric. Ecosyst. Environ.* **29**, 111–114.

Dumas, E., Tahiri-Alaoui, A., Gianinazzi, S. and Gianinazzi-Pearson, V. (1990). In *Endocytobiology IV*, (P. Nardon, V. Gianinazzi-Pearson, A. M. Grenier, L. Margulis and D. C. Smith, eds), pp 153–157. INRA, Paris.

Gay, J. L. (1984). In *Plant Diseases, Infection Damage and Loss*, (R. K. S. Wood and G. J. Jellis, eds), pp. 49–59. Blackwell Scientific, Oxford.

Gianinazzi, S. (1984). In *Plant–Microbe Interactions* (T. Kosuge and E. W. Nester, eds), pp. 321–342. Macmillan, New York.

Gianinazzi, S. (1991). *Agric. Ecosyst. Environ.* **35**, 105–119.

Gianinazzi, S. and Gianinazzi-Pearson, V. (1990). In *Endocytobiology IV* (P. Nardon, V. Gianinazzi-Pearson, A. M. Grenier, L. Margulis and D. C. Smith, eds), pp. 83–90. INRA, Paris.

Gianinazzi, S., Gianinazzi-Pearson, V. and Dexheimer, J. (1979). *New Phytol.* **82**, 127–132.

Gianinazzi-Pearson, V. (1984). In *Genes Involved in Microbe–Plant Interactions* (E. S. Dennis, B. Hohn, Th. Hohn, P. King, I. Schell and D. P. S. Verma, eds), pp. 225–253. Springer-Verlag, Vienna and New York.

Gianinazzi-Pearson, V. and Gianinazzi, S. (1976). *Physiol. Veg.* **14**, 833–841.

Gianinazzi-Pearson, V. and Gianinazzi, S. (1978). *Physiol. Plant Pathol.* **12**, 45–53.

Gianinazzi-Pearson, V. and Gianinazzi, S. (1983). *Plant Soil* **71**, 197–209.

Gianinazzi-Pearson, V. and Gianinazzi, S. (1988). In *Cell to Cell Signals in Plant, Animal and Microbial Symbiosis* (S. Scannerini, D. C. Smith, P. Bonfante-Fasolo and V. Gianinazzi-Pearson, eds), NATO ASI Series, Vol. H17, pp. 73–84. Springer-Verlag, Berlin.

Gianinazzi-Pearson, V. and Gianinazzi, S. (1989a). *Génome* **31**, 336–341.

Gianinazzi-Pearson, V. and Gianinazzi, S. (1989b). In *Nitrogen, Phosphorus and Sulphur Utilization by Fungi* (L. Boddy, R. Marchant and D. J. Read, eds), pp. 227–241. Cambridge University Press, Cambridge.

Gianinazzi-Pearson, V., Morandi, D., Dexheimer, J. and Gianinazzi, S. (1981). *New Phytol.* **88**, 633–638.

Gianinazzi-Pearson, V., Dexheimer, J., Gianinazzi, S. and Jeanmaire, C. (1984). *Z. Pflanzenphysiol.* **114**, 201–205.

Gianinazzi-Pearson, V., Bonfante-Fasolo, P. and Dexheimer, J. (1986). In *Recognition in Microbe Plant Symbiotic and Pathogenic Interactions* (B. Lugtenberg, ed.), pp. 273–282. Springer-Verlag, Berlin.

Gianinazzi-Pearson, V., Gianinazzi, S., Dexheimer, J., Morandi, D., Trouvelot, A., Dumas, E. (1988). *Cryptogamie-Mycologie* **9**, 201–209.

Gianinazzi-Pearson, V., Gianinazzi, S. and Brewin, N. J. (1990). In *Endocytobiology IV* (P. Nardon, V. Gianinazzi-Pearson, A. M. Grenier, L. Margulis and D. C. Smith, eds), pp. 127–131. INRA, Paris.

Gianinazzi-Pearson, V., Smith, S. E., Gianinazzi, S. and Smith, F. A. (1991). *New Phytol.* **117**, 61–76.

Giovannetti, M. and Mosse, B. (1980). *New Phytol.* **84**, 489–500.

Hall, J. L. (ed.) (1978). *Electron Microscopy and Cytochemistry of Plant Cells.*

138     S. GIANINAZZI and V. GIANINAZZI-PEARSON

Elsevier/North Holland Biomedical Press, Amsterdam, Oxford and New York.
Hayat, M. A. (ed.) (1989). *Colloidal Gold: Principles, Methods and Applications*, Vols 1 and 2. Academic Press, New York.
Holgate, C. S., Jackson, P., Cowen, P. N. and Bird, C. C. (1983). *J. Histochem. Cytochem.* **31**, 938–944.
Huet, M., Benchimol, S., Berlinguet, J. C., Castonguay, C. and Cantin, M. (1974). *J. Microsc.* **21**, 147–158.
Jabaji-Hare, S. H., Thérien, J. and Charest, P. M. (1990). *New Phytol.* **114**, 481–496.
Jacquelinet-Jeanmougin, S., Gianinazzi-Pearson, V. and Gianinazzi, S. (1987). *Symbiosis* 3, 269–286.
Jeanmaire, C., Dexheimer, J., Marx, C., Gianinazzi, S. and Gianinazzi-Pearson, V. (1985). *J. Plant Physiol.* **119**, 285–293.
Jensen, W. A. (1962). *Botanical Histochemistry, Principles and Practice.* Freeman, San Francisco.
Kough, J., Malajczuk, N. and Linderman, R,. G. (1983). *New Phytol.* **94**, 57–62.
Kough, J. L., Gianinazzi-Pearson, V. and Gianinazzi, S. (1987). *New Phytol.* **106**, 707–715.
Kuc, J. (1982). In *Active Defense Mechanisms in Plants* (R. K. S. Wood, ed.), NATO ASI Series, Vol. A37, pp. 157–178. Plenum Press, New York.
Lemoine, M. C., Gianinazzi-Pearson, V., Gianinazzi, S. and Straker, C. J. (1990). *Abstract of IVth International Mycological Congress*, Regensburg, FRG, 28 August–3 September.
MacDonald, R. M. and Lewis, M. (1978). *New Phytol.* **80**, 135–141.
MacGonigle, T. P., Miller, M. H., Evans, D. G., Fairchild, G. L. and Swan, J. A. (1990). *New Phytol.* **115**, 495–501.
Marx, C., Dexheimer, J., Gianinazzi-Pearson, V. and Gianinazzi, S. (1982). *New Phytol.* **90**, 37–43.
Mueller, W. C., Tessier, B. J. and Englander, L. (1986). *Can. J. Bot.* **64**, 718–723.
Nemec, S. (1981). *Can. J. Bot.* **59**, 609–617.
Ocampo, J. A. and Barea, J. M. (1982). In *Les Mycorhizes, Partie Intégrante de la Plante: Biologie et Perspectives d'Utilisation*, Les Colloques de l'INRA no. 13, (S. Gianinazzi, V. Gianinazzi-Pearson and A. Trouvelot, eds), pp. 267–271. INRA, Paris.
Pearse, A. G. E, (1968). *Histochemistry, Theoretical and Applied*, Vol. I, pp. 1–759. Churchill Livingstone, Edinburgh and London.
Pearse, A. G. E. (1972). *Histochemistry, Theoretical and Applied*, Vol. II, pp. 760–1515. Churchill Livingstone, Edinburgh and London.
Phillips, J. M. and Hayman, D. S. (1970). *Trans. Br. Mycol.Soc.* **55**, 158–161.
Ravolanirina, F. (1990). *L'endomycorhization VA des plantes ligneuses (vigne, pommier et poirier) micropropagées: techniques d'inoculation, analyse de la morphogenèse racinaire et approches biochimique et immunologique.* Thesis, l'Université de Bourgogne, France. 131pp.
Reynolds, E. S. (1983). *J. Cell. Biol.* **17**, 208–212.
Roland, J. C., Lembi, C. A. and Morré, D. J. (1972). *Stain Technol.* **47**, 195–200.
Scannerini, S. and Bonfante-Fasolo, P. (1979). *New Phytol.* **83**, 87–94.

Scannerini, S. and Bonfante-Fasolo, P. (1983). *Can. J. Bot.* **61**, 917–943.
Serrigny, J. (1985). *Synthèse mycorhizienne* in vitro. *Etude cytologique comparée entre deux mycorhizes naturelles et artificielles et leur champignon symbiote isolé.* Thesis, Université de Nancy, France. 149pp.
Serrigny, J. and Dexheimer, J. (1985). *Cytologia* **50**, 779–788.
Smith, S. E. and Gianinazzi-Pearson, V. (1990). *Austral. J. Plant Physiol.* **17**, 177–188.
Smith, S. E., McGee, P. A. and Smith, F. A. (1990). In *Endocytobiology IV* (P. Nardon, V. Gianinazzi-Pearson, A. M. Grenier, L. Margulis and D. C. Smith, eds), pp. 91–98. INRA, Paris.
Spanu, P. and Bonfante-Fasolo, P. (1988). *New Phytol.* **109**, 119–124.
Spanu, P., Boller, T., Ludwig, A., Wiemken, A., Faccio, A. and Bonfante-Fassalo, P. (1989). *Planta* **177**, 447–455.
Straker, C. J., Gianinazzi-Pearson, V., Gianinazzi, S., Cleyet-Marel, J. C. and Bousquet, N. (1989). *New Phytol.* **111**, 215–221.
Swift, J. A. (1968). *J. Roy. Miscrosc. Soc.* **88**, 449–460.
Tahiri-Alaoui, A., Dumas, E. and Gianinazzi, S. (1990). *Plant Mol. Biol.* **14**, 869–871.
Thiéry, J. P. (1967). *J. Microscopie* **6**, 987–1018.
Tinker, P. B. (1978). *Physiol. Veg.* **16**, 743–751.
Tisserant, B. (1990). *L'endomycorhization VA chez le platane: influence sur le développement de la plante et recherche de marqueur biochimique de l'activité fonctionnelle de la symbiose.* DEA. Université de Bourgogne, France. 25 pp.
Trouvelot, A., Kough, J. L. and Gianinazzi-Pearson, V. (1986). In *Physiological and Genetical Aspects of Mycorrhizae* (V. Gianinazzi-Pearson and S. Gianinazzi, eds), pp. 217–221. INRA, Paris.
Valentines, R. C. (1961). *Adv. Vir. Res.* **8**, 287–290.
Wang, T. L. (ed.) (1986). *Immunology in Plant Science.* Cambridge University Press, Cambridge.
Wells, B. (1985). *Micron Microsc. Acta* **16**, 49–53.
Woods, A. M. and Gay, J. L. (1987). *Physiol. Mol. Plant Pathol.* **30**, 167–185.

# 8

# Pathogenic and Endomycorrhizal Associations

PAOLA BONFANTE-FASOLO

Dipartimento di Biologia Vegetale dell'Università, Viale Mattioli 25, 10125
Torino, Italy

PIETRO SPANU

Abteilung Pflanzenphysiologie Botanisches Institut der Universität Basel,
Hebelstrasse 1, CH 4056 Basel, Switzerland

METHODS IN MICROBIOLOGY
VOLUME 24  ISBN 0-12-521524-X

Copyright © 1992 by Academic Press Limited
All rights of reproduction in any form reserved

# I.  Introduction

Higher plants interact with a wide diversity of soil micro-organisms ranging from necrotrophic to biotrophic pathogens, from plant surface colonizers to mycorrhizal fungi, the association formed depending on the nutritional strategy of the microbe (Jeffries, 1987).

Comparative experiments using a mycorrhizal and a pathogenic fungus are of interest for two reasons. Firstly, although pathogenic and mycorrhizal associations differ as far as the nutrient flux and the effects on the colonized plants are concerned, the analysis of some cellular interactions in both groups, suggests that mycorrhizal and biotrophic pathogenic fungi have some features in common (Bonfante-Fasolo and Perotto, 1990). During their life-cycle, endomycorrhizal fungi show a strong dependency on the host for their metabolic processes, regulate the organization of their cell surfaces following the infection process and cause both the fungal and host contact surfaces to increase in order to improve nutrient exchanges. Similar events occur in many pathogenic fungi, where modifications of their surface take place during the adhesion and the infective phases and where haustorium development leads to an increased exchange surface. A common background for the penetration strategies in mycorrhizal and pathogenic fungi can therefore be suggested. Moreover, vesicular-arbuscular mycorrhizal fungi have so far resisted attempts at cultivation *in vitro*, in a similar way to some rust fungi (Bonfante-Fasolo and Perotto, 1990).

Secondly, it has been suggested that mycorrhizal fungi can protect roots against pathogen attack. For ectomycorrhiza this protection may be due to production of antibiotics, physical protection afforded by the fungal mantle, competition with pathogens for sugars and other substances released by the host root, stimulation of an antagonistic microflora associated with the mantle, and elicitation of the production of antimicrobial metabolites by the root (Marx, 1982). Vesicular-arbuscular mycorrhiza are also considered to protect roots from pathogen infections (Schonbeck and Dehne, 1989). However, other experimental results seem to lead to the opposite conclusion: vesicular-arbuscular mycorrhizal fungi may facilitate pathogen penetration, caused by fungi or nematodes (Ilag *et al.*, 1987).

The establishment of a symbiosis between plant and micro-organism involves the following events: attraction; adhesion; penetration; reaction of the plant; establishment of the interface; exchange of metabolites; and breakdown of the association. In this chapter we report protocols and provide a basic reference list, aiming to shed some light onto some aspects shared by both mycorrhizal and pathogenic fungi.

## II.   Adhesion of fungi to the plant surface

Adhesion of pathogenic fungi to the plant surface is essential to the successful establishment of pathogenesis (Nicholson and Epstein, 1990). Attachment is mediated by adhesive and non-adhesive compounds secreted by fungi and plants in the extracellular environment (Longman and Callow, 1987; Nicholson *et al.*, 1988). One model proposed to explain the molecular basis of recognition and adhesion is that they are determined by specific binding of lectins and haptens on the surfaces of host and pathogen. For example, *Magnaporthe grisea* produces a lectin-binding mucilage at the spore tip prior to germination; as concanavalin A (Con A binds glucose and mannose residues) inhibits adhesion, it has been suggested that these sugar haptens are responsible for this first step of infection (Hamer *et al.*, 1988).

Monoclonal antibodies raised against extracellular material of *Phythium aphanidermatum* allowed the identification of the glycoproteins involved in the adhesion of the fungal propagules as well as the secretion pathway (Estrada-Garcia *et al.*, 1990a). Fungal adhesion and zoospore encystment are influenced by root mucilage. In *P. aphanidermatum* they are triggered by some components of cress mucilage: uronic acids are the most effective, but commercial pectin as well as some lectins are also active inducers (Estrada-Garcia *et al.*, 1990b).

Very little information is available on the adhesion characteristics of endomycorrhizal fungi, even though such features have often been hypothesized (Bonfante-Fasolo, 1988). Attention has been devoted to ericoid mycorrhizal fungi. In pure culture, some fungal strains related to the species *Hymenoschyphus ericae* produce a loose weft of extracellular fibrillar material. When fungi are inoculated into host plants, the amount of extracellular material increases and it becomes organized into a sheath which radiates out from the hyphal wall. When the amount of fibrillar material was quantified in various strains by using computer-aided image analysis, infectivity correlated positively with production of the fibrillar sheath (Gianinazzi-Pearson *et al.*, 1986). The composition of this extracellular material has been analysed by cytochemical and biochemical methods: polysaccharides were located on the fibrillar sheath and on the outer layer of the cell wall by using lectin binding or the PATAg test (Protocols A and B). Con A bound weakly to non-infective strains and strongly to the infective ones (Bonfante-Fasolo *et al.*, 1987). Wheat-germ agglutinin (WGA), the lectin which binds to *N*-acetylglucosamine residues, only bound to the inner cell wall layers (Figs 1–4). Observations on salt-extracted fractions from intact myce-lium, sodium dodecyl sulphate-polyacrylamide gel electrophoresis (SDS-PAGE) and Con A labelling on Western blot demonstrated the pres-ence of surface glycoproteins (Perotto *et al.*, 1990a). Cross-reaction between some of the cell surface antigens could be seen by using immunofluorescence techniques (Protocol C) with monoclonal antibo-dies raised against *H. ericae* (Perotto *et al.*, 1988) and against a cell surface mannoprotein isolated from *Candida albicans* and important in the adhesion process between the yeast and epithelial cells (Cassone *et al.*, 1988).

On the roots of *Calluna vulgare*, host to *Hymenoschyphus ericae*, some mucilaginous and cell wall components were analysed using cytochemical (Figs 5 and 6) and biochemical methods. Soluble and insoluble molecules (glucose, mannose, *N*-acetyl-glucosamine, pectins, glucans and proteins) showed a heterogeneous surface distribution, suggestive of a role in recognition and binding between the two partners (R. Peretto *et al.*, 1990).

While it was possible to correlate the presence of hapten residues on the fungal surface with infectivity by using lectin-binding experiments, the quest for lectins on the host surface by using corresponding neoglycoproteins was not successful (P. Bonfante-Fasolo, unpubl. res.).

---

**Fig. 1.**   Con A–FITC binds to the cell surface of an infective *Hymenoschyphus ericae* strain. × 500.

**Fig. 2.**   WGA–FITC does not bind to the cell surface: only septa (S) are fluorescent. × 800.

**Fig. 3.**   Thin section of *Hymenoschyphus ericae* seen at ultrastructural level after the PATAg reaction. A Loose weft of fibrillar material (arrows) originates from the cell wall. Glycogen (G) is also PATAg positive N, nucleus. × 29 000.

**Fig. 4.**   Thin section of *H. ericae* after WGA–gold treatment. Gold granules are only present on the septum (S, arrow) and in the inner layer of the wall (W). × 18 000.

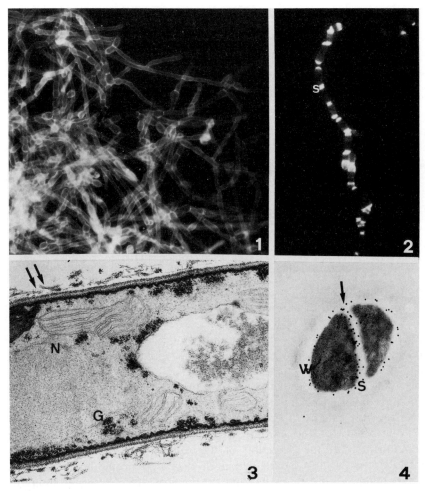

**Figs 1–4.**   Lectin binding to the cell surface of *Hymenoschyphus ericae*, an ericoid mycorrhizal fungus.

## III.   Enzymes that degrade plant cell walls

Penetration of the plant tissues requires the ability to degrade primary, secondary and modified cell walls (Cooper, 1984). A wealth of literature exists reporting extensive analysis of pathogen-derived enzymes whose substrates are cell wall components: cellulose, hemicelluloses, pectic polymers, proteins, polyesters. The role of these enzymes is thought not

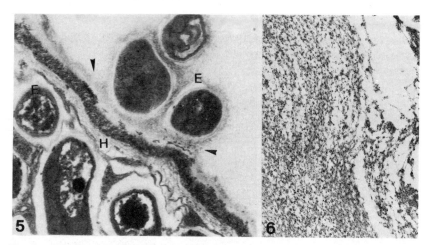

**Fig. 5.** Micrograph showing the contact point between *Hymenoschyphus ericae* hyphae (E) and the cell surface of a *Vaccinium* root (H). Fibrillar material (arrow) is adherent to the epidermal cell wall. × 22000.

**Fig. 6.** Higher magnification of the mucilage material covering the cell surface of the host plant after PATAg reaction. × 31000.

to be restricted to allowing penetration, but is also implicated in the release of oligomeric fragments which can regulate various physiological events including elicitation of the plant's defence response (Cervone *et al.*, 1989). Elucidation of the function of cell wall degrading enzymes in endomycorrhizal symbioses will need to consider the following: (1) the ability of the fungus to produce the enzyme (*in vitro*, where possible); (2) the ability of the enzyme to attack and degrade plant cell wall components; (3) the detection of the cell wall degrading enzymes in infected tissues; (4) the detection of degraded or modified cell wall components in infected tissue; and (5) the correlation between ability to produce the enzymes and to penetrate. It will also be necessary to mimic the effects in plants by application of purified enzymes and inhibitors, to engineer the expression of the enzymes. Such a task is evidently daunting and has indeed been carried out only for a few better studied plant–pathogen interactions. The path leading to the analysis of the requirements is fraught with difficulties, among which are the inability to grow many endomycorrhizal fungi *in vitro*, the presence of interfering substances (e.g. inhibitors) *in vivo*, and the lack of information regarding the genetics of the endosymbionts. Nevertheless, a few steps have already been made. We present in Section VII a protocol that may be used for the assay of polygalacturonase, an enzyme that catalyses the degradation of one of the many cell wall components (Protocol D).

The possibility of growing the fungus, *Hymenoschyphus ericae, in vitro*, and the existence of strains which exhibit different abilities to infect, has made it possible to address some of the points listed above. The results show that polygalacturonase was measurable in the culture medium, albeit at a much lower level and at a later stage compared to some pathogens; interestingly culture medium from strains with an increased mycorrhizal ability showed lower and delayed accumulation of polygalacturonase (Cervone *et al.*, 1988). With regards to vesicular-arbuscular mycorrhiza, the only biochemical data available are restricted to measurement of pectolytic enzymes in spore extracts and indirect observations on the ability of the fungus to penetrate roots in the presence of various pectic substrates (Garcia-Romera *et al.*, 1990).

Ultrastructural observations of plant cell walls provide indirect proof that endomycorrhizal fungi produce hydrolytic enzymes. The middle lamella has a loosened texture when vesicular-arbuscular mycorrhizal fungi grow between cortical cells in many different plants (Bonfante-Fasolo, 1984). At the moment of penetration of the cortical cell, the host cell walls seem to be plastic and to invaginate rather than break and gross alterations of the structure are not evident. More refined methods of analysis of the cell wall texture based on chemical extractions and on *in situ* labelling show that there are differences between cell walls of infected and non-infected roots (Figs 7–9): the former appear more swollen and laminated after solvent treatment because the matrix components disappear and the fibrillar components are exposed (Bonfante-Fasolo and Vian, 1989). Pathogenic fungi, such as *Rhizoctonia solani*, quickly destroy host cell walls, while other hemibiotrophic fungi (for example *Colletotrichum lindemuthianum*) reveal the fibrillar texture of the host wall simply by their presence (O'Connel and Bailey, 1986): solvent extractions are not needed.

## IV. The plant's response

Plants are endowed with the ability to respond to invading organisms in a variety of ways which are aimed at limiting infection of the invading organism. Components are produced which modify the cell walls, making them more difficult to penetrate; examples are callose, cell wall bound phenolics, lignin, cell wall proteins (e.g. extensins) and cell wall peroxidases. Small molecular weight toxic compounds phytoalexins, accumulate on pathogen attack and are released at the sites of infection

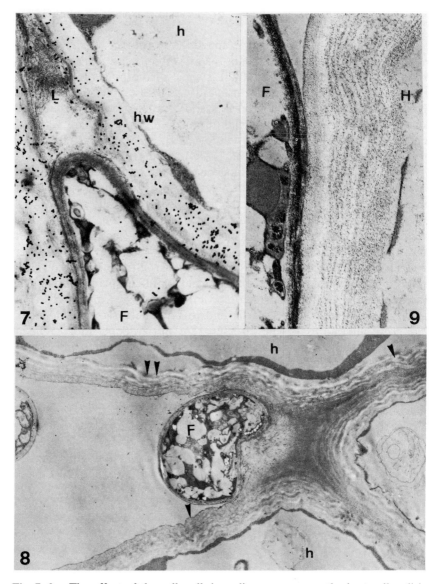

**Figs 7–9.** The effect of the cell wall degrading enzymes on the host cell wall in a vesicular-arbuscular mycorrhiza.

**Fig. 7.** An intercellular hypha of *Glomus versiforme* (F) is proceeding along an intercellular junction. Remnants of the middle lamella (L) are present, while the fibrillar component of the primary wall (hw) is labelled by gold granules after cellobiohydrolase I-gold treatment. h, host cell. × 31 500.

(Hahlbrock *et al.*, 1986). Hydrolytic enzymes such as chitinases and glucanases are induced (Boller, 1985); these are able to degrade fungal and bacterial cell walls and to inhibit fungal growth *in vitro* (Schlumbaum *et al.*, 1986; Mauch *et al.*, 1988). The ethylene biosynthetic pathway is induced with the consequent evolution of this gaseous hormone (Spanu and Boller, 1989).

Some of these events have been analysed in endomycorrhizal associations. Cell wall bound phenolics in roots of leek and *Ginkgo biloba* do not change after infection with *Glomus versiforme* (Codignola *et al.*, 1989) and production of callose has never been demonstrated (P. Bonfante-Fasolo, unpubl. res.). The accumulation of the main phytoalexin of soybean, glyceollin, could not be detected during the first 30 days after mycorrhization, while there was an evident increase of the compound in roots infected by *Rhizoctonia solani* (Wyss *et al.*, 1991). A higher phytoalexin content was detected at later stages of the symbiosis (Morandi *et al.*, 1984). In leek, both cell wall bound peroxidase and chitinase (see Protocol E) increased above control levels during the first days of mycorrhization, only to return to or fall below those of non-mycorrhizal roots at later stages of established symbiosis (Spanu and Bonfante-Fasolo, 1988; Spanu *et al.*, 1989). Moreover, comparative analysis of tobacco roots infected with a pathogen and mycorrhizal fungi showed that in the latter the usual complement of pathogenesis-related proteins, including chitinase and glucanase, were not induced (Dumas *et al.*, 1990). No increase in ethylene biosynthesis could be measured in mycorrhizal soybean roots (P. Wyss, pers. commun.) which, on the other hand, produced greatly increased amounts of the gas on attack by *Phytophthora megasperma* var. *soyae* (Reinhardt *et al.*, 1991).

Immunofluorescence and gold localization of chitinase activity by using a polyclonal serum (see Protocols F and G) demonstrated that the enzyme is localized around the fungus both in the intercellular spaces and in the vacuole of the infected cells (Figs 10 and 11), but it does not seem to get into direct contact with the fungus *in vivo*. However, when treated at 100 °C to eliminate soluble components, chitinase bound to the *Glomus* cell walls (Spanu *et al.*, 1989). The results of this experiment are consistent with the finding that chitin is protected by other cell wall components (Bonfante-Fasolo *et al.*, 1990). When a pathogenic

---

**Fig. 8.**   Cell walls of cortical cells (h) appear swollen and polylamellate (arrows) in leek roots, after fungal passage (F) and dimethylsulphoxide extraction. × 14 000.

**Fig. 9.**   Fungal (F) wall–host (H) wall contact in a leek root after methylamine extraction. The fibrillar texture of the host wall is clearly seen. × 33 500.

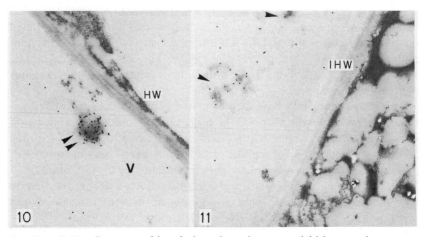

**Figs 10 and 11.** Immunogold technique by using an antichitinase polyserum on thin sections of *Allium porrum* infected by *Glomus versiforme*. Chitinase activity is localized in the vacuole (V) (Fig. 10) or in the intercellular spaces around the fungus, but never onto the fungus (IHW) (Fig. 11). HW, host cell wall. × 15 500.

fungus–plant interaction was analysed, Benhamou *et al.* (1990b) obtained opposite results: here an intense plant chitinase activity accumulated on the fungal surface, suggesting an *in vivo* action.

Hydroxyproline-rich glycoproteins (HRGPs) are structural glycoproteins, involvement of which in plant defence is suggested by their increase after pathogenic infection. When a polyclonal antiserum against melon HRGPs was used on thin sections of infected tissues, a strong accumulation of gold granules was observed indicating the presence of the antigen around pathogenic fungi (Benhamou *et al.*, 1990b). A regular distribution of gold granules was also observed around the arbuscular hyphae of vesicular-arbuscular mycorrhizal fungi after using the same antibody (Bonfante-Fasolo *et al.*, 1991).

However, the experiments carried out to date have not always been of a comparative nature and are limited to a few types of host; a more systematic approach considering both compatible and incompatible mycorrhizal and pathogenic associations is needed. This could shed some light on the mechanism of recognition in the different interactions.


## V.    The interface

Interfaces are defined as specialized areas where nutrient exchanges between plant and fungi take place (Bracker and Littlefield, 1973). The establishment of a succession of different interfaces is a common feature in walled organisms and represents a critical site for interactions between hosts and their symbionts (Smith, 1990). Many original papers and reviews devoted to such topics offer an up-to-date physiological and structural interpretation of the different types in relation to nutrient exchange in mutualistic and pathogenic symbiotic associations (Bonfante-Fasolo and Scannerini, 1991; Smith and Smith, 1990).

The best known interface is that found at the haustoria of the biotrophic pathogens and is composed of the intracellular fungal structure and the extrahaustorial membrane. In endomycorrhiza, and particularly in vesicular-arbuscular mycorrhiza, the situation is very similar. The interface is created by the fungal branch penetrating a root cortical cell, by the interfacial material and by the surrounding host membrane.

From an experimental point of view, interfaces with all their components (membranes, cell walls and cell wall-like materials) can be investigated using ultrastructural, cytochemical and electrophysiological methods.

## A.   The extrahaustorial matrix and the interfacial material

In rusts, the extrahaustorial matrix varies greatly in appearance, depending on haustorial age, degree of host compatibility and preparation and staining procedures (O'Connel, 1987; Harder, 1989). Ultrastructural affinity techniques (Fig. 12), which are important tools in identifying molecules present in complex compartments, such as interface areas, provided useful information. A variety of complex substances have been located: protease-sensitive polysaccharides, cellulose-like molecules, $\alpha$-linked glucans or mannans (Harder, 1989), as well as arabinogalactan-proteins (Rohringer et al., 1989). All these molecules may be of host or fungal origin.

In vesicular-arbuscular mycorrhiza the interfacial material is morphologically continuous with the peripheral host cell wall and changes during the infection process: it is thick and electron dense around the fungal coils in the outer root layers and at the penetration points; it is loose and ill-defined around the thin arbuscular branches and becomes thicker around the collapsed arbuscular branches (Bonfante-Fasolo, 1984). The affinity techniques used (Protocols G and H) demonstrate that in different plants host cell wall molecules, such as $\beta$-glucans and polygalacturonic acids, occur (Fig. 13). However, the architecture of the interfacial material, as well as its staining properties, differs from that of the peripheral wall. The results suggest that the plant reacts to the fungal colonization by activating the synthesis of cell wall-related

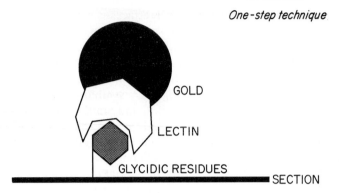

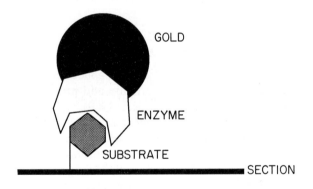

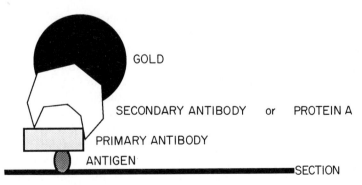

**Fig. 12.**   Ultrastructural affinity techniques.

molecules, but the assembly process is altered, leading to the creation of a "cell wall-like envelope" but not to a true cell wall (Bonfante-Fasolo *et al.*, 1990b; Perotto *et al.*, 1990b).

The interface zone should allow apoplastic transfer owing to the hydrophilic properties of its components. However, in the "neck ring" zone of some rust haustoria, hydrophobic molecules produce an apoplastic block, similar to the Casparian strips of vascular plants (Heath, 1976). This leads to important differences in the hypotheses of solute transport, since the haustorial matrix is seen as a compartment maintained closed by the seal of the neckband (Gay and Woods, 1987). No such structures can be observed in vesicular-arbuscular mycorrhizal arbuscules.

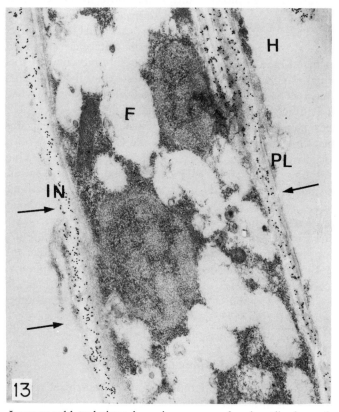

**Fig. 13.** Immunogold technique by using a monoclonal antibody against polygalacturonic acid demonstrates the presence of such molecules in the interface material of a vesicular-arbuscular mycorrhiza (pea infected by *G. versiforme*). F, fungus; H, host; PL, plasma membrane; IN (with arrow), interface material.
× 15 000.

## B. The extrahaustorial membrane and the periarbuscular membrane

A common feature of biotrophic infections is the presence of a membrane which surrounds the intracellular fungal component. The origin, structure and function of this membrane pose similar questions in both pathogenic and mycorrhizal interactions.

In most fungal infections the morphological continuity between the host plasmalemma and the invaginated membrane hints at its host origin, even though in the presence of a neck ring, in pathogens, the membrane continuity is not clearly defined (Harder, 1989). Many cytochemical findings have indicated extensive structural differences between the extrahaustorial membrane and the host plasmamembrane. Freeze-fracture replicas indicate differences both in particle distribution and in the sterol composition in *Uromyces* infections (Harder and Mendgen, 1982), while no information is available for mycorrhizal associations. However, the most important differences between extra-haustorial membrane and periarbuscular membrane concern their functional activities. Smith and Smith (1990) exhaustively discussed the distribution, the meaning and the possible role of the membrane-bound ATPases. Pathogenic associations show a membrane-associated ATPase on the peripheral plasma membrane, but not on the extrahaustorial membrane, while symbiotic associations are characterized by an ATPase activity occurring on the periarbuscular membrane, too (Gianinazzi-Pearson *et al.*, 1991). This ATPase may be involved in an active transport mechanism, and its disappearance from the extrahaustorial membrane is consistent with the efflux of solutes in pathogens; the cytochemical distribution of the ATPase activity in mycorrhizal roots is consistent with the double flux in this symbiosis. The different ATPase distribution is correlated to different values of potential difference (Protocol I): in pathogenic associations a membrane depolarization is usually observed (Pelissier *et al.*, 1986), while vesicular-arbuscular mycorrhiza show a stable hyperpolarization of the membrane (Scannerini *et al.*, 1990) (Fig. 14).

## C. Haustorial cell wall and arbuscular cell wall

After host penetration, both pathogenic and mycorrhizal fungi undergo extensive modifications of their cell wall. Cytochemical studies on *Albugo* and many rusts show a predominant glycoprotein composition as well as the presence of specific lectin binding sites (Harder, 1989). As the haustorium matures the cell wall changes cytochemically: for example it acquires the ability to bind WGA, indicating the presence of *N*-acetylglucosamine residues (Chong *et al.*, 1985). Changes in lectin

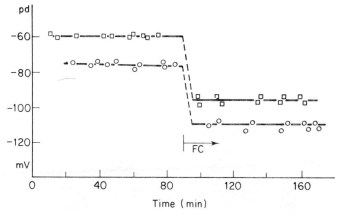

**Fig. 14.** Transmembrane electrical potentials (pd) in onion roots in the presence (○) and absence (□) of *Glomus* sp. before and after addition of 10 μm
Fusicoccin (FC). The conditions of the experiment are described in the Protocol
I. The membrane in vesicular-arbuscular mycorrhizal roots is constantly hyper-
polarized, in comparison with that in uninfected roots. The addition of FC,
which activates the protonic pump only in the plant membrane, causes the
expected hyperpolarization, both in the uninfected and mycorrhizal roots
(M. Fieschi, M. Bellando and S. Scannerini, unpubl.).

binding occur at the *Uromyces* surface during the infection process
(Kapooria and Mendgen, 1985).

Similar events can be observed in vesicular-arbuscular mycorrhiza: the
fungal cell wall thins out and becomes more simplified passing from the
extraradical to the intracellular arbuscular hyphae. In the intracellular
phase the fungal cell wall is thin and amorphous. The ability to bind
WGA (Protocol J) is constant along all the different types of hyphal cell
walls from the outer thick fibrillar (Fig. 15) to the inner thin amorphous
ones (Fig. 16) (Bonfante-Fasolo, 1988: Grandmaison *et al.*, 1988;
Bonfante-Fasolo *et al.*, 1990a).


## VI.  Conclusions

We have compared some findings obtained on plant–pathogen interactions and endomycorrhiza, tracing the common features between the
two as well as describing some of the methods used to investigate these
interactions.

The process of adhesion has been analysed using spores where data
regarding hyphae are not available; adhesion in mycorrhiza still remains
hypothetical. The mechanisms of tissue penetration, extensively studied

156 P. BONFANTE-FASOLO and P. SPANU

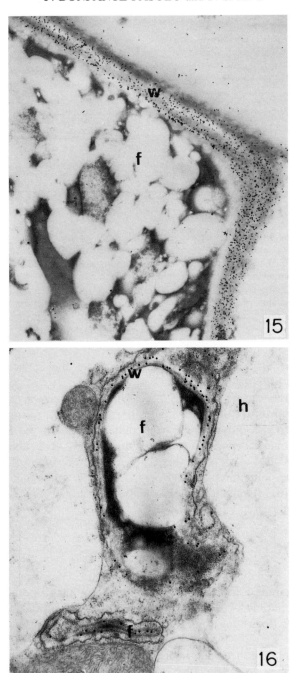

in pathogens have only been approached in endomycorrhiza: present knowledge is limited to that obtained from ultrastructural studies of infected tissues and from measurements of hydrolytic activity in extra-radical structures. The plants' response to infection has been studied mostly by looking for responses analogous to those found upon pathogen attack. In general, in endomycorrhiza, the extensive alteration of gene expression seen in pathogens has not been found, even though generalizations must be avoided due to the limited number of data available. This, together with the absence of a clear endomycorrhiza-specific response, seems to indicate that these fungi can penetrate and colonize roots without eliciting defence responses. The question therefore arises whether plants do "recognize" the presence of the symbiont at all, or rather that the defence responses are repressed by the fungus in order to allow a compatible interaction. The interface is probably the most studied compartment: here the main common feature is the presence of structures which increase the contact surface and reduce the passive barriers to exchange; differences concern the properties of the host membrane: in endomycorrhiza it maintains, if not enhances, its features aimed at sustaining bidirectional flow, while extrahaustorial membranes are depolarized and leaky, allowing unhindered nutritional flow to the colonizing fungus.

It is clear that the data are patchy and lack a truly comparative nature except in a few cases; such experiments are needed if anything is to be said regarding differences in recognition between plant–pathogen and plant–endomycorrhizal associations and concerning the reciprocal influence during simultaneous infection.

## VII.  Protocols

### A.  Lectin binding test

1. Grow fungal colonies in liquid medium or on microculture slides.
2. Wash in buffer, a different one for each lectin, following the manufacturer's instructions. The most common buffers are 10 mM

---

**Fig. 15.** WGA–gold binds to the thick and fibrillar cell walls (w) of the external hyphae of a vesicular-arbuscular mycorrhizal fungus, *G. versiforme* (f). Gold labelling only occurs on the fungal cell wall (W). × 13 000.

**Fig. 16.** WGA–gold binds to the thin and amorphous cell walls of the arbuscular hyphae (f) of *G. versiforme*, colonizing cortical cells (h) of a leek root. × 28 000.

(N-[2'-Hydroxyethyl]    piperazine-N'-[2-ethanesulphonic    acid])
(HEPES) or 10 mM phosphate-buffered saline (PBS), pH range:
7–8.

3. Incubate with the fluorescein isothiocyanate (FITC)-lectin in the
   dark for 30 min. Lectin concentration ranges from $10-100~\mu g\,ml^{-1}$
   in the appropriate buffer.
4. Wash the sample in buffer and rinse in distilled water.
5. Observe using photomicroscope equipped for fluorescence, with a
   485 nm excitation filter and a 520 nm barrier filter for all the
   FITC-labelled chemicals.

*Control experiments*

1. Inhibitory sugars are added at 300 mM concentration to the labelled
   lectin before incubation (Table I).
2. Samples are treated with FITC without lectins.
3. Samples are directly observed to check their autofluorescence.
4. Samples can be treated with a pronase solution for 10–20 h, and
   then treated with the labelled lectin, in order to reveal whether the
   glycidic determinants are bound to proteins.

**B.   Ultrastructural polysaccharide localization with the PATAg reaction
(periodic acid, thiocarbohydrazide, Ag proteinate)**   (Roland, 1978)

1. Cut thin sections from glutaraldheyde–osmium-fixed, Araldite-
   embedded samples; collect them on gold grids or on plastic rings.
2. Treat the sections with 1% periodic acid for 20–30 min.
3. Wash carefully in distilled water.
4. Float the sections on 0.2% thiocarbohydrazide in 20% acetic acid
   or 1% thiosemicarbazide in 10% acetic acid. Sections are kept at
   room temperature for 3–4 h or overnight.
5. Wash the sections with an aqueous solution of acetic acid with
   decreasing concentration for 30–60 min.

TABLE I

| Lectins | Inhibitory sugars |
| --- | --- |
| Concanavalin A (Con A) | $\alpha$-methyl-D-mannoside |
| Wheat-germ agglutinin (WGA) | $N$, $N'$, $N''$-triacetyl-chitotriose and ovomucoid |
| *Ricinus communis* ($RCA_{120}$) | D-galactose |
| *Ricinus communis* ($RCA_{60}$) | $N$-acetyl-D-galactosamine |
| *Ulex europeaus* agglutinin (UEA) | L-fucose |
| Soybean agglutinin (SBA) | $N$-acetyl-D-galactosamine |

6. Wash with distilled water for 30 min.
7. Treat the sections with 1% aqueous silver proteinate (prepared and used in the dark) for 30 min.
8. Wash carefully in distilled water for 10 min.
9. Observe dry sections without further staining.

## C. Immunofluorescence test for localization of cell surface antigens

1. Grow fungal colonies in liquid medium or on microculture slides.
2. Wash samples in 0.05 M phosphate buffer with 0.9% NaCl (PBS) pH 7.2.
3. Wash samples with PBS containing 0.5% bovine serum albumin (BSA).
4. Incubate fungi with the first antibody (monoclonal antibodies are used undiluted) for 3–4 h or overnight at 4° C.
5. Wash 4–5 times with PBS + 0.5% BSA buffer.
6. Incubate the samples for 1 h at 4° C with the secondary antibody diluted from 1:10 to 1:100 in 0.05 M PBS + 0.5% BSA. The secondary antibody can be an FITC-labelled goat anti-rat immunoglobulin or an FITC-labelled goat anti-rabbit immunoglobulin, depending on the origin of the first antibody.
7. Thoroughly wash with the buffer and rinse with distilled water.
8. Observe under a fluorescence microscope at 485 nm.

*Control experiments*

1. Samples are treated with the secondary antibody alone.
2. Samples are directly observed to check their autofluorescence.
3. First antibody is pre-absorbed with the purified antigen—when known—before reaction.

## D. Polygalacturonase assay   (Collmer *et al.*, 1988)

This assay is based on the determination of oligosaccharides released by the enzyme from polygalacturonic acid, measured with copper–arsenomolybdate reagent (Somogyi, 1952).

1. Preparation of the copper reagent:
   (a) Dissolve 12 g of sodium potassium tartrate and 24 g $Na_2CO_3$ in 250 ml of water.
   (b) Add in sequence 4% (w/v) $CuSO_4.5(H_2O)$ in water, 16 g $NaHCO_3$, 500 ml 36% (w.v) $Na_2SO_4$ degassed by boiling, bring to 1 litre with water.
   (c) Filter away the precipitate that will form during the first week.

2. Preparation of the arsenomolybdate reagent:
    (a) Dissolve 25g $(NH_4)_2 MoO_4.4H_2O$ in 450 ml.
    (b) Add 21 ml of 96% $H_2SO_4$.
    (c) Add 25 ml of 12% (w.v) $Na_2HAsO_4.7H_2O$ in water.

Incubate the final solution at 37° C for 24 h and store in the dark.

3. Preparation of the substrate stock:
    (a) Mix vigorously, 20 ml 0.6 M NaCl and 80 ml 75 mM sodium
        acetate buffer pH 5.3, 7.5 mM EDTA, 0.3% (w/v) polygalac-
        turonic acid. Adjust the final pH to 5.3 with NaOH. Add
        0.02% $NaN_3$ and store at 4° C.

4. Assay:
    (a) Mix 2.5 ml of substrate stock with 0.5 ml enzyme sample.
    (b) Incubate at 30° C.
    (c) Remove 0.5 ml aliquots at timed intervals and mix with 0.5 ml
        copper reagent.

Incubate the samples at 100° C for 10 min. Cool to room temperature,
add 1 ml of arsenomolybdate, and centrifuge to remove the precipitate.
Measure extinction at 500 nm. One unit of enzyme activity forms 1 $\mu$mol
of oligogalacturonate in 1 min.

*Notes*
This assay is directly applicable to enzyme preparations obtained from
culture filtrates or fungal homogenates; caution should be observed
when attempting determinations in infected tissues because of the
possible presence of pectic enzyme inhibitors (Collmer and Keen, 1986)
and other substances which interfere with the assay.

### E.   Chitinase assay

There are three types of assays available to measure chitinolytic activity:
a colorimetric one based on determination of N-acetylglucosamine
released (Boller and Mauch, 1988), a viscosimetric assay based on the
reduction is viscosity of a solution of water-soluble chitin analogues
(Ohtakara, 1988), and a radiochemical one based on the release of
soluble radioactive fragments from regenerated tritiated chitin (Boller,
1990). These different methods have advantages and disadvantages but
under appropriate conditions yield similar results. We will therefore
present here the protocol for the radiochemical assay, which is the most
rapid and simple one if facilities for work with radioactive isotypes are
available (Boller, 1991).

1. Prepare collodial radioactive regenerated chitin.

2. Extract plant tissues with 0.1 M sodium phosphate buffer pH 7.0 or 0.1 M sodium acetate buffer pH 5.0; clarify by centrifugation.
3. Set up reaction mixtures in 1.5 ml polypropylene microcentrifuge tubes on ice: 0–150 µl extract and 150–0 µl extraction buffer (to a final volume of 150 µl) and 100 µl [$^3$H]chitin.
4. Close tubes, mix and incubate at 37° C for 30 min.
5. Stop the reaction by adding 250 µl 10% trichloroacetic acid, mix and centrifuge at 3000 g for 10 min in a swing-out rotor.
6. Remove carefully 250 µl supernatant so as not to resuspend the sediment.
7. Mix with 4 ml scintillation liquid and measure radioactivity in a liquid scintillation counter.

*Notes*

The chitinase activity must be determined by comparison with a calibration curve constructed by serial dilution of the extract after subtracting the appropriate substrate blanks. A new calibration curve should be constructed for chitinases of different plant origins and for each new batch of [$^3$H]chitin prepared.


**F.   Immunofluorescence test on cryotome-cut sections for localization of intra/extracellular anitgens in tissues**   (Knox *et al.*, 1989)

1. Fix in 4% (w/v) formaldehyde in 50 mM PIPES buffer pH 6.9 containing 5 mM MgSO$_4$ and 5 mM ethylene glycol bis ($\beta$-aminoethylether) tetraacetic acid. Tissue is fixed for at least 1 h, washed briefly in water and mounted in OCT compound (Miles) to bind tissues to the specimen block.
2. Cut tissue at −10° C to −20° C depending on cryomicrotome type.
3. Collect sections of 7–10 µm on poly-L-lysine-coated multi-well slides and store them dry until use.
4. Incubate sections for 5 min with 5% calf serum in 50 mM phosphate saline buffer, pH 7.2.
5. Incubate sections with the antiserum or monoclonal antibody from 1 h to overnight at 4° C.
6. Wash sections 3–5 times with PBS.
7. Incubate sections with FITC-labelled secondary antibody diluted from 1:10 to 1:100 in 5% calf serum in PBS for 1 h. The secondary antibody can be an FITC-labelled goat anti-rat immunoglobulin or an FITC-labelled goat anti-rabbit immunoglobulin, depending of the origin of the first antibody.
8. Wash thoroughly with buffer and rinse with distilled water.
9. Mount the samples in an anti-fade compound such as Citifluor.
10. Observe under a fluorescence microscope at 485 nm.

## G.  Immunogold labelling of pectin and chitinase with mono- or polyclonal antibodies

The immunogold technique is a two-step, post-embedding method.

1. Cut sections from samples which have been fixed in glutaraldehyde alone and embedded in LR White or Lowicryl resins. Mount sections onto gold or nickel grids or handle as floating sections.
2. Incubate for 20 min in normal goat serum diluted 1:30 in 50 mM Tris-HCl with 0.5 M NaCl (TBS, pH 7.5) and 0.1% BSA.
3. Incubate overnight with (a) the cell culture supernantant of JIM 5, an anti-pectin monoclonal antibody (Vandenbosh *et al.*, 1989), or (b) polyclonal antiserum against chitinase diluted 1:1500 in TBS with BSA (Spanu *et al.*, 1989).
4. Wash the grids with TBS for 30 min.
5. Incubate for 1 h with the secondary antibody diluted 1:20 in 20 mM Tris pH 8.2 with 0.1% BSA. The secondary antibody can be (a) a 15 nm colloidal gold–goat anti-rat immunoglobulin complex when the first antibody is a monoclonal, or (b) a 15 nm colloidal gold–goat anti-rabbit immunoglobulin complex when the first antibody is a polyclonal antibody raised in rabbit.
6. Wash the grids with TBS, rinse twice in water and dry.
7. Counterstain the grids with uranyl and lead citrate.

*Note*
Control sections are run (a) by omitting the first antibody, (b) by its substitution with irrelevant antibodies, (c) by antibody pre-absorption with 1 mg ml$^{-1}$ solutions of pectin, polypectate or Na-polygalacturonate, 12 h before incubation (for JIM 5), and (d) using pre-immune serum for the experiments with polyclonals.

## H.  Ultrastructural labelling with cellobiohydrolase–immunogold complex for $\beta$-1,4-glucan localization

*Preparation of colloidal gold* (Bendayan, 1985)

Colloidal gold (CG) is obtained by reducing choloroauric acid according to Frens method (1973). With this method gold particles have a size of 15 nm. Alternative methods are listed in Roth (1983).

1. Prepare a 0.01% tetrachloroauric acid double-distilled water solution.
2. Add 4 ml of a 1% aqueous trisodium citrate solution to a 100 ml of the boiling tetrachloroauric solution, until the reduction process is over, as indicated by a stable wine red colour.

3. Cool the suspension and adjust the suspension to the desired pH with 0.2 M potassium carbonate or with 0.1 M HCl. The pH of the suspension must be near to the isoelectric point (p*I*) of the macromolecule, since absorption of macromolecules to colloidal gold occurs at the basic side of the p*I* value.

*Preparation of enzyme–colloidal gold solution* (Bendayan, 1985; Berg *et al.*, 1988)

1. Purified enzymes must be used.
2. For each enzyme the minimum amount necessary to stabilize the colloidal gold must be determined by preparing constant volumes (5 ml) of the CG solution and mixing them with 0.1 ml of serial dilutions of each enzyme. After 3–5 min, 0.5 ml of 10% NaCl solution is added and the colour changes (from wine red to dark blue) are observed. The lowest dilution before the colour changes (i.e. sign of flocculation) gives the minimum amount of enzyme for gold stabilization.
3. In the case of cellobiohydrolase (CBH I, EC 3 2.1.91) a 2 mg ml$^{-1}$ stock solution is prepared.
4. A 100 $\mu$l aliquot of the solution is added to 10 ml of CG solution (pH 4.5) while stirring continuously.
5. After 5 min, 500 $\mu$l of 1% polyethylene glycol 20000 (PEG) are added.
6. The enzyme–gold mixture is centrifuged at 14000 rpm for 1 h at 4° C.
7. The clear supernatant containing the free enzyme is discarded, as is the black solid pellet near the bottom of the tube corresponding to the free metallic gold. The mobile, dark red pellet corresponding to the enzyme complex is collected in 5 ml of 50 mM citrate–phosphate buffer (CPB), pH 4.9, to which 0.02% (w/v) PEG is added.

*Cytochemical labelling* (Bonfante-Fasolo *et al.*, 1990b)

The enzyme–gold technique is a direct one-step, post-embedding method.

1. Cut sections from samples which have been fixed in glutaraldehyde-OsO$_4$ or glutaraldehyde alone and embedded in Araldite, Epon or LR White. Mount sections onto gold or nickel grids or handle as floating sections.
2. Float the grids on 0.05 M CPB for 10 min.

3. Treat the grids for 30 min with the enzyme–gold solution (dilutions in the incubation buffer from 1:10 to 1:50 may be used).
4. Wash the grids with buffer, rinse twice in water and dry.
5. Counterstain the grids with uranyl and lead citrate.

*Note*

Control sections are treated with an enzyme gold solution to which carboxymethylcellulose has been added at 1 mg ml$^{-1}$ final concentration, 1 h before use.

## I. Membrane potential measurements in vesicular-arbuscular mycorrhizal and uninfected roots (Scannerini *et al.*, 1990)

1. Grow uninfected and mycorrhizal plants on a mineral substrate in controlled conditions.
2. Check the degree of mycorrhization and sample the roots at different times, preparing a time course of the infection process.
3. Excise 1.5 cm long apical segments from mycorrhizal and uninfected roots.
4. Wash in aerated 0.4 mM lithium succinate buffer, pH 5.5 containing 0.5 mM CaSO$_4$ and 0.1 mM KCl.
5. Measure transmembrane potential differences (pd) under flowing buffer with glass microelectrodes, filled with 1 mM KCl. Microelectrodes are inserted in cells of the second and third root layers 0.5–0.8 cm from the root apex. Measurements are performed for periods not exceeding 80 min.
6. Measure membrane potential after addition of either:
   (a) 10 $\mu$M Fusicoccin (FC) to activate plant membrane ATPases; or
   (b) 0.1 M 2,4-dinitrophenol (DNP) to inhibit ATPase activity and the production of pd differences both in plant and fungus.

Measurements of pd are performed for at least 60 min.

Alternatively, the collected roots (step 3) may be maintained for 24 h in an aerated 0.5 mM CaSO$_4$ solution, in order to lower the ionic intracellular concentration and facilitate the membrane ionic exchanges. Follow from step 4.

*Comments*

1. Each experiment must be repeated three times for each sampling.
2. Each measurement series (step 5 and 6) must include a number of pd measurements (at least 10) in order to define the stable cell potential.
3. Values are analysed statistically.

## J. Labelling with wheat-germ agglutinin–gold complex for ultrastructural localization of N-acetylglucosamine residues

*Preparation of colloidal gold* (Bendayan, 1985)
Details for the preparation of colloidal gold [CG] are given in Protocol H.

*Preparation of wheat-germ agglutinin–colloidal gold solution* (Roberts *et al.*, 1983)

1. Wheat-germ agglutinin (WGA) is commercially available.
2. A 2.5 mg ml$^{-1}$ stock solution in 2.5 M HCl is prepared.
3. A 500 $\mu$l aliquot of the solution is added to 10 ml of CG solution; the pH is adjusted to 9.7.
4. After 5 min, 500 $\mu$l of 1% polyethylene glycol (PEG) are added.
5. The WGA–gold mixture is centrifuged at 14000 rpm for 1 h at 4° C.
6. The clear supernatant containing the free protein is discarded, as is the black solid pellet near the bottom of the tube corresponding to the free metallic gold. The mobile, dark red pellet corresponding to the lectin complex is collected in 5 ml of 20 mM Tris buffer, pH 7.0, to which 0.04% (w/v) PEG is added.

*Cytochemical labelling* (Bonfante-Fasolo *et al.*, 1987)

The WGA–gold technique is a direct one-step, post-embedding method; two-step protocols can also be used (Benhamou, 1989).

1. Cut sections from samples which have been fixed in glutaraldehyde (GA)–OsO$_4$ or GA alone and embedded in Araldite, Epon or LR White. Mount sections onto gold or nickel grids or handle as floating sections.
2. Float the grids on a drop of 4% Na metaperiodate for 30 min and wash.
3. Treat the grids for 40 min with the lectin-gold solution (dilutions in Tris buffer from 1:1 to 1:10 may be used).
4. Wash the grids with the buffer, rinse twice in water and dry.
5. Counterstain the grids with uranyl and lead citrate.

*Notes*
Control sections are treated with a lectin–gold solution to which $N$, $N'$, $N''$-triacetyl-chitotriose (0.3 M) has been added 1 h before use.
    WGA–gold complexes are now commercially available.

# References

Bendayan, M. (1985). In *Techniques in Immunocytochemistry* (G. R. Bullock and P. Petrusz, eds), Vol. 3, pp. 179–201. Academic Press, London.

Benhamou, N. (1989). In *Colloidal Gold: Principles, Methods and Applications*, Vol. 1, pp. 95–143. Academic Press, London.

Benhamou, N., Joosten, M. H. A. J. and de Wit, P. J. G. M., (1990a). *Plant Physiol.* **92**, 1108–1120.

Benhamou, N., Mazau, D. and Esquerre-Tugaye, M. T. (1990b). *Phytopathology* **80**, 163–173.

Berg, R. H., Erdos, G. W., Gritzali, M. and Brown, R. D. Jr (1988). *J. Electron Microsc. Tech.* **8**, 371–379.

Boller, T. (1985). In *Cellular and Molecular Biology of Plant Stress* (A. Riss, ed.), pp. 247–262. R. Lyss, New York.

Boller, T. (1991). *Molecular Plant Pathology: A Practical Approach* (D. Bowers, ed.). Oxford University Press (in press).

Boller, T. and Mauch, F. (1988). *Meth. Enzymol.* **161**, 426–430.

Bonfante-Fasolo, P. (1984). In *VA Mycorrhizas* (C. L. Powell and D. J. Bagyaraj, eds), pp. 5–33. CRC Press, Boca Raton, FL.

Bonfante-Fasolo, P. (1987). *Symbiosis* **3**, 249–268.

Bonfante-Fasolo, P. (1988). In *Cell to Cell Signals in Plant, Animal, and Microbial Symbiosis* (S. Scannerini, D. C. Smith, P. Bonfante-Fasolo and V. Gianinazzi-Pearson, eds), pp. 219–235. Springer-Verlag, Berlin and Heidelberg.

Bonfante-Fasolo, P. and Perotto, S. (1990). In *Electron Microscopy Applied in Plant Pathology* (K. Mendgen and D. E. Lesemann, eds). Springer-Verlag, Berlin (in press).

Bonfante-Fasolo, P. and Scannerini, S. (1991). In *Functioning in Mycorrhizae* (M. Allen, ed.), pp. 265–275. Academic Press, London.

Bonfante-Fasolo, P. and Vian, B. (1989). *Ann. Bot.* **10**, 97–109.

Bonfante-Fasolo, P. and Perotto, S., Testa, B. and Faccio, A. (1987). *Protoplasma* **137**, 25–35.

Bonfante-Fasolo, P., Faccio, A., Perotto, S. and Schubert, A. (1990a). *Mycol. Res.* **94**, 157–165.

Bonfante-Fasolo, P., Vian, B., Perotto, S., Faccio, A. and Knox, J. P. (1990b). *Planta* **180**, 537–547.

Bonfante-Fasolo, P., Tamagnone, L., Peretto, R., Esquére-Tugaye, M. T., Mazau, D., Mosiniak, M. and Vian, B. (1991). *Protoplasma* (in press).

Bracker, C. E. and Littlefield, L. J. (1973). In *Fungal Pathogenicity and the Plant's Response* (R. J. W. Byrde and C. V. Cutting, eds), pp. 159–317. Academic Press, London.

Cassone, A., Torosantucci, A., Boccanera, M., Pellegrini, G., Palma, C. and Malavasi, F. (1988). *J. Med. Microbiol.* **27**, 233–238.

Cervone, S., Castoria, R., Spanu, P. and Bonfante-Fasolo, P. (1988). *Trans. Br. Mycol. Soc.* **91**, 537–539.

Cervone, S., Hahn, M. G., de Lorenzo, G., Darvill, A. and Albersheim, P. (1989). *Plant Physiol.* **90**, 542–548.

Chong, J., Harder, D. E. and Rohringer, R. (1985). *Can. J. Bot.* **63**, 1713–1724.

Codignola, A., Verotta, L., Maffei, M., Spanu, P., Scannerini, S. and Bonfante-

Fasolo, P. (1989). *New Phytol.* **112**, 221–228.
Collmer, A. and Keen, N. T. (1986). *Ann. Rev. Phytopathol.* **24**, 383–409.
Collmer, A., Ried, J. L. and Mont, M. J. (1988). *Meth. Enzymol.* **161**, 329–335.
Cooper, R. M. (1984). In *Plant Diseases: Infection, Damage and Loss* (R. K. S. Wood and G. J. Jellis, eds), pp. 13–27. Blackwell Scientific, Oxford.
Dumas, E., Tahiri-Alaoui, A., Gianninazzi, S. and Gianinazzi-Pearson, V. (1990). In *Endocytobiology IV* (P. Nardon, V. Gianinazzi-Pearson, A. M. Grenier, L. Margulis and D. C. Smith, eds), pp. 153–157. INRA, Paris.
Estrada-Garcia, T., Callow, J. A. and Green, J. R. (1990a). *J. Cell Sci.* **95**, 199–206.
Estrada-Garcia, T., Ray, T. C., Green, J. R., Callow, J. A. and J. F. Kennedy (1990b). *J. Exp. Bot.* **41**, 693–699.
Frens, G. (1973). *Nature Phys. Sci.* **241**, 20–21.
Garcia-Romera, I., Garcia-Garrido, J. M., Martinez-Molina, E. and Ocampo, J. A. (1990). *Soil Biol. Biochem.* **22**, 149–152.
Gay, J. L. and Woods, A. M. (1987). In *Fungal Infection of Plants* (G. F. Pegg and P. G. Ayres, eds), pp. 79–91. Cambridge University Press, Cambridge.
Gianinazzi-Pearson, V. Bonfante-Fasolo, P. and Dexheimer, J. (1986). In *Recognition in Microbe–Plant Symbiotic and Pathogenic Interactions* (B. Lugtenberg, ed.), NATO ASI Series, Vol. H4, pp. 273–282. Springer-Verlag, Berlin.
Gianinazzi-Pearson, V., Smith, S. E., Ginaninazzi, S. and Smith, F. A. (1991). *New Phytologist* **117**, 61–74.
Grandmaison, J., Bemamou, N., Furlan, V. and Visser, S. A. (1988). *Biol. Cell* **63**, 89–100.
Hahlbrock, K., Cuypers, B., Douglas, C., Fritzemeier, K. H., Hoffman, H., Rohwer, F., Scheel, D. and Schultz, W. (1986). In *Recognition in Microbe–Plant Symbiotic and Pathogenic Interactions* (B. Lugtenberg, ed.), NATO ASI Series, Vol. H4, pp. 311–323. Springer-Verlag, Berlin.
Hamer, J. E., Howard, R. J., Chumbley, F. G. and Valent, B. (1988). *Science*, **239**, 288–290.
Harder, D. E. (1989). *Can. J. Plant Pathol.* **11**, 91–99.
Harder, D. E. and Mendgen, K. (1982). *Protoplasma* **112**, 46–54.
Heath, M. C. (1976). *Can. J. Bot.* **54**, 2484–2489.
Ilag, L. L., Rosales, A. M. and Mew, T. W. (1987). In *Mycorrhizae in the Next Decade, Proceedings of the 7th North American Conference on Mycorrhizae* (D. M. Sylvia, L. L. Hung and J. H. Graham, eds), p. 201. Institute of Food and Agriculture Sciences. University of Florida, Gainsville, FL, USA.
Jeffries, P. (1987). In *Fungal Infection of Plants* (G. F. Pegg and P.G. Ayres, eds), pp. 60–78. Cambridge University Press, Cambridge.
Kapooria, R. G. and Mendgen, K. (1985). *Phytopathol. Z.* **113**, 317–323.
Knox, J. P., Day, S. and Roberts, K. (1989). *Development* **106**, 47–56.
Longman, D. and Callow, J. A. (1987). *Physiol. Mol. Plant Pathol.* **30**, 139–150.
Marx, D. H. (1982) *Ann. Rev. Phytopathol.* **10**, 429–454.
Mauch, F., Mauch-Mani, B. and Boller, T. (1988). *Plant Physiol.* **88**, 936–942.
Morandi, D., Bailey, J. A. and Gianinazzi-Pearson, V. (1984). *Physiol. Plant Pathol.* **24**, 357–364.
Nicholson, R. L. and Epstein, L. (1990). In *Fingal Spore and Disease Initiation*

*in Plants and Animals* (S. T. Cole and H. C. Hoch, eds), pp. 3–23. Plenum Press, New York.

Nicholson, R. L., Hoshioka, H., Yamaoka, N. and Kunoh, H. (1988). *Exp. Mycol.* **12**, 336–349.

O'Connell, R. J. (1987) *New Phytol.* **107**, 725–735.

O'Connell, R. J. and Bailey, J. A. (1986). In *Biology and Molecular Biology of Plant–Pathogen Interactions* (J. A. Bailey, ed.), NATO ASI Series, Vol. H1, p. 39. Springer-Verlag, Berlin.

Ohtakara, A. (1988). *Meth. Enzymol.* **161**, 426–430.

Pelissier, B., Thibaud, J. B., Grignon, C. and Esquerre-Tugaye, M. T. (1986). *Plant Sci.* **46**, 103–109.

Peretto, R., Perotto, S., Faccio, A. and Bonfante-Fasolo, P. (1990). *Protoplasma* **155**, 1–18.

Perotto, S. Faccio, A. Malavasi, F. and Bonfante-Fasolo, P. (1988). *Giorn. Bot. Ital.* **122**, 60–61.

Perotto, S. Peretto, R., Moré, D. and Bonfante-Fasolo, P. (1990a). *Symbiosis* **9**, 167–172.

Perotto, S., Vandenbosch, K. A. Brewin, N. J., Faccio, A., Knox, J. P. and Bonfante-Fasolo, P. (1990b). In *Endocytobiology IV* (P. Nardon, V. Gianinazzi-Pearson, A. M. Grenier, M. Margulis and D. C. Smith, eds), pp. 114–117. INRA, Lyon.

Reinhardt, D., Wiemken, A. and Boller, T. (1991) *J. Plant Physiol.* (in press).

Roberts, R. L., Bowers, B., Slater, M. and Cabib, E. (1983). *Mol. Cell Biol.* **3**, 922–930.

Rohringer, R., Chong, J., Gillespie, R. and Harder, D. E. (1989). *Histochemistry* **91**, 383–393.

Roland, J. C. (1978). In *Electron Microscopy and Cytochemistry of Plant Cells* (J. L. Hall, ed.), pp. 1–62. Elsevier, Amsterdam.

Roth, J. (1983). In *Techniques in Immunocytochemistry* (G. R. Bullock and P. Petrutz, eds), pp. 217–284. Academic Press, London.

Scannerini, S., Fieschi, M., Alloatti, G., Sacco, S. and Berta, G. (1990). *Proceedings of the 8th North American Conference on Mycorrhizae*. Jackson, Wyoming, USA.

Schonbeck, F. and Dehne, H. W. (1989). In *Interrelationships Between Microorganisms and Plants in Soil* (V. Vancura and F. Kunc, eds), pp. 83–92. Elsevier, Amsterdam.

Schlumbaum A., Mauch, F., Vogeli, U. and Boller, T. (1986). *Nature* **324**, 365–367.

Smith, D. C. (1990). In *Endocytobiology IV* (P. Nardon, V. Gianinazzi-Pearson, A. M.Grenier, L. Margulis and D. C. Smith, eds), pp. 29–35. INRA, Paris.

Smith, S. E. and Smith, F. A. (1990). *New Phytol.* **114**, 1–38.

Somogyi, M. (1952). *J. Biol. Chem.* **195**, 19–23.

Spanu P. and Boller T. (1989). *J. Plant Physiol.* **134**, 533–537.

Spanu, P. and Bonfante-Fasolo, P. (1988). *New Phytol.* **109**, 119–124.

Spanu, P., Boller, R., Ludwig, A. Wiemken, A., Faccio, A. and Bonfante-Fasolo, P. (1989). *Planta* **177**, 447–455.

Vandenbosch, K. A., Bradley, D. J., Knox, J. P., Perotto, S., Butcher, G. W. and Brewin, J. B. (1989). *EMBO J.* **8**, 335–342.

Wyss, P., Boller, T. and Wiemken, A. (1991). *Experientia* **47**, 395–399.

# 9
# Isozyme Analysis of Mycorrhizal Fungi and their Mycorrhiza

SØREN ROSENDAHL

*Institute for Sporeplanter, University of Copenhagen, Øster Farimagsgade 2 D, DK-1353 Copenhagen K, Denmark*

ROBIN SEN*

*Department of Botany, University of Helsinki, Unioninkatu 44, SF-00170 Helsinki, Finland*

## I. Introduction

Electrophoretic analysis of isozyme (isoenzyme) variation has long been a valuable tool in systematic and population genetic studies of bacteria

*Present address: Department of Forest Protection, Finnish Forest Research Institute, PO Box 18, SF-01301 Vantaa, Finland.

METHODS IN MICROBIOLOGY
VOLUME 24   ISBN 0-12-521524-X

Copyright © 1992 by Academic Press Limited
All rights of reproduction in any form reserved

(Selander *et al.*, 1986), plants (Tanksley and Orton, 1983) and fungi (Micales *et al.*, 1986). The first applications of electrophoretic techniques in mycorrhizal studies enabled the separation of fungal proteins in *Eucalyptus–Pisolithus* ectomycorrhiza (Seviour and Chilvers, 1972) and the detection of fungus and plant-specific acid and alkaline phosphatase isozymes within vesicular-arbuscular mycorrhizal roots (Gianinazzi-Pearson and Gianinazzi, 1976). More recently, isozyme techniques have featured in taxonomic and population genetic studies of vesicular-arbuscular (Sen and Hepper, 1986; Hepper, 1987; Hepper *et al.*, 1988a; Rosendahl, 1989) and ectomycorrhizal fungi (Ho and Trappe, 1987; Ho, 1987; Zhu *et al.*, 1988; Sen, 1989, 1990a). The methodology has also been further extended to enable the identification of different fungal symbionts in endo- (Hepper *et al.*, 1986, 1988b; Rosendahl and Hepper, 1987) and ectomycorrhiza (Sen, 1990b). These preliminary investigations indicate that isozyme methodology has considerable potential in the study of mycorrhizal host–plant interrelationships at both the biochemical and ecological levels.

This chapter describes isozyme methods for the analysis of vesicular-arbuscular and ectomycorrhizal fungi and for the identification and quantification of mycorrhizal fungi within plant roots. Present areas of application are reviewed briefly.

## II.   Methodology

### A.   Vesicular-arbuscular mycorrhizal sample preparation

As vesicular-arbuscular mycorrhizal fungi cannot be cultured axenically, fungal material in the form of resting spores or external and/or internal root mycelium, together with surface-cleaned infected root systems, are routinely analysed. These samples can be prepared and extracted as follows.

### 1.   Spore extracts

Spores of vesicular-arbuscular mycorrhizal fungi are the easiest structures to recover and extract. It is recommended that spores of a reasonable size ($> 80\ \mu$m diameter) are analysed initially, as they are much simpler to handle than those of certain species which form spore clusters. In the case of the larger *Glomus* species (e.g. *G. mosseae* and *G. caledonium*) only a few spores per sample are required, providing that electrophoretic separation is carried out in thin polyacrylamide gels

(70 mm × 70 mm × 0.8 mm) with small well capacities (13 $\mu$l). Immediately before electrophoresis, spores can be subjected to either of two extraction procedures.

(1) Spores (5–10 per sample) are collected in a 500 $\mu$l Eppendorf centrifuge tube to which is added 10–15 $\mu$l cold extraction buffer (EB: 10 mM Tris-HCl, 10 mM NaHCO$_3$, 10 mM MgCl$_2$, 0.1 mM Na-ethylenediaminetetraacetic acid (NaEDTA) and 10 mM $\beta$-mercaptoethanol, pH 8.0) containing 150 g litre$^{-1}$ sucrose (SEB) and 1 ml litre$^{-1}$ Triton X-100 (STEB). The spores are crushed with a glass pestle made from a flamed Pasteur pipette tip and the tubes centrifuged at 10 000 $g$ for 20 min (at 4 °C). A 10 $\mu$l aliquot of the supernatant is then ready for loading into a well. If the spores are crushed in a micro-glass homogenizer before transfer to the Eppendorf tube for centrifugation, maximum recovery of extract may not be possible.

(2) Spores are transferred to an ice-cold 5 $\mu$l drop of bromophenol blue (BPB) (20 mg litre$^{-1}$) amended SEB on a chilled Petri dish lid. The spores are crushed using fine forceps and the extract mixed with a further 5 $\mu$l of ice-cold STEB (Sen and Hepper, 1986).

Method 1 gives reproducible results, although it is difficult to ensure that all spores are broken, while in Method 2 some loss of extract is inevitable and spore wall fragments remain in the extract prior to loading into the wells.

The spores of cluster-forming species (e.g. *G. deserticola* and *G. etunicatum*) may be more difficult to extract. Clusters recovered from soil sievings should be washed in sterile tap water and examined for contaminating fungi and detritus before transfer to Eppendorf tubes. After centrifugation for 15 min at 2500 $g$ and removal of the supernatant, the spore clusters remaining can be subjected to the Eppendorf extraction procedure described (Method 1).

## 2. Extraction of infected roots

Surface-cleaned and washed whole roots or segments representing fresh, deep-frozen (−75 °C) or freeze-dried material can be used.

(1) Fresh roots are extracted by grinding in an ice-chilled mortar with cold STEB (1 ml g$^{-1}$ roots) containing PVPP (polyvinylpolypyrrolidone, Sigma) or Polyclar AT (insoluble polyvinylpyrrolidone, BDH) (8 mg ml$^{-1}$). Roots can also be similarly extracted in a

glass homogenizer with a larger volume of SEB. After homogenization the slurry is transferred to a centrifuge tube and centrifuged at 17 000 $g$ for 30 min (at 4 °C). In both cases concentration of these crude extracts may be necessary (see below).

(2) More efficient extraction of pre-weighed fresh or deep-frozen root samples can be achieved by grinding to a powder in a mortar containing liquid nitrogen. The resulting powder is transferred to a centrifuge tube and suspended in cold EB (5 ml g$^{-1}$ roots) containing Polyclar AT (8 mg ml$^{-1}$). After centrifugation for 20 min at 23 000 $g$ (at 4 °C) the supernatant is concentrated (Hepper et al., 1986, 1988b).

(3) Freeze-dried root material also retains enzyme activities, but as both proteins and salts are concentrated during the drying procedure it is important to pre-wash the roots in distilled water. Samples are then ground in a mortar. For quantitative studies EB is added at a rate of approximately 10 $\mu$l mg$^{-1}$ of sample, before centrifugation at 17 000 $g$ for 30 min (at 4 °C). Extracts prepared by this method may contain 3–4 $\mu$g protein $\mu$l$^{-1}$ and can be run directly without the need for a protein concentration step. In all cases it is advisable to wash the pellet with aliquots of EB to ensure maximum protein recovery. The root pellets should be retained, stored at −75 °C and later subjected to a glucosamine assay (Hepper, 1977), enabling a direct correlation between specific isozyme activity and total infection to be obtained (Rosendahl et al., 1989a).

## 3. Concentration of proteins

Where protein levels in fresh and deep-frozen root extracts are low, concentration of the extracts is essential. Freeze-drying of the extracts cannot be recommended, as the procedure causes further increases in salt concentration. High salt concentrations in the extracts disturb both enzyme mobility during electrophoresis and the staining procedure.

Proteins in the root extracts can be concentrated (100 to 300-fold) to the low volumes (30–50 $\mu$l) needed for electrophoresis using two membrane-based systems.

(1) Static concentration of large volumes of extract (10–15 ml) can be made in a Minicon B15 or CS15 concentrator unit containing a 15 kD cut-off membrane (Amicon Corp., USA). Chambers are filled and maintained overnight at 4 °C. The resulting concentrated extract (30–50 $\mu$l) is then removed from each chamber using a finely drawn out Pasteur pipette and either frozen in liquid

nitrogen and stored at −75 °C or immediately prepared for electrophoresis (Hepper *et al.*, 1986, 1988a).
(2) A second, more rapid, system (1 h) involves concentration by centrifugation. These concentrators (Sartorius, Germany) are available with different molecular weight cut-off membranes. A 20 kD cut-off is recommended.

Both methods, however, are expensive and time-consuming. Some loss of sample is inevitable which may result in an underestimation of total enzyme activities.

### 4.  Recovery of internal mycelium

Internal mycelium is recovered from roots using a modification of earlier enzymic digestion procedures (Capaccio and Callow, 1982; Smith *et al.*, 1985) in order to obtain the maximum amount of active fungal tissue with the least amount of plant contamination.

Infected root segments are incubated in 10 mM 2(*N*-morpholino) ethanesulphonic acid (MES) containing 10% sorbitol, 2% (w/v) cellulase R-10 (Yakult Honsha Co.), 1% (w/v) pectinase (Sigma) and 0.5% (w/v) hemicellulase (Sigma) (pH 5.0) at 7 °C for 16 h. After incubation, the roots are washed in 10 mM MES containing 10% sorbitol and maintained at 4 °C. Internal mycelium dissected from the roots using fine needles and forceps is then prepared directly for extraction or frozen in liquid nitrogen and stored at −75 °C. Efficient recovery of internal mycelium from leek (*Allium porrum*) (Hepper *et al.*, 1986) has been possible, but recovery from other plant species such as maize (*Zea mays*) and cucumber (*Cucumis sativus*) has not been as successful.

### 5.  Recovery of external mycelium

Infected roots are submerged in cold water (4 °C) and the external mycelium emanating from the root surface is freed of soil particles, sporocarps and resting spores under a binocular microscope using fine forceps before transfer to cold EB. After removal of excess buffer on a filter paper, the mycelium is weighed and either directly prepared for enzyme extraction or stored at −75 °C after a preliminary freezing in liquid nitrogen.

### 6.  Extraction of external and internal mycelium

Fresh or deep-frozen mycelium is transferred to a 100 µl glass homogenizer and extracted with EB (1 µl mg$^{-1}$ mycelium) containing Polyclar

AT (30 mg ml$^{-1}$) on ice. After extraction the cell debris can be removed by centrifugation and the extract stored at $-75\,°C$, following pre-freezing in liquid nitrogen. Before electrophoresis 5 $\mu$l samples of extract are mixed with an equal volume of STEB containing BPB (20 mg litre$^{-1}$).

## B. Standardization of extracts

In order to standardize the amount of extracts for quantitative studies, two methods should be considered:

(1) *Dry weight.* Vesicular-arbuscular mycorrhizal infection is known to alter the percentage dry weight of the host plant, as mycorrhizal roots are often heavier than uninfected roots. As this difference may not be apparent from fresh weight determinations, it is advisable to quantify a representative sample of the root material on a dry weight basis. This is achieved most simply with freeze-dried root material.

(2) *Protein content.* The samples can also be quantified on the basis of their protein content (Lowry *et al.*, 1951; Bradford, 1976). The method measures the total protein content colorimetrically using bovine serum albumin (BSA) as a standard. The protein content obtained will be a sum of all proteins, including non-enzymic structural proteins. As most isozyme electrophoresis studies are made on crude extracts, it is important to bear this in mind when trying to quantify specific isozyme activities. Protein concentration of the extract can also be used as a measure of the efficiency of the extraction procedure.

## C. Ectomycorrhizal sample preparation

Unlike the fungi forming vesicular-arbuscular mycorrhizal associations, it is possible to culture many of the ectomycorrhizal fungi, particularly members of many Ascomycete and Basidiomycete families. The production of taxonomically recognizable fruiting structures (also a source of extractable tissue) makes it possible to isolate dikaryotic and, in certain cases, monokaryotic mycelium from individuals of the same species. Ectomycorrhiza are also more easily recognizable, although amounts of infected root material may vary considerably depending on the host species (see Agerer, 1987).

## 1. Recovery of sporocarp tissue

Where ectomycorrhizal cultures are not available or field studies are being made it is possible to extract sporocarp tissues as follows.

Young fleshy sporocarps (e.g. mushroom-forming basidiomycetes) should be harvested from the field and immediately brought to the laboratory. Overnight storage at 4 °C is possible but rapid processing is advisable. Cap or stipe tissues should be exposed in a sterile laminar flow cabinet and samples contaminated by larvae, etc. rejected. Tissues should be cut out using a sterile scalpel and a portion set aside for culture. The remaining tissue should be weighed and then frozen in liquid nitrogen before storage at −75 °C. Basidiospores should also be collected and cultured where possible.

## 2. Mycelial culture

Mycelium can be grown in either (1) agar or (2) liquid culture.

(1) Autoclaved cellophane discs (Courtaulds British Cellophane Ltd) are placed on the surface of the solidified media (e.g. Modified Melin-Norkrans of Hagems medium; Molina and Palmer, 1982) and inoculated with agar plugs of the test fungus. After incubation at 25 °C (mycelium should be harvested before the cessation of radial growth) the central portion of mycelium, containing the inoculum, should be removed with a scalpel. The remaining colony is separated from the cellophane, weighed and either frozen in liquid nitrogen and stored at −75 °C or immediately extracted.

(2) Liquid media (see Molina and Palmer, 1982) contained in flasks or bottles should be inoculated with agar plugs of inocula and incubated on a rotary shaker. After sufficient growth has occurred (not longer than 4 weeks) the culture is vacuum filtered through a membrane (Millipore) and the trapped mycelium washed with distilled water. Mycelium gently removed from the membrane surface is then treated as in Method 1.

Although the second method of culture produces large amounts of mycelium, not all ectomycorrhizal fungi can grow in submerged culture. Agar culture using cellophane membranes would therefore seem to be the method of choice for most fungi studied.

## 3. Extraction of sporocarp tissues and mycelium

Fresh or frozen samples (40–200 mg) are ground to a fine powder in a mortar containing liquid nitrogen and 20 mg Polyclar AT before transfer to a chilled Eppendorf tube. The powder is suspended in STEB (see above) (2 mg fresh wt $\mu l^{-1}$) at 4 °C and centrifuged for 1 min at 16 000 $g$ in an Eppendorf centrifuge. Following assay of the soluble protein content (Bradford, 1976) of the supernatant, aliquots containing 5, 10 and 15 $\mu g$ soluble protein are frozen in liquid nitrogen and stored at −75 °C. The remaining fungal pellet can also be deep-frozen (−75 °C) and later subjected to glucosamine determination (Hepper, 1977). For isozyme analysis, standard amounts of protein are always loaded onto the polyacrylamide gels (Sen, 1990a).

### 4.  Preparation and extraction of host roots

Isozyme analysis for identification studies has only been carried out on ectomycorrhiza of *Pinus sylvestris* L. (Sen, 1990b) and *Pinus cembra* L. (G. Keller, pers. commun.). The following procedures may therefore require some modification for ectomycorrhiza of other host species.

Fresh or frozen (−75 °C) non-mycorrhizal root material, harvested from sterile growth media (e.g. nutrient agar, liquid media, peat and artificial substrates), is processed either as (1) whole root samples or (2) after fractionation into component parts.

(1) Whole roots are washed, extracted and concentrated as described for vesicular-arbuscular mycorrhizal roots.
(2) Bulked samples of excised short roots and long root tips from individual plants are ground in a glass homogenizer with STEB (approx. 1 mg fresh wt 10 $\mu l^{-1}$). Aliquots containing 2–5 $\mu g$ protein (Bradford, 1976) are loaded onto the gels for electrophoresis.

### 5.  Preparation and extraction of whole ectomycorrhiza

Synthesized or natural ectomycorrhizal roots are thoroughly washed free of growth substrate and submerged in cold water under a binocular microscope. Ectomycorrhiza, identified visually, are excised (from the surface of the long roots in *Pinus* spp.) and individually transferred to Eppendorf tubes, rapidly frozen in liquid nitrogen and stored at −75 °C. Each ectomycorrhiza is ground in a 100 $\mu l$ glass homogenizer (Uniform, Jencons) containing 30 $\mu l$ STEB and 5 $\mu g$ Polyclar AT at 0 °C. The extract is removed from the homogenizer with a finely drawn out Pasteur pipette into an Eppendorf tube and centrifuged for 1 min at 16 000 $g$. A 10 $\mu l$ or 15 $\mu l$ aliquot (0.8–2.5 $\mu g$ protein) is then prepared for gel electrophoresis.

*Pinus cembra* ectomycorrhiza have also been successfully extracted in an Eppendorf tube using a tight-fitting pestle (G. Keller, pers. commun.). The procedure described above has been developed for individual ectomycorrhiza of *P. sylvestris* where the amount of infected tissue is low (0.3–0.8 mg fresh wt). Other host-fungus combinations are known to produce extensively ramified ectomycorrhiza (Agerer, 1987) which will produce larger volumes of extract per mycorrhiza for analysis.

### 6.   *Dissection and extraction of ectomycorrhiza tissue fractions*

Individual *P. sylvestris-Suillus* ectomycorrhiza have also been separated into component sheath and core (host/Hartig net tissue) fractions for isozyme analysis (Sen, 1990b). In similar studies the core tissue of *Picea abies* and *Fagus sylvatica* ectomycorrhiza has been further dissected into vascular cylinder and cortical tissue fractions (Chalot *et al.*, 1990). In both studies the component tissues were dissected under a binocular microscope.

For *P. sylvestris-Suillus* ectomycorrhiza sheath tissue is removed from individual ectomycorrhiza submerged in cold SEB (0 °C) using fine forceps and needles. A sheath tissue fraction is then examined for host contamination and the samples immediately ground in the micro-homogenizer with 30 $\mu$l of STEB containing 5 $\mu$g Polyclar AT at 4 °C. Sheath and core tissue extracts each containing 1.5 $\mu$g protein are loaded into individual wells.

## D.   Electrophoresis

### 1.   *Supporting medium*

Protein extracts can be subjected to electrophoretic separation in a number of gel formats in which starch or polyacrylamide are routinely used as the supporting medium. In fungal population studies, starch gel electrophoresis has been popular due to the ease of gel preparation and potential for analysis of many isolate/enzyme combinations (Micales *et al.*, 1986). However, polyacrylamide gel electrophoresis (PAGE) is favoured by us because:

(1) Polyacrylamide gels produce better resolved banding patterns as proteins are separated not only on the basis of net charge, a feature common to both media, but also on the basis of molecular weight.

(2) Enzyme activities in polyacrylamide gels are quantifiable through the loading of extracts of known volume or protein concentration,

unlike in starch gels where extracts absorbed onto filter paper wicks are analysed.

These features of PAGE are particularly important for identification and quantification studies of mycorrhizal infections within host roots as it is important to separate the host and fungal activities clearly for diagnosis.


## 2.  The electrophoretic system

Zonal electrophoresis is the most accessible system. Proteins can be separated in either a gel containing a fixed concentration, or a composite gel with two different concentrations of acrylamide. Moreover, composite gels with different buffer systems (discontinuous systems) can be used. The discontinuous buffer system is recommended for extracts with low protein concentration, as the proteins are first concentrated in a stacking gel. For mycorrhizal studies a vertical discontinuous system is recommended in which proteins migrating anodally are first concentrated at low pH in a large pore (low strength) acrylamide stacking gel before separation in a supporting higher pH, small pore running gel. A number of manufacturers (e.g. LKB, Pharmacia and Hoeffer) produce vertical electrophoresis systems that accept small sized gels (approx. 70 mm × 70 mm) containing 6–15 wells. The gel is cast in a cassette made of glass plates separated by spacers (0.8, 1.5 or 3.0 mm) which determine the final gel thickness. Large numbers of gels (10 or 15) can also be cast from the same acrylamide solutions in specially designed casting boxes which saves time and improves reproducibility.


## 3.  Preparation of polyacrylamide gels and electrode buffers

Two different gel and electrode buffer combinations are routinely used.

(a) Combination 1. In most of our mycorrhizal studies stacking gels containing 3.0% acrylamide or running gels containing 6.0 or 7.5% acrylamide in 70.6 mM Tris-HCl (pH 7.8) are freshly prepared as follows:

Stacking gel. 1.5 g acrylamide, 0.375 g bis-acrylamide and 0.515 g Trisma base are dissolved in 30 ml distilled water. 35 $\mu$l TEMED are added and the solution adjusted to pH 7.8 with conc. HCl. The solution is made up to 45 ml and stored at 4 °C for 1 h.

*Running gel* (6.0% acrylamide). 6.95 g acrylamide, 0.25 g bis-acrylamide and 1.03 g Trisma base are dissolved in 70 ml distilled water. 44 μl TEMED are added and the solution adjusted to pH 7.8 with conc. HCl. The resulting solution is made up to 90 ml. The solution is de-aerated under vacuum and 84 mg ammonium persulphate and 30 ml cold 1.6% Triton X-100 immediately added and gently mixed. This solution is then poured directly into a gel casting box containing 10 gel cassettes (65 mm × 80 mm × 1.5 mm) to a height of 55 mm. The surface of each gel is then layered with isobutanol and the acrylamide allowed to polymerize for 1 h (for 7.5% acrylamide gels use 8.85 g acrylamide, and 0.274 g bis-acrylamide).

After polymerization the isobutanol is removed and the gel surfaces washed in several changes of distilled water and drained. The cold stacking gel solution is then de-aerated and amended with 42 mg ammonium persulphate and 15 ml cold Triton X-100 (1.6%) before being poured over the gels. A comb containing 15 wells is immediately inserted into each gel and the acrylamide is then allowed to polymerize for 1 h. Gels can be stored for up to two weeks at 4 °C. It is important to note that acrylamide is a potent neurotoxin so extreme care should be taken when handling this chemical.

*Electrode buffer.* Gels are run submerged in 8.3 mM Tris-barbitone buffer (pH 7.45). 5 g Trisma base and 27.6 g barbitone (BDH) are dissolved in 4500 ml distilled water on a magnetic stirrer and made up to 5000 ml. The buffer should be stored at 4 °C before use.

*(b) Combination 2.* A discontinuous buffer system that has been found to be useful in studies of several fungi including vesicular-arbuscular mycorrhizal fungi is given below.

Gels are prepared from stock solutions which can be stored at 4 °C for several weeks, thus limiting the exposure to acrylamide.

(i) Acrylamide (Stock A): 30 g acrylamide and 0.8 g bis-acrylamide are dissolved in 100 ml distilled water and the solution filtered.

(ii) Running gel buffer (Stock B): 36.3 g Trisma base are dissolved in 70 ml distilled water, adjusted to pH 8.8 with conc. HCl and made up to a volume of 100 ml.

(iii) Stacking gel buffer (Stock C): 6 g Trisma base are dissolved in 70 ml distilled water, adjusted to pH 6.8 with conc. HCl and made up to 100 ml.

*Running gel* (7.5% acrylamide, 375 mM Tris-HCl). 7.5 ml Stock A, 3.75 ml Stock B and 17.5 ml distilled water are mixed together and de-aerated. 30 mg ammonium persulphate, 1.25 ml Triton X-100 (2%) and 30 $\mu$l TEMED are added to the solution and mixed gently before pouring over the gel cassettes. Space should be left above for the stacking gel. The best concentration of proteins is obtained if the stacking gel is twice the height of the sample loaded in the well. The acrylamide surface is overlaid with 40% ethanol and allowed to poly- merize for 1 h, after which time the ethanol is poured off and the gel surfaces washed in distilled water.

*Stacking gel* (3.8% acrylamide, 125 mM Tris-HCl). 2.5 ml Stock B, 5 ml Stock C and 11 ml distilled water are mixed together and de-aerated. A further 20 mg ammonium persulphate, 20 $\mu$l TEMED and 1.5 ml Triton X-100 (2%) are added, poured over the gels, and the combs are inserted. After polymerization the gels are ready for use and should not be kept longer than 24 h.

*Electrode buffer* (24.8 mM Tris-glycine (0.19 M), pH 8.3). 3 g Trisma base and 14.4 g glycine are dissolved in 1000 ml distilled water and stored at 4 °C.

Between 1 and 4 small format gels can be run in most vertical electrophoresis systems. Gels in their glass cassettes, with the exposed wells facing upwards, are then inserted into the electrophoresis tank so that the bottom half is submerged in a large volume (e.g. 3500 ml Tris-barbitone) of cold buffer (approx. 4 °C). The upper half of the gel, which is isolated from the lower anode buffer tank, is then filled until the buffer level is about 8–10 mm above the top edge of the gel.

## 4. Gel loading

A small volume (2–4 $\mu$l) of bromophenol blue (BPB) (20 mg litre$^{-1}$ STEB) tracking dye, which will follow the buffer front, can either be incorporated with each extract or pipetted into empty wells of each gel just before loading. Extracts are then loaded into the submerged wells using a 10–20 $\mu$l automatic pipette or syringe.

## 5. Gel running

After the gels are loaded the tank is connected to a power supply and a constant current of 15 mA per gel is applied. The choice of running time (2.5–6.0 h), based on the distance of migration of the BPB dye front

(15–60 mm), is dependent on the fungi being studied, the pH and pore size of the gel and the enzyme to be stained. When large numbers of samples are being analysed on two or more gels, standardization of gel runs can be achieved by loading reference extracts on all gels and by ensuring that the BPB tracking dye migrates the same distance in each gel before completion of the run. After electrophoresis the gel is removed from the glass cassette and submerged in an enzyme staining solution.

## E.  Enzyme staining

Gel stains have been developed for the detection of the most central and secondary metabolic enzyme activities (Harris and Hopkinson, 1977; Shaw and Prasad, 1970). In our mycorrhizal investigations the following enzyme activities have been regularly visualized: malate dehydrogenase (MDH) (Figs 1 and 2), esterase (EST) (Fig. 3), peptidase (PEP)

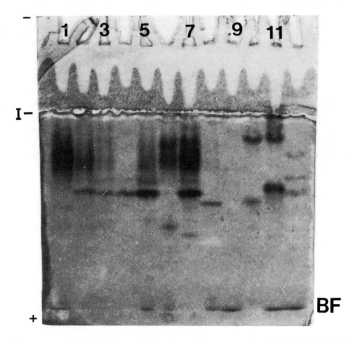

**Fig. 1.** Extracts of spore cluster-forming *Glomus* spp. stained for MDH. Lanes 1–5 and lane 7 are morphologically similar isolates showing highly conserved banding patterns. The remaining lanes show banding patterns of morphologically dissimilar isolates. The isolate in lane 6 was used for inoculation of cucumber shown in Fig. 2. I, interface between stacking and running gels; BF, buffer front.

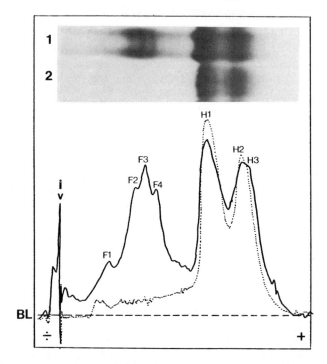

**Fig. 2.** Lanes showing infected (lane 1) and uninfected cucumber roots (lane 2) stained for MDH and the densitometric tracings of these lanes. The solid line is the tracing of the root system of cucumber infected with a *Glomus* sp. (lane 1). The dotted line represents similar non-inoculated cucumber roots (lane 2). Peaks F1–F4 and H1–H3 are the respective fungal and plant MDH activites. BL, baseline; I, gel interface.

(Fig. 4), glutamate oxaloacetate transaminase (GOT), hexokinase (HK), phosphoglucomutase (PGM), malic enzyme (ME), acid phosphatase (ACP), peroxidase (PER), glucose 6-phosphate dehydrogenase (Gd) and glutamate dehydrogenase (GDH). In most cases the stain solution is a buffer, at optimum pH, containing a natural or artificial substrate, cofactors, and a dye. On incubation coloured bands appear in the gel at positions of isozyme activity. Recipes for making sufficient staining solution for a single gel are as follows (all chemicals from Sigma):

*Malate dehydrogenase* (EC 1.1.1.37). 150 mg malic acid and 2.5 ml unbuffered 0.5 M Trisma base are dissolved in 20 ml distilled water and adjusted to pH 8.0 with 1 M KOH. The solution is made up to 25 ml and 5 mg NAD$^+$, 5 mg nitroblue tetrazolium (NBT) or MTT and 3 mg phenazine methosulphate (PMS) are added.

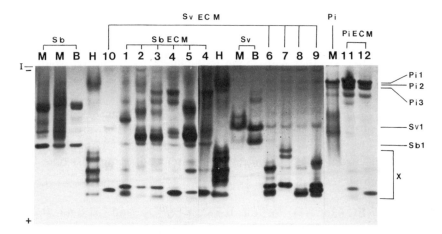

**Fig. 3.** Esterase isozyme activities of individual *P. sylvestris* ectomycorrhiza (ECM) synthesized between *Suillus bovinus* (Sb), *S. variegatus* (Sv) and *Paxillus involutus* (Pi). Lanes have been loaded with: bulked host radicle extracts (H) (8 μg protein), mycelial extracts (M) (1.5–3.0 μg protein), basidiocarp tissue extracts (B) (1.5 μg protein), extracts of individual *S. bovinus* (ECM1-5), *S. variegatus* (ECM6-10) and *P. involutus* (ECM11-12) ectomycorrhiza (0.8–2.5 μg protein). Conserved species diagnostic bands (Pi1, 2 and 3; Sv1; Sb1) are indicated. Additional induced isolate-specific bands in *S. bovinus* ectomycorrhiza (ECM1-5) visible in the upper half of the gel and common bands (group X) in all ECM extracts were accountable in sheath tissue extracts (Sen, 1990b). I, gel interface.

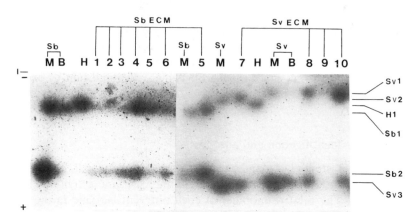

**Fig. 4.** Extracts from the same individual *S. bovinus* (Sb) and *S. variegatus* (Sv) ectomycorrhiza of *P. sylvestris* stained for peptidase (PEP). Extract designations are as in Fig. 3. Upper diagnostic Sv (Sv1 and Sv2) and Sb (Sb 1) bands are separated by the single host (H1) band. The lower bands (Sb2 and Sv3) are also diagnostic at the species level. I, gel interface.

*Esterase* (EC 3.1.1.1). To 2.5 ml of 0.1 M Tris-HCl (pH 7.1) 22.5 ml distilled water are added, 10 mg α-naphthyl acetate and/or β-naphthyl acetate are dissolved in 1 ml 50% acetone and suspended in the Tris-HCl buffer together with 25 mg Fast blue BB salt and the solution is filtered.

*Glutamate oxaloacetate transaminase* (EC 2.6.1.1). 73 mg α-ketoglutaric acid, 266 mg PVP-40, 100 mg EDTA and 2.84 g $Na_2HPO_4$ are dissolved in 100 ml distilled water and pH is adjusted to 7.4. 12.5 ml of this solution are mixed with 12.5 ml distilled water and 50 mg Fast blue BB salt.

*Peptidase* (EC 3.4.1.1). 4.7 ml 0.2 M $KH_2PO_4$ and 31.8 ml 0.2 M $Na_2HPO_4$ are mixed and the solution is diluted to 100 ml. 3.5 ml 0.1 M $MnCl_2$ are added to this solution, which is then filtered. To 12.5 ml of the filtrate 10 mg L-leucylglycine, 5 mg *Bothrops atrox* snake venom and 10 mg peroxidase are added. 19 mg 3-amino-9-ethyl carbazole in 0.5 ml *N,N*-Dimethyl formamide are mixed and added to the substrate solution. 200 mg agar are dissolved in 12.5 ml boiling distilled water and maintained at 55–60 °C.

The first three stains (MDH, EST and GOT) are poured onto the gel immediately after preparation and incubated at room temperature. All stains containing PMS and MTT or NBT (e.g. MDH) must be incubated in the dark. For peptidase staining the warm agar solution is rapidly mixed with the substrate/dye solution and poured over the gel. After the agar has set the gel is incubated at 37 °C until peptidase bands have developed in the agar overlay. Most stained gels can be fixed in 7% acetic acid and stored in 50% glycerol. Again it is important to exercise care when handling staining chemicals as many are toxic.

## F.  Isozyme identification and analysis

For each gel stained a number of bands of activity will be detected representing different isozymes with a common enzyme activity (see Figs 1, 3 and 4). The relative mobility of each different isozyme in the gel is dependent on its net charge, size and shape. As these variables are directly related to the amino acid composition of the molecule, the relative mobility can be directly correlated with variations in the nucleotide sequence of the coding DNA. Differences between individuals can thus be identified in the variations of their banding patterns.

*1. Mobility and activity measurements*

The relative mobilities of each isozyme band in the gel can be recorded either manually or by densitometry.

*(a) Manual measurement.* The distances of migration of all bands in each lane, representing the isozyme activities of a single isolate, should be measured as accurately as possible from the gel interface between the stacking and running gel. Bands can be seen more clearly if the gels are placed on top of a ground glass screen above a bright light source. For permanent records the gels are photographed.

*(b) Densitometric measurement.* The gels can also be scanned by a densitometer (Rosendahl *et al.*, 1989a) or image analyser. These instruments both determine the mobilities and measure intensities of the bands, providing information on specific isozyme activities (Fig. 2).

*2. Numerical analysis*

For analysing isozyme variation in fungal populations a number of numerical methods can be applied to the banding data (Sneath and Sokal, 1973). All possible band positions should be numbered and the presence or absence of bands in each lane recorded from the mobility measurements. Matrices (e.g. squared Euclidean distance matrix) can then be constructed from the absence/presence data and similarities between isolates then assessed using unpaired group mean average (UPGMA) cluster analysis. Relationships between the isolates are normally displayed in the form of dendograms (Hepper *et al.*, 1988a; Rosendahl, 1989; Sen 1990a).

*3. Genetic analysis*

Further information on the genetic and nuclear condition of a fungal isolate can be inferred from certain characteristic isozyme banding patterns (Micales *et al.*, 1986). These patterns result from isozyme expression that is dependent on one or a combination of three different phenomena: (1) multiple allelism at a single genetic locus; (2) multiple loci coding for single enzymes; and (3) post-translational modifications and the formation of secondary isozymes (Harris and Hopkinson, 1976).

Genetic interpretations of banding patterns are normally made with reference to enzyme subunit compositions (Darnall and Klotz, 1972).

Multiple allelism at a single locus can be detected as patterns of closely migrating bands. As each allele codes for a structurally distinct polypeptide subunit, the banding pattern expressed will be dependent on both the subunit composition of the functional isozyme and the number of alleles present at each locus. The latter property will be dependent on the nuclear condition of the fungus (i.e. monokaryotic or dikaryotic), ploidy number and whether the fungus is homozygous or heterozygous. Fungi that are diploid and heterozygous expressing two separate alleles will produce more complex banding patterns than those that are haploid or homozygous. Based on the banding patterns observed, vesicular-arbuscular mycorrhizal fungi are considered to be haploid (S. Rosendahl, unpubl. res.). Ectomycorrhizal fungi studied have exhibited more complicated dikaryotic (or diploid equivalent) banding patterns (Zhu *et al.*, 1988; Sen, 1990a).

Multiple loci, originating from gene duplications, often code for isozymes which show greater structural variation and thus appear in very different regions of the gel. The expression of these loci is also related to the development state of the fungus. The phenomenon occurs if the nuclear conditions are not similar in all parts of the fungus, e.g. monokaryotic and dikaryotic stages, but has also been observed between different parts of the basidiocarp in *Coprinus cinereus* (Moore and Jirjis, 1981) and *Agaricus bisporus* (Paranjpe *et al.*, 1970). In order to interpret isozyme variation conclusively, the genetic background for the isozyme pattern should be clarified (Scandalios, 1974; Burdon *et al.*, 1986; Roux and Labarere, 1991). Mycorrhizal examples of multiple loci expression include a putative monomorphic PEP locus expressed in internal and external mycelium of infected roots, but not in spores of *G. caledonium* and *G. mosseae* (Hepper *et al.*, 1986), and basidiocarp- and mantle-specific monomorphic EST and ACP loci in *Suillus bovinus* and *S. variegatus* (Sen, 1990a,b).

Secondary isozymes can arise from non-genetic post-translational modifications or result from proteolytic activity during enzyme extraction. Bands resulting from non-genetic modifications are easily identifiable as groups of several closely migrating bands with lower activities than the original unmodified band.

Due to space restrictions no fuller treatment of this important approach is given, but the reader is directed to the review by Micales *et al.* (1986) and fungal source papers (May *et al.*, 1979; Royse and May, 1982) for details of the banding patterns that are detectable and the genetic nomenclature used.

### III. Applications

To date the isozyme analysis of vesicular-arbuscular and ectomycorrhizal fungi has been restricted to only a few studies but potentially important information has been obtained on the taxonomic and ecological relationships between individuals. Our ability to differentiate and identify fungal and host activities within mycorrhiza, a relatively novel application of the technique, is also of value for taxonomic, biochemical and ecological studies of the mycorrhizal host–plant relationship.

#### A.  An aid to taxonomy

Morphological characteristics are traditionally used in the species designation of fungi. However, identification problems persist in certain fungal species because many morphological characters show different degrees of environmental variability.

In an effort to overcome problems of morphological instability and lack of definable traits (e.g. the perfect state) other chemotaxonomic methods have been developed. Isozyme analysis has been one of these methods used in taxonomic studies of several controversial groups of fungi. As the banding patterns represent direct gene products they reflect the fungal genotype more closely than most morphologically displayed traits. Relationships between and within groups of fungi have been characterized in plant pathogens such as *Botrytis* (Backhouse *et al.*, 1984), *Pseudocercosporella* (Julian and Lucas, 1990), *Erysiphe* (Koch and Kohler, 1990) and *Armillaria* (Morrison *et al.*, 1985).

The identification of vesicular-arbuscular mycorrhizal fungi (Glomales) has been further confounded by their unculturability and the considerable environmental plasticity that can be displayed by the key resting spore structures.

Isozyme analysis has now been applied in taxonomic studies of vesicular-arbuscular mycorrhizal fungi, particularly the *Glomus* spp. (Sen and Hepper, 1986; Hepper, 1987; Hepper *et al.*, 1988a; Rosendahl and Hepper, 1987; Rosendahl, 1989). An important feature of this analysis was the stability of the EST, MDH, PEP and PGM isozyme patterns displayed by *G. caledonium* spores taken from different host and soil combinations. Similar stability of spore morphological characters has been shown by Morton (1985). In these preliminary studies we were able to detect clear interspecific variation between certain species groupings. Morphologically similar isolates forming the species grouping *G. clarum/G. manihotis* were clearly separated from another grouping containing *G. mosseae/G. monosporum*. However, high levels of isozyme variation within the *G. mosseae* grouping were also detected. Species designation can still be very difficult, even though morphological characters of the Glomales have been subjected to cladistic analysis (Morton, 1990). Several epithets may cover several distinct species. Walker and Koske (1987) have redescribed *G. fasciculatum*, but the epithet is still used for almost all spore cluster-forming *Glomus* spp. Similar descriptions of *G. mosseae* have not been made, although several distinct species are covered under this epithet. Almost all laboratories involved in mycorrhizal research are in possession of an isolate termed *G. mosseae* but isozyme analysis has revealed that some of the isolates bearing this particular epithet are genetically very distant (Hepper *et al.*, 1988a).

Fortunately, the taxonomy of the higher ectomycorrhizal fungi is on a sounder footing, although many of these fungi are also unculturable and of unknown affinities. Ho and Trappe (1987) used acid phosphatase isozyme analysis to separate sections and species of *Rhizopogon*. They found that isozyme mobilities differed greatly between host-related sections but were more conserved within species. In a comparison of the isozyme banding patterns of seven enzymes in *Suillus bovinus* and *S. variegatus*, the respective species clusters showed considerable genetic dissimilarity (11% similarity), whereas intraspecific similarity in both species was greater than 65% (Sen, 1990a). It is interesting to compare these results with those of the *Glomus* studies. Although high levels of intraspecific variability do exist in fungi the results suggest that certain *G. mosseae* isolates of the study may in fact have represented different taxonomic species. In order to use isozyme analysis in taxonomic studies, particularly in the case of the vesicular-arbuscular mycorrhizal fungi, it is of major importance that the morphological characters of the specimens are carefully studied. It is also very important to study a large population of the taxon, including isolates from a number of particular habitats. Isozyme analysis, of over 20 enzyme systems, can then reveal whether a taxon is homogenous, or whether it clusters into two or more groups which could be biovars or new taxa. Alternatively, this analysis can also be used for lumping taxa together if the variation between two taxa is less than the variation found within the taxon.

In his review, Morton (1988) stressed the problems still inherent in vesicular-arbuscular mycorrhizal taxonomy and pointed to the importance of combining morphological and chemotaxonomic criteria for answering these questions.

### B.  Population structures of mycorrhizal fungi

As has been pointed out, the uncertainties of species classification in vesicular-arbuscular mycorrhizal fungi makes population studies of these fungi difficult. Isolate level variation was detected in *G. mosseae* but this may have in fact represented interspecific variability for reasons described. However, clear intraspecific variation of isozyme expression has been detected in the ectomycorrhizal fungi *Laccaria laccata* (Ho, 1987), *Suillus tomentosus* (Zhu *et al.*, 1988), *Suillus bovinus* and *S. variegatus* (Sen, 1990a). In all these studies, those isolates showing the greatest degree of isozyme homology could be directly related to a particular tree species or geographical location. Polymorphism at specific loci was also correlated with host species (Zhu *et al.*, 1988) and habitat location (Sen, 1990a). In mycorrhizal associations the host must exert considerable selection pressure on the fungal symbiont, which over time may result in genetic divergence within a population. This situation has arisen within the root pathogen *Heterobasidion annosum*, where inter-sterility groups specific for pine (P-strains), spruce (S-strains) and fir (F-strains) now exist (Capretti *et al.*, 1990). The individuality of different mycelia has also been correlated by isozyme and somatic incompatibility methods (Sen, 1990a) which provide an opportunity to detect genetic relatedness of strains within a forest stand (Dahlberg and Stenlid, 1990).

## C. Identification and quantification of root infections

Vesicular-arbuscular mycorrhizal fungi can be easily identified within roots through their arbuscular infections of root cortical cells. However, further identification to the species level is complicated by the near identical infection morphologies exhibited by different vesicular-arbuscular mycorrhizal fungi. No host-related infection patterns are detectable. Subtle differences in the anatomy of the intercellular hyphae and vesicles can be identified and used as criteria for species level identification (Abbott, 1982), but morphological characterization requires much training and patience.

Identification of individual and combined infections of *G. mosseae*, *G. caledonium* and *G. fasiculatum* in the roots of leek (*Allium porrum*) and maize (*Zea mays*) (Hepper *et al.*, 1986) and of *G. deserticola* and *G. fasciculatum* in cucumber (*Cucumis sativus*) (Rosendahl and Hepper, 1987) has now been made possible through isozyme analysis. Host and fungal PEP, GOT and EST isozyme have been differentially identified by comparison with those of uninoculated roots (Fig. 2) and fungal controls (spores, external and internal mycelium).

Several methods for quantifying vesicular-arbuscular mycorrhizal infections in plant roots have been described. These are mostly based on a quantitative estimate of the biomass of the fungus. Per cent root length infected (Giovanetti and Mosse, 1980) is most commonly used, whereas colorimetric assays such as that determining glucosamine (Hepper, 1977) are used less frequently. Assessment of the amount of metabolically active mycelium has been used to determine the fungitoxic effect of benomyl (Kough et al., 1987).

Estimation of metabolic activity of vesicular-arbuscular mycorrhizal hyphae within roots has also been possible by relating total hyphal infection (as measured by the glucosamine assay) to visual assessment of the fungal isozyme activities (Hepper et al., 1988b) and to absolute activities measured densitometrically (Rosendahl et al., 1989a). In this latter study the infection was quantified from a standard curve made by diluting infected roots with uninfected roots.

Using the same principle, morphologically similar synthesized *Pinus sylvestris* L. ectomycorrhiza of *S. bovinus* and *S. variegatus* have been individually identified (Sen, 1990b). This list has been extended to include Scots pine mycorrhiza of *S. luteus*, *Paxillus involutus* (Fig. 3), *Piloderma croeceum*, *Pisolithus tinctorius* and *Lactarius rufus* (R. Sen, unpubl. res.). Conserved species-specific diagnostic EST (Fig. 3) and PEP (Fig. 4) fungal isozymes were detected against a background of little or no host activity. By comparison vesicular-arbuscular mycorrhizal roots exhibited large amounts of host activity (Fig. 2). The *S. bovinus* ectomycorrhiza also exhibited induced fungal isolate-specific EST isozymes (Fig. 3). Natural ectomycorrhiza of *Pinus cembra* have also been subjected recently to isozyme analysis, enabling the clear identification of the fungal symbiont *S. plorans* (G. Keller, pers. commun.).

## D.  Ecological studies

Isozyme patterns can be used to distinguish morphologically similar infection types, and they therefore provide a valuable technique for ecological studies of mycorrhizal fungi. Competition between introduced and indigenous pairs of *G. caledonium*, *G. mosseae* and *G. fasciculatum* for infection of leek has been studied in a microcosm experiment (Hepper et al., 1988b). The experiment showed that root colonization by an introduced fungus effectively prevented entry by a competing indigenous fungus. However, the ecological unit of vesicular-arbuscular mycorrhizal fungi is not well defined, thus creating problems for identification of root infections under field conditions. In a preliminary study of the diversity of vesicular-arbuscular mycorrhizal fungi in a semi-natural grassland, isozyme analysis indicated that different plant species did not share the same endophytes. Moreover, several similar, but still different, strains of vesicular-arbuscular mycorrhizal fungi could be isolated from the same host plant (Rosendahl *et al.*, 1989b).

Identification of ectomycorrhiza at the species and isolate level provides an opportunity to relate above-ground host and below-ground fungal diversities. These types of study will provide important information on the dynamics of succession, which until now has been mainly studied through analysis of sporocarp distribution.

### E.  Biochemical and genetic studies

The influence of vesicular-arbuscular mycorrhiza on the metabolism of the host plant has been the subject of several studies but in most cases the plant and the internal mycelium of the fungus have not been separated (Dehne, 1986; Dodd *et al.*, 1987). We have shown that the two components can be separated by electrophoresis, enabling the measurement of host-specific enzyme activities (S. Rosendahl, unpubl. res.) (Fig. 2). This will provide further biochemical information on the host–fungus relationship, complementing earlier work on host- and fungus-specific acid and alkaline phosphatases (Gianinazzi-Pearson and Gianinazzi, 1976, 1978).

In ectomycorrhizal research, the detection of little or no host isozyme activity in individual *Pinus sylvestris* ectomycorrhiza (Sen, 1990b) is in accordance with a recent two-dimensional polypeptide analysis of *Eucalyptus–Pisolithus* ectomycorrhiza in which over 80% of the host-related polypeptides were found to be repressed (Hilbert and Martin, 1988). Electrophoretic analysis of nitrogen cycling enzymes in *Picea excelsa* and *Fagus sylvatica* ectomycorrhiza has also shown that expression of fungus- and host-specific isozyme activities is strongly dependent on the host–fungus species combination (Dell *et al.*, 1989; Chalot *et al.*, 1990). These studies should therefore allow a better understanding of the contributions of host and fungus to the mycorrhizal symbiosis.

### IV.  Conclusions

Isozyme analysis has been shown to be a valuable tool in mycorrhizal research. The well-established importance of this method in the separation of species and strains should be appreciated and, along with recent molecular methods, used as additional characters in mycorrhizal taxonomy. The separation and identification of host and fungal enzyme activities in mycorrhiza should further their biochemical characterization and help in the analysis of mycorrhizal species and strain interactions under ecologically relevant conditions.

## Acknowledgements

We are very grateful for having had the opportunity of working with Dr Christine Hepper, whose untimely retirement from active research is much regretted. SR receives funding from the Danish Natural Science Research Council and RS thanks the Maj and Tor Nessling Foundation and the Finnish Natural Resources Research Foundation for financial support in Finland.

## References

Abbott, L. K. (1982). *Austral. J. Bot.* **30**, 485–499.
Agerer, R. (1987). *Colour Atlas of Ectomycorrhizae.* Einhorn-Verlag, Munich.
Backhouse, D., Willetts, H. J. and Adam, P. (1984). *Trans. Br. Mycol. Soc.* **82**, 625–630.
Bradford, M. (1976). *Anal. Biochem.* **72**, 248–254.
Burdon, J. J. and Marshall, D. R. (1983). *Isozymes in Plant Genetics and Breeding*, Part A (S. D. Tanksley and T. J. Orton, eds), pp. 401–412. Elsevier Science Publishers, Amsterdam.
Burdon, J. S., Roelfs, A. P. and Brown, A. H. D. (1986). *Can. J. Genet. Cytol.* **28**, 171–175.
Capaccio, L. C. M. and Callow, J. A. (1982). *New Phytol.* **91**, 81–91.
Capretti, K., Korhonen., K., Mugnai, L. and Romagnoli, C. (1990). *Eur. J. For. Pathol.* **20**, 231–240.
Chalot, M., Brun, A., Khalis, A., Dell, B., Rohr, R. and Botton, B. (1990). *Can. J. Bot.* **68**, 1756–1762.
Dahlberg, A. and Stenlid, J. (1990). *New Phytol.* **115**, 487–493.
Darnal, D. W. and Klotz, I. M. (1972). *Arch. Biochem. Biophys.* **149**, 1–14.
Dehne, H. W. (1986). In *Physiological and Genetical Aspects of Mycorrhizae* (V. Gianinazzi-Pearson and S. Gianinazzi, eds), pp. 431–435. INRA, Paris.
Dell, B., Botton, B., Martin, F. and Le Tacon, F. (1989). *New Phytol.* **111**, 683–692.
Dodd, J. C., Burton, C. C., Burns., R. G. and Jeffries, P. (1987). *New Phytol.* **107**, 163–172.
Gianinazzi-Pearson, V. and Gianinazzi, S. (1976). *Physiol. Veg.* **14**, 833–841.
Gianinazzi-Pearson, V. and Gianinazzi, S. (1978). *Physiol. Plant Pathol.* **12**, 45–53.
Giovanetti, M and Mosse, B. (1980). *New Phytol.* **84**, 489–500.
Harris, H. and Hopkinson, D. A. (1976). *Handbook of Enzyme Electrophoresis in Human Genetics (with supplements).* North-Holland, Amsterdam.

Hepper, C. M. (1977). *Soil Biol. Biochem.* **9**, 15–18.
Hepper, C. M. (1987). In *Mycorrhizae in the Next Decade, Practical Applications and Research Priorities* (D. M. Sylvia, L. L. Hung and J. H. Graham, eds), pp. 308–310. IFAS, Gainesville, FL.
Hepper, C. M., Sen, R. and Maskall, C. S. (1986). *New Phytol.* **102**, 529–539.
Hepper, C. M., Sen, R., Azcon-Aguilar, C. and Grace, C. (1988a). *Soil Biol. Biochem.* **20**, 51–59.
Hepper, C. M., Azcon-Aguilar, C., Rosendahl, S. and Sen, R. (1988b). *New Phytol.* **93**, 401–413.
Hilbert, J. L. and Martin, F. (1988). *New Phytol.* **110**, 339–346.
Ho, I. (1987). *Can. J. For. Res.* **17**, 855–858.
Ho, I. and Trappe, J. M. (1987). *Mycologia* **79**, 553–558.
Julian, A. M. and Lucas, J. A. (1990). *Plant Pathol.* **39**, 178–190.
Koch, G. and Kohler, W. (1990). *J. Phytopathol.* **129**, 89–101.
Kough, J. L., Gianinazzi-Pearson, V. and Gianinazzi, S. (1987). *New Phytol.* **106**, 707–715.
Leuchtmann, A. and Clay, K. (1989). *Can. J. Bot.* **67**, 2600–2607.
Lowry, O. H., Rosenburg, N. J., Farr, A. L. and Randall, R. J. (1951). *J. Biol. Chem.* **193**, 265–275.
May, B., Roberts, D. W. and Soper, R. S. (1979). *Exp. Mycol.* **3**, 289–297.
Micales, J. A., Bonde, M. R. and Peterson, G. L. (1986). *Mycotaxon* **27**, 405–449.
Molina, R. and Palmer, J. G. (1982). In *Methods and Principles of Mycorrhizal Research* (N. C. Schenck, ed.), pp. 115–130. The American Phytopathological Society, St. Paul, MN.
Moore, D. and Jirjis, R. I. (1981). *New Phytol.* **87**, 101–113.
Morrison, D. J., Thomson, A. J., Chu, D., Peet, F. G. and Sahota, T. S. (1985). *Can. J. Microbiol.* **31**, 651–653.
Morton, J. B. (1985). *Mycologia* **77**, 192–204.
Morton, J. B. (1988). *Mycotaxon* **32**, 267–324.
Paranjpe, M. S., Chen, P. K. and Jong, S. C. (1979). *Mycologia* **71**, 469–478.
Rosendahl, S. (1989). *Opera Botanica* **100**, 215–223.
Rosendahl, S. and Hepper, C. M. (1987). In *Mycorrhizae in the Next Decade, Practical Applications and Research Priorities* (D. M. Sylvia, L. L. Hung and J. H. Graham, eds), p. 319. IFAS, Gainesville, FL.
Rosendahl, S., Sen, R., Hepper, C. M. and Azcon-Aguilar, C. (1989a). *Soil Biol. Biochem.* **21**, 519–522.
Rosendahl, S., Rosendahl, C. N. and Søchting, U. (1989b). *Agric. Ecosyst, Environ.* **29**, 329–336.
Roux, P. and Labarère, J. (1991). *Mycol. Res.* **95**, 851–860.
Royse, D. J. and May, B. (1982). *Mycologia* **74**, 93–102.
Scandalios, J. G. (1974). *Ann. Rev. Plant Physiol.* **25**, 225–258.
Selander, K. R., Caugant, D. A., Ochman, H., Musser, J. A., Gilmour, M. N. and Whittam, S. T. (1986). *Appl. Environ. Microbiol.* **51**, 873–884.
Seviour, R. J. and Chilvers, G. A. (1972). *New Phytol.* **71**, 1107–1110.
Sen, R. and Hepper, C. M. (1986). *Soil Biol. Biochem.* **18**, 29–34.
Sen, R. (1989). *Agric. Ecosyst. Environ.* **28**, 463–468.
Sen, R. (1990a). *New Phytol.* **114**, 607–616.
Sen, R. (1990b). *New Phytol.* **114**, 617–626.
Shaw, C. R. and Prasad, R. (1970). *Biochem. Genet.* **4**, 297–320.

Smith, S. E., St John, B. J., Smith, F. A. and Nicholas, D. J. D. (1985). *New Phytol.* **99**, 211–227.

Sneath, P. H. A. and Sokol, R. R. (1973). *Numerical Taxonomy: The Principles and Practice of Classification.* Freeman, San Francisco.

Tanksley, S. D. and Orton, T. J. (eds) (1983). *Isozymes in Plant Genetics and Breeding,* Part A. Elsevier Science Publishers, Amsterdam.

Tooley, P. W., Fry, W. E. and Villarreal-Gonzalez, M. J. (1985). *J. Hered.* **76**, 431–435.

Walker, C. and Koske, R. E. (1987). *Mycotaxon* **30**, 253–262.

Zhu, H., Higginbotham, K. O., Dancik, B. P. and Navratil, S. (1988). *Can. J. Bot.* **66**, 588–594.

# 10

# Enzymic Separation of Vesicular-arbuscular Mycorrhizal Fungi from Roots: Methods, Applications and Problems

P. A. McGEE

*School of Biological Sciences, A12, University of Sydney, 2006 NSW, Australia*

S. E. SMITH

*Department of Soil Science, Waite Agricultural Research Institute, University of Adelaide, Glen Osmond, 5064 SA, Australia*

## I.  Introduction

The physiology of vesicular-arbuscular mycorrhiza has been studied in some detail (see Smith and Gianinazzi-Pearson, 1988) and we know much about the influence of infection on the physiology of mycorrhizal plants. However, we know relatively little about the functioning of the fungi in vesicular-arbuscular mycorrhizal roots. Electron microscope studies have shown that the fungal structures inside the root differ from extramatrical hyphae, germ tubes and spores (e.g. Bonfante-Fasolo and Grippiolo, 1982; Bonfante-Fasolo *et al.*, 1986). Furthermore, specialized intraradical structures like arbuscules are not found externally. It is also clear that the establishment of symbiosis alters the physiology of the

METHODS IN MICROBIOLOGY
VOLUME 24   ISBN 0-12-521524-X

Copyright © 1992 by Academic Press Limited
All rights of reproduction in any form reserved

germ tubes, so that nuclear division can occur in the fungus (Burggraaf and Beringer, 1989) and changes in the pattern of protein synthesis occur as infection is established (Dumas *et al.*, 1990). In view of these findings it is not reasonable to expect major advances in our knowledge of the physiology and biochemistry of the intraradical fungus to follow from studies of the fungus grown in culture (even if this were possible) and alternative methods of distinguishing the activities of the fungus from that of the host root have been sought. One of these methods involves dissection of the fungus from roots following digestion of the cell walls with cellulase and pectinase. The separated fungus can then be used for cytochemical and chemical studies and for extraction and assay of enzymes.

## II.  Methods

The methods used to digest the walls of the plant root cells are based on those developed for the production of plant protoplasts. The enzymes pectinase and cellulase hydrolyse the pectic layer between cells and then remove the cellulosic cell walls. In theory, fungal walls composed of chitin are not affected, so that fungal structures should remain intact. The principle was first applied by Capaccio and Callow (1982). Roots of onion infected with *Glomus mosseae* were placed in an unbuffered, aqueous solution of cellulase (0.1%) and pectinase (1%) for 16 h. Intercellular hyphae and arbuscules were separated from the root tissue and utilized for assays of enzymes involved in polyphosphate interconversion.

The solution used by Capaccio and Callow (1982) was modified by Gianinazzi-Pearson *et al.* (1984) to include an osmoticum, sorbitol at $100 \, \text{mg} \, \text{ml}^{-1}$, in a solution buffered by 0.1 M 2-(N-morpholino) ethane sulphonic acid (MES). The aim of these changes was to plasmolyse the tissues and therefore prevent bursting of the fungal structures, particularly arbuscules which have very thin walls, and to keep the pH at around 6.5, optimum for the cell wall degrading enzymes. Following reports of damage to the fungus by the digestion treatment (see Gianinazzi-Pearson *et al.*, 1984) and apparent loss in activity (Smith *et al.*, 1985), McGee and Smith (1990) investigated the effects of the composition of the digestion solution on the activity of the extracted fungus (determined by the reduction of nitroblue tetrazolium to formozan in the presence of succinate, i.e. succinate dehydrogenase activity) and the effectiveness of the digestion process. They confirmed damage to the fungus, but did not improve the method.

Following incubation of short segments of root in the enzyme solution for the appropriate period of time, the fungus has to be removed from the macerated tissue by hand. Recovery of the fungus is a slow process. Under a dissecting microscope, sharp forceps and mounted needles are used to remove the undigested root material (stele and epidermal and hypodermal layers) and, to pull the hyphae and arbuscules out. In many investigations this dissection stage has been carried out over ice. Separated fungus is then transferred to an appropriate solution, depending on the use that will be made of it. A reasonable estimate of the amount of material that can be harvested from heavily infected root segments would be about 15–20 mg wet weight per hour (S. E. Smith, pers. commun.). Care must be taken to ensure that host cells are removed from the preparation.

### III.   Applications and limitations of the method

The methods oulined above have been used to remove several species of vesicular-arbuscular mycorrhizal fungi from the roots of *Allium cepa* (Capaccio and Callow, 1982; Gianinazzi-Pearson *et al.*, 1984; Kough and Gianinazzi-Pearson, 1986; Smith *et al.*, 1985), *A. porrum* (McGee and Smith, 1991) and *Ornithogallum umbellatum* (Casana and Bonfante-Fasolo, 1982; Gianinazzi-Pearson *et al.*, 1984), but have not been successful with *Trifolium subterraneum* (P. A. McGee and S. E. Smith, unpubl. res.), *Ginko biloba*, *Citrus*, grapevine or raspberry (Gianinazzi-Pearson *et al.*, 1984). The plants so far found to be susceptible to digestion are all in the Liliflorae. Reasons for unsuccessful digestion are unclear, but are likely to be related to the anatomy of the roots and the chemical composition of cell walls. Deposition of suberin, which occurs, for example, in the hypodermis of many roots (see Peterson *et al.*, 1978; Shishkoff, 1987) would be expected to protect the pectin and cellulose from the activity of the enzymes. Indeed, in *Allium* digestion does not lead to separation of the suberized hypodermal cells, so that a "sleeve" of cells derive from this layer remains surrounding the remnants of the cortical cells and fungal tissue. This sleeve has to be stripped away from the fungus. It is likely that modifications of the composition of the digestion solutions will be required to extend the methods to other plant species.

It has been assumed that species of vesicular-arbuscular mycorrhizal fungi have similar responses to enzymic digestion. Variation in ease of removal from roots of different fungal species has not been examined. Choice of fungal species may influence the results obtained.

The structure of the mycelium separated from roots by enzymic means has been studied under the electron microscope and the material also used to determine the activity of several enzymes, using both *in vitro* and cytochemical methods. Chemical determination of the presence and concentration of polyphosphate and lipid in vacuoles and cytoplasm and the concentration and distribution of carbohydrates and chitin in fungal walls has also been made. Table I gives the details of these investigations and references.

Observations in a number of the experiments suggest that the digestion process may lead to damage to the fungus and therefore to artefacts in the experimental results.

At the structural level, the main features of the fungus appeared normal (Casana and Bonfante-Fasolo, 1982). However, in another study the plasmalemma appeared to be fragmented and mitochondria altered structurally (Gianinazzi-Pearson *et al.*, 1984).

Total protein and nitrogen (but not total phosphate) concentrations of the extracted fungus were low (Smith and Gianinazzi-Pearson, 1987). Loss of protein could occur if hyphae and arbuscules burst, losing their

TABLE I

Analyses carried out on fungal mycelium and arbuscules following digestion in solutions containing cellulase and pectinase.

| Analyses | Reference |
|---|---|
| *Enzymes* | |
| Exopolyphosphatase | Capaccio and Callow (1982) |
| Alkaline phosphatase | Gianinazzi-Pearson and Gianinazzi (1983) |
| | Gianinazzi-Pearson *et al.* (1984) |
| | Smith and Gianinazzi-Pearson (1987) |
| | Kough and Gianinazzi-Pearson (1986) |
| Succinate dehydrogenase | Kough and Gianinazzi-Pearson (1986) |
| | McGee and Smith (1991) |
| Glutamine synthetase | Smith *et al.* (1985) |
| | Smith and Gianinazzi-Pearson (1987) |
| Lipase | Kough and Gianinazzi-Pearson (1986) |
| | |
| *Other analyses* | |
| Fresh weight/dry weight | Smith and Gianinazzi-Pearson (1987) |
| Total phosphate | Smith and Gianinazzi-Pearson (1987) |
| Polyphosphate | Capaccio and Callow (1982) |
| | Smith and Gianinazzi-Pearson (1987) |
| Total nitrogen | Smith and Gianinazzi-Pearson (1987) |
| Total protein | Smith and Gianinazzi-Pearson (1987) |
| Acridine orange fluorescence | McGee and Smith (1990) |

contents, while low nitrogen levels could result both from this and low chitin concentration (see below).

Glutamine synthetase activity which could be attributed to fungal enzymes, appeared to be lower after digestion (Smith et al., 1985) and the deposition of formozan as the result of activity of succinate dehydrogenase was clearly reduced in fungal structures after roots had been subjected to the digestion process (Gianinazzi-Pearson et al., 1984; McGee and Smith, 1990). Effects on mitochondrial membranes (shown in electron micrographs) may explain the reduction in activity of succinate dehydrogenase. Other reductions in enzyme activity, such as glutamine synthetase, have not been linked to structural changes, but are nevertheless significant. Furthermore, McGee and Smith (1990) observed a reduction in fluorescence with acridine orange, suggesting changes in DNA structure.

Although chitin is known to be reduced in the walls of arbuscules, compared with intercellular or extraradical hyphae (Bonfante-Fasolo and Grippiolo, 1982), it is certainly not absent from infected roots and the concentration of chitin has been used to estimate the extent of mycorrhizal infection (Hepper, 1977; Bethlenfalvay et al., 1982). It is therefore surprising that chitin assays on digested fungus have given negative results (Casana and Bonfante-Fasolo, 1982). It is possible that chitin is removed from fungal walls during digestion, due to the action of chitinase which is present in commercial preparations of cellulase (Hamlyn et al., 1981; Peberdy, 1985). Considerable variation exists in the capacity of commercial cellulases to degrade fungal cell walls (Hamlyn et al., 1981). Either a source of cellulase which is free of chitinase or swamping existing chitinase with chitin during digestion could be worthwhile.

The role of different components in the digestion solution in causing reductions in fungal activity (determined cytochemically as succinate dehydrogenase activity and acridine orange fluorescence) was the subject of an investigation by McGee and Smith (1990). They varied the concentrations of enzymes, buffer, osmoticum and added salts of either or both $Mg^{2+}$ and $Ca^{2+}$. All treatments which permitted successful separation of fungus from root tissues of Allium porrum, also reduced the activity of succinate dehydrogenase in hyphae and arbuscules. The presence and concentration of cellulase and pectinase had the greatest effect on the reduction in activity. Presence of MES buffer was significantly correlated with increasing activity of the enzyme, but as the change in pH was less than 0.23 units, it is hard to see how MES could have affected enzyme activity. Presence of $Mg^{2+}$ and $Ca^{2+}$ had no significant effect on enzyme activity, suggesting that membrane function was not compromised. Also of significance was the osmolarity of the solution. Osmolarity was negatively correlated with enzyme activity as well as the plasmolysis of young vesicles. Given that plasmolysis is induced to assist separation of plant and fungal tissue and to prevent loss of cytoplasm from fungal structures, consideration of the use of isotonic solutions is warranted in future.

## IV. Conclusions

At present the utility of the method discussed in this chapter is limited. It is time-consuming and has now been shown to have deleterious effects on enzyme activity and other characteristics of the extracted fungus. It is clear that it could be used for qualitative determination of the presence of a particular enzyme and possibly for some chemical analyses, but quantitative data and data suggesting the absence of a particular enzyme must be viewed critically.

There is scope for research into improving the method. This could include screening sources of cellulase for chitinase activity and correlating this with effects on fungi, determining the best source and concentration of osmoticum for the solution and modifying methods to suit different species of plant roots. In the meantime, the method can be useful but should always be employed in conjunction with an independent method and appropriate controls.

## References

Bethlenfalvay, G. J., Pacovsky, R. S. and Brown, M. S. (1982). *Phytopathology* **72**, 894–897.
Bonfante-Fasolo, P. and Grippiolo, R. (1982). *Can. J. Bot.* **60**, 2303–2312.
Bonfante-Fasolo, P., Marzachi, C. and Testa, B. (1986). In *Recognition in Plant–Microbe Symbiotic and Pathogenic Interaction* (B. Lugtenberg, ed.), pp. 283–286. Springer-Verlag, Berlin.
Burggraaf, A. J. P. and Beringer, J. E. (1989). *New Phytol.* **111**, 25–33.
Cappaccio, L. C. M. and Callow, J. A. (1982). *New Phytol.* **91**, 81–91.
Casana, M. and Bonfante-Fasolo, P. (1982). *Allionia* **25**, 17–25.
Dumas, E., Tahiri-Alaoui, A., Gianinazzi, S. and Gianinazzi-Pearson, V. (1990). In *Endocytobiology IV* (P. Nardon, V. Gianinazzi-Pearson, L. Margulis and D. C. Smith, eds), pp. 91–98. INRA, Paris.
Gianinazzi-Pearson, V. and Gianinazzi, S. (1983). *Plant Soil* **71**, 197–209.
Gianinazzi-Pearson, V., Bonfante-Fasolo, P., Dexheimer, J. and Gianinazzi, S. (1984). In *Proceedings of the 6th North American Conference on Mycorrhizae,*

(R. Molina, ed.), p. 378. Forest Research Laboratory, Oregon State University, Corvallis, OR.

Hamlyn, P. F., Bradshaw, R. E., Mellon, F. M., Santiage, C. M., Wilson, J. M. and Peberdy, J. F. (1981). *Enzyme Microbiol. Technol.* **3**, 321–325.

Hepper, C. M. (1977). *Plant Soil* **9**, 15–18.

Kough, J. L. and Gianinazzi-Pearson, V. (1986) In *Mycorrhizae: Physiology and Genetics* (V. Gianinazzi-Pearson and S. Gianinazzi, eds), pp. 223–226. INRA, Paris.

McGee, P. A. and Smith, S. E. (1990). *Mycol. Res.* **94**, 305–308.

Peberdy, J. F. (1985). In *Fungal Protoplasts: Applications in Biochemistry and Genetics* (J. F. Peberdy and L. Ferenczy, eds), pp. 31–44. Marcel Dekker, New York.

Peterson, C. A., Peterson, R. L.. and Robards, A. W. (1978). *Protoplasma* **96**, 1–21.

Shishkoff, N. (1987). *Ann. Bot.* **60**, 1–15.

Smith, S. E. and Gianinazzi-Pearson, V. (1987). In *Mycorrhizae in the Next Decade* (D. M. Sylvia, L. L. Hung and J. H. Graham, eds), p. 221. Institute of Food and Agricultural Sciences, University of Florida, Gainsville, FL.

Smith, S. E. and Gianinazzi-Pearson, V. (1988). *Ann. Rev. Plant Physiol. Mol. Biol.* **39**, 221–244.

Smith, S. E., St John, B. J., Smith, F. A. and Nicholas, D. J. D. (1985). *New Phytol.* **99**, 211–227.

# 11
# Axenic Culture of Arbuscular Mycorrhizal Fungi

P. G. WILLIAMS

*School of Biological Science, University of New South Wales, PO Box 1, Kensington, NSW 2033, Australia*

## I.  Introduction

### A.  Scope

The development of a method which would enable the culturing of arbuscular mycorrhizal fungi is widely perceived as being highly desirable but such a method remains as elusive in the last decade of the twentieth century as it was in the first. However, advances made in culturing other biotrophs such as rust fungi (Williams *et al.*, 1967) may

METHODS IN MICROBIOLOGY
VOLUME 24   ISBN 0-12-521524-X

Copyright © 1992 by Academic Press Limited
All rights of reproduction in any form reserved

be usefully applied to studies of arbuscular fungi. The development and applicability of such methods is described here and their merits are discussed from both practical and theoretical standpoints. For a broader treatment of attempts to devise an axenic culture technique, and of all of the possible approaches to this question the reader is referred to Millner (1988) and Hepper (1984).

The text inevitably uses a series of terms "axenic", "monoxenic", etc., many of which have been repeatedly misused in the past. Accurate definitions, as used in this chapter, are provided in Appendix I.


## B. Background

Coming to research on arbuscular endophytes after two decades in the field of physiology of rust fungi, it was suggested to me that techniques for the axenic culture of cereal rust fungi might be applied to good effect to the long-standing problem of culturing arbuscular mycorrhizal fungi. Clearly, despite the very different biology of the two groups of fungal biotrophs, the parasitic and mutualistic symbioses have many features in common; enough in common to justify the application of the "principles" of rust culture, if such there were, to the culture of arbuscular mycorrhizal fungi. With the passage of time and the increase in knowledge of the mutualistic endophytes the soundness of this proposal is more compelling than ever. Especially striking is the remarkable similarity of the infection process in arbuscular endophytes and in species of rust fungi that enter by direct penetration of the epidermis, e.g. *Puccinia graminis* Pers. f.sp *tritici* Eriks. and E. Henn. basidiospores on barberry (Waterhouse, 1921) or *Phakopsora pachyrhizi* Syd. urediospores on soybean (Bonde *et al.*, 1976). In both mutualistic and parasitic symbionts infection involves adhesion, penetration, formation of an intracellular chamber and finally differentiation of the inter- and intracellular structures representing the primordia of the assimilative state, respectively, the distributive hyphae and arbuscules and the haustorium mother cells and haustoria.

Probably the most valuable general insight to emerge from the discovery of techniques for culturing some rust fungi is the recognition that culturing these biotrophs involves finding solutions to two problems, not one. The culture of rust fungi had always been regarded solely as a problem in fungal nutrition. It was thought that the fungi failed to grow on complex nutrient media because the media lacked an "essential growth factor" or because the urediospore germ tubes had a "biochemical lesion", a "metabolic block". However, the lesson from hindsight is that axenic culture is a dual problem. Firstly, it is a morphogenetic problem, the difficulty being to obtain differentiated fungal cells with the capacity for assimilation and growth, i.e. haustorium mother cells. Secondly, it is a physiological problem, involving the need to devise a nutrient medium which provides what the haustorium mother cells need for their growth. Urediospore germlings can be induced to form haustorium mother cells but the results are unpredictable. It is much simpler to use the cells which already exist in young rust infections, as described below. The nutrients, in all cases to date, are to be found in commercial animal or plant extracts.

Work on the axenic culture of arbuscular mycorrhizal fungi has been handicapped by the same narrow and simplistic thinking that blocked progress toward culturing rust fungi for so long. Culturing the mutualistic symbionts *in vitro* has also been seen exclusively in terms of fungal nutrition. Hyphae from germinated spores or extramatrical hyphae from monoxenic cultures were cut off and placed on a medium. Failure of the fungus to grow indicated that the medium lacked an essential metabolite. No alternative conclusion for example, that the germ tubes and extramatrical hyphae were specialized for finding and infecting roots and therefore inappropriate forms of the fungus to be offered nutrients, was considered.

The idea of applying to arbuscular endophytes the beautifully simple procedures of Lane and Shaw (1974) that had proved effective in bringing many species of rust fungi into axenic culture was attractive from the start. In this technique pieces of host tissue containing young rust infections are surface sterilized and placed on a nutrient medium. As is typical of many biotrophs, the mycelia grow very slowly but after incubation for 2–6 weeks tufts of fungal hyphae appear. On longer incubation, sometimes for up to 6 months, the mycelia achieve sufficient mass for them to be successfully cut off and cultured axenically. Recent examples of the application of the technique include the culture of two pine rust fungi, *Endocronartium harknessii* (J. P. Moore) Y. Hiratsuka (Allen *et al.*, 1988) and *Cronartium quercuum* Miyabe and Shirai (Yamazaki and Katsuya, 1987).

In empirical terms, this experimental approach sets up a monoxenic culture of the test fungus and its symbiotic partner on a medium which promotes the outgrowth of mycelia. During the incubation, the host cells senesce and the nutrition of the fungus shifts progressively from biotrophic to saprotrophic. Eventually the mycelia can be established by themselves in axenic culture.

The technique succeeds because the haustorium mother cells and intercellular hyphae in the infections represent the assimilation and growth phenotype of the rust fungus. When these cells are offered suitable nutrients from an external source the mycelia proliferate in the tissue and eventually grow out. The mycelia are then excised and transferred to fresh medium where they achieve the status of an axenic culture. This procedure is remarkably like that used in the classical method for isolating plant pathogens!

To apply the Lane and Shaw method for rust fungi to arbuscular mycorrhizal fungi a method for establishing the latter fungi in mono-xenic culture was needed. At that time there were at least half a dozen papers describing methods for monoxenic culture of arbuscular endo-pytes, beginning with Mosse and Phillips (1971). The methods all involved bringing together disinfected, germinated spores and aseptically grown seedlings or root organ cultures. The large number of papers, of itself, was a hint that the methods were unreliable. This indeed was the case in my experience and in that of St. John et al. (1981). The method described recently by Bécard and Fortin (1988 and Bécard and Piché (Chapter 6, this volume)) for monoxenic culture of Gi. margarita in transformed carrot roots promises to be more widely repeatable than any of its predecessors. The procedure for disinfecting the spores is especially rigorous and the inoculation is arranged with unique preci-sion: a single root explant is placed on the medium with its most infection-prone region in the path of an approaching germ tube.

When application of published methods to the problem of obtaining monoxenic cultures failed, the possibility of achieving the same end by applying disinfectants to mycorrhizal roots was investigated.* In 1986 Strullu and Romand published a method for disinfecting strawberry mycorrhiza to generate large numbers of pathogen- and pest-free propagules. A technique subsequently developed for disinfecting arbus-cular mycorrhiza in clover and onion (Williams, 1990) was intended to supply material for axenic culture experiments on the lines of the Lane and Shaw procedure for rust fungi.

The disinfection technique for clover and onion mycorrhiza consists of three steps: selecting the mycorrhiza, disinfecting them, and incubating them on a nutrient medium. These steps provide a framework for the remainder of this chapter. For each step there is a discussion of general principles; then the technical details of the published method are contrasted with recent variations and future directions.

---

*"Surface sterilization" properly describes a treatment given to above-ground plant tissues, but "disinfection" is the apt term for soaking delicate roots carrying young mycorrhiza in dilute household bleach.

## II.   Choosing the mycorrhiza

### A.   Theory

The techniques recommended here are based on the proposition, which has growing support (Bécard and Piché, 1989), that the distributive hyphae and arbuscules of the endomycorrhizal fungi are indispensable to the symbiosis because they express the assimilation and growth phenotype of these organisms. These vital structures are therefore homologous with the haustorium mother cells and haustoria of the rust fungi. As such, the distributive hyphae (with or without arbuscules) become the most favoured subjects for treatments to induce the formation of saprotrophic mycelia and from them, axenic cultures.

Distributive hyphae and arbuscules undergo developmental changes during the life of an infection unit. The nature of the changes and the rate at which they take place are likely to affect their effectiveness as sources of saprotrophic mycelia. Arbuscules exhibit visible deterioration in physical integrity in a matter of days or months depending, apparently, on whether they are in crop plants (Alexander et al., 1989) or woodland plants (Brundrett and Kendrick, 1990), respectively. There is no precise information relating to the developmental changes in distributive hyphae. Observations indicate several possible alternatives: they may encyst, they may form vesicles or intraradical spores or they may become vacuolate and partitioned by adventitious septa (Bonfante-Fasolo, 1984). Tests with vital stains showed that the viability of intraradical hyphae declined with increasing age of infection units (Hamel et al., 1990). Other than this, nothing is known about the maturation dynamics of these hyphae. These considerations of the quality of distributive hyphae and arbuscules emphasize how important is the choice of mycorrhizal roots for use as a source of inoculum.

Different sources of distributive hyphae and arbuscules have their advantages and disadvantages. Observations of Williams, (unpubl.) on roots agree with those of Tommerup and Kidby (1980) on spores: as a rule field material carries more contaminants than pot culture material. But field material has the advantage that it contains diverse genotypes, thus increasing the chance that an easily cultured strain of endophyte will be included in an inoculum (the endophytes are heterokaryotic and ease of culture is bound to be a heritable character, as it is in rust fungi). Unless they happen to have spores associated with them, the numerous endophytes in field material are usually of unknown identity; pot cultures have the advantage that they normally contain only one arbuscular endophyte of known provenance.

## B. Practice

The first part of the published technique, choosing the mycorrhiza and preparing them for disinfection, is described below.

The procedure uses roots taken from pot cultures of onion (*Allium cepa* L.) or white clover (*Trifolium repens* L.). The roots are collected carefully by first shaking the soil away to expose them, then cutting them 10–15 cm from the apex. After the bulk of the adhering soil has been removed in running water the roots are examined with a dissecting microscope. Only white or pale coloured, sound roots are collected. Lengths of root with mature infections are identified by the attached extramatrical hyphae, the presence of vesicles or both of these. Recent infections lack vesicles and have scant extramatrical hyphae but can sometimes be recognized by a slightly water-soaked appearance under favourable illumination with a fibre optic. Apparently uninfected regions are trimmed off.

The trimmed, infected roots (3–6 cm) are brushed in sterile distilled water with a pair of fine camel hair artists' brushes to remove soil particles and debris. After rinsing in several changes of sterile water the roots are observed with a dissecting microscope at 100 × magnification. While holding a root with a yoke made from a wooden toothpick, pieces about 1 mm in length are cut with a surgical blade. For a modest experiment 100–150 root pieces are collected in a sieve in which they are transferred through the various solutions and rinses. The sieve is made by cementing a piece of nylon mesh (0.2 mm openings) to a cut down polyethylene cap 20 mm in diameter. Following vacuum infiltration in sterile water for 2 min the sample is ready for disinfection.

This routine for choosing roots and preparing them for disinfection was worked out mainly with pot culture mycorrhiza of *Glomus fasciculatum* (Thaxter sensu Gerd.) Gerd. & Trappe in onion. Mycorrhiza of onion and *Glomus mosseae* (Nicol. & Gerd.) Gerd. & Trappe, *Acaulospora laevis* (Gerd. & Trappe) and *Gigaspora decipiens* Hall & Abbott have also been successfully disinfected using the technique. The method has also been successfully applied to roots of white and red clovers collected from a pasture (P. G. Williams, unpubl. res.).

The principal weakness of the published method is the vagueness of the directions for selecting roots. Ames *et al.* (1982) recommended fluorescence microscopy as an elegantly simple, non-destructive method for locating arbuscular mycorrhiza in a root system. These authors reported that arbuscules gave a greenish-yellow autofluorescence with blue light excitation (450–490 nm) of whole roots; but Jabaji-Hare *et al.* (1984) were unable to confirm this. Numerous observations (P. G.

Williams, unpubl.) of intact roots of several mono- and dicotyledons agree with those of Ames *et al.*, but indicate that autofluorescence is only given by senescent arbuscules. Arbuscules in *Agrostis* primary roots can be viewed directly, without staining. Intact arbuscules in young infection units in *Agrostis* (see below) do not autofluoresce with either blue or violet light excitation; only degraded arbuscules show fluorescence. The fluorescent substances are probably phenolic compounds deposited by the plant cell on the collapsed arbuscular branches (A. E. Ashford, pers. commun.). The rather non-specific autofluorescence of endomycorrhizal structures reported by Jabaji-Hare *et al.* (1984) is not relevant in this context because the emission can be seen only in sections of roots.

A practice has recently been adopted which allows non-destructive distinction between mycorrhizal and non-mycorrhizal roots and discrimination between intact and senescent arbuscules to be obtained; it also affords a supply of young infection units of such immaturity as to be ripe for producing saprotrophic mycelia. The practice, which owes much to Brundrett *et al.* (1985), is described only in outline here since a firm schedule has not yet been worked out. Seeds of *Agrostis* sp. are implanted 48 h after germination in a network of extramatrical hyphae in a pot culture or an undisturbed core from the field. After about 5 days (the incubation period must be determined empirically) the seedlings are disentangled from the matrix using a dissecting microscope and the primary root examined, preferably with plan-apochromat objectives. The *Agrostis* primary root (0.10–0.13 mm diameter) consists only of an epidermis, an exodermis and one, or at most two, cortical layers. Arbuscular infection units can therefore be readily identified with a good dissecting microscope; the status of arbuscules can be determined by inspection with a research microscope. After selection, the mycorrhizal roots are brushed and rinsed in preparation for the disinfection step described below.

The published procedure for preparing roots (Williams, 1990) describes cutting the explants into lengths of about 1 mm after washing and brushing and before disinfection. It is likely that an endophyte suffers less damage from the chlorine if the bleach is applied to long pieces of root rather than short ones. A case can also be made to limit cutting to a minimum on the grounds that these fungi are especially sensitive to mechanical damage. The classical demonstration of the peripheral growth zone of a mycelial fungus (Trinci, 1971) involves cutting across a sector of a fungal colony on agar and observing which hyphae continue to grow without interruption and which ones suffer non-fatal or fatal injury. Trinci's measurements showed that the peripheral growth zone of four aseptate fungi was significantly greater than that of four septate fungi. While this suggests that the distributive hyphae of arbuscular endophytes may be especially sensitive to damage by cutting, the fungi are known to be well adapted to repairing damaged hyphae (Gerdemann, 1955a). The subject deserves investigation in relation to the manipulations necessary for establishing axenic cultures.

## III. Disinfection

### A. Theory

It has long been known that cutting mature arbuscular mycorrhiza causes the resting structures within them, i.e. vesicles, intraradical spores, encysted hyphae, to become briefly active and produce "regrowth hyphae". Such regrowth hyphae which are also generated from dead roots can cause new infections (Tommerup and Abbott, 1981; Strullu and Romand, 1986). Efforts by workers early in the century to obtain the outgrowth of these hyphae after treating lengths of mycorrhizal roots with disinfectants failed (see Peyronel, 1923 for references). Neither were Jones (1924), Niell (1944) or Magrou (1946) successful; all reported that disinfectant treatments that effectively destroyed unwanted micro-organisms also killed the endophytes. It was not until antibiotics became available that Tolle (1958) succeeded in obtaining a short regrowth hypha under aseptic conditions. Regrettably her achievement was not followed up, probably largely because the findings of Mosse (1953) and Gerdemann (1955b) changed the basis of axenic culture research from roots to spores. The recent publication by Strullu and Romand (1986) of a technique for disinfecting strawberry mycorrhiza was a welcome sign that the domination of axenic culture research by experiments on spores was coming to an end.

The method of Strullu and Romand is only suitable for robust mycorrhiza that can withstand an especially rigorous treatment: 10 min of ultrasound, 2–3 min in 95% ethyl alcohol, 1–2 min in 6% calcium hypochlorite and 5–20 min in a solution containing 200 mg litre$^{-1}$ streptomycin and 20 g litre$^{-1}$ Chloramine T. The authors give no data on the yield of disinfected root segments that are obtainable using this method. It appears to be ill-suited for fine absorbing roots.

A superior method, described below, is suitable for disinfecting mycorrhiza in delicate, absorbing roots. The method employs dissolved chlorine gas for the differential killing of contaminant and mycorrhizal fungi (Tommerup and Kidby, 1980) and antibiotics to eliminate surviving bacteria. If large numbers (50–100) of small (1–2 mm) sections of mycorrhiza are disinfected together and then incubated individually it becomes possible to modulate the disinfection so that an acceptable proportion of the treated root pieces contains living endophyte and is uncontaminated.

The effectiveness of the technique depends heavily on the quality of the mycorrhizal roots: a minimum load of propagules of unwanted fungi, bacteria, actinomycetes, algae and protozoa and a maximum content of endophyte (see Section II). The concentration of chlorine in the bleach solution and the immersion time are so adjusted that propagules of common moulds such as *Mucor, Pythium, Fusarium, Aspergillus* and *Rhizoctonia* spp. on or in some root pieces survive. Under these conditions there is a high probability that the endophyte will also survive in a proportion of those pieces that are not contaminated by moulds. The experimental data in Table I show that longer soaking of a sample of onion mycorrhiza in dilute bleach solution progressively reduced the fraction contaminated by moulds and bacteria, and the fraction which was contaminant-free and contained endophyte structures able to form the common regrowth hyphae.

## B. Practice

That part of the published procedure (Williams, 1990) concerned with disinfection runs as follows: after selection, cleaning, trimming and cutting, the root pieces are transferred to a sterile work station and immersed in a solution of household bleach (e.g. "Xixo", nominally 4% available chlorine): 2 ml diluted to 100 ml with distilled water just before use. The tissues are briefly agitated immediately and at 30 s intervals for

**TABLE I**

Percentages of explants contaminated by fungi or bacteria and percentages of explants with and without regrowth hyphae of *Glomus fasciculatum* 2 weeks after treatment with 2% household bleach for indicated times

| Treatment time (min) | Contaminated (%) | With regrowth hyphae (%) | Without regrowth hyphae (%) | Total number of explants |
|---|---|---|---|---|
| 2.5 | 40 | 51 | 9 | 45 |
| 5 | 28 | 38 | 34 | 68 |
| 10 | 18 | 30 | 52 | 50 |

2 min, then rinsed in three changes of sterile water. Next, each explant is placed in a drop (about 0.03 ml) of freshly prepared, filter sterilized incubation medium. The medium (see below for recent changes) contains penicillin (500 mg litre$^{-1}$), streptomycin (500 mg litre$^{-1}$) and bovine albumin (20 g litre$^{-1}$) pH 6.4. The drops of medium are arranged on the inside of the inverted lid of a sterile plastic Petri dish. The bottom of the dish contains plain agar (8 g litre$^{-1}$) to provide water vapour. When all the drops (about 30 per dish) have been seeded with a root piece the dishes are sealed with wax film and incubated in an inverted position at 23 °C in the dark.

Probably any common household bleach, except perhaps one of the "lemon scented" products, is a suitable source of chlorine gas for the first stage of disinfection. The concentration of sodium hypochlorite stated on the label of some lines is not a reliable guide to the concentration of chlorine because sodium chloride is sometimes added to increase the solubility of chlorine through the common ion effect. Because of the product's instability it is a sensible practice to keep a container of bleach solution only for 8–10 weeks and then replace it. In my experience Chloramine T is to be avoided. The stability of the compound is unpredictable and a product of its decomposition is very toxic to regrowth hyphae.

Since the disinfection method was published (Williams, 1990) the procedure has changed significantly. The mycorrhiza are no longer infiltrated under vacuum and bovine albumin is no longer included in the antibiotics solution. Instead, after rinsing to remove the bleach solution, the explants are soaked in a solution containing only penicillin and streptomycin (500 mg litre$^{-1}$ of each) for 3 h with periodic agitation. The tissues are then transferred to a solution containing bovine albumin (20 g litre$^{-1}$) and various test substances that may promote saprotrophic growth (see Section IV). When disinfecting *Agrostis* primary roots, whole roots (15–20 mm) are cut into three approximately equal lengths during the period of soaking in antibiotics.

Several aspects of the disinfection schedule deserve investigation in the future. For example, it may be possible to protect the arbuscule-forming distributive hyphae from damage by chlorine by pre-soaking in reducing agents (ascorbic acid, cysteine, etc.). Disinfection with silver nitrate and rinsing in sodium chloride is another alternative worthy of study. Instead of relying solely on wide spectrum chemical agents like chlorine, it might be possible to couple their diminished use to more narrowly specific fungicidal compounds (see Trappe *et al.*, 1984); i.e. develop a selective medium for arbuscular endophytes (Tsao, 1970). Yet another option could be to combat mould contamination by assiduous brushing and washing, which has been effective for isolating the endophytes from the very fine absorbing roots of ericacious plants (Pearson and Read, 1973; Reed, 1989).


## IV.  Incubation

### A.  Theory

The final step in disinfecting arbuscular mycorrhiza is the incubation. Here the first objective is to separate the contaminated from the uncontaminated root pieces. To expedite the separation, substances are added to the incubation medium that will promote the growth of the contaminating micro-organisms.

The second objective of the incubation is to switch the nutrition of the distributive hyphae and arbuscules from a biotrophic to a saprotrophic mode. If the incubation medium contains suitable nutrients it is expected that these hyphae will grow out of the explant and eventually can be cut off and become the inoculum for an axenic culture. In all likelihood only one incubation medium will be necessary because the compounds which promote saprotrophic growth of the endophyte are bound to stimulate the growth of the contaminants as well.

The contaminated explants are mostly identified and eliminated during the first weeks of an incubation. On longer incubation the uncontaminated root pieces separate into two groups, those that never produce regrowth hyphae and those that sooner or later do so. An explant may fail to produce regrowth hyphae for the simple reason that it contained no fungus. Another reason could be that the fungus was fatally damaged by cutting or disinfection. A third reason for the failure of an endophyte to make an appearance could be that although the fungus was present, viable and in an appropriate form, the incubation medium lacked suitable nutrients for saprotrophic growth. When testing possible nutrients for these fungi it will be essential to be able to discriminate between the latter two explanations. A means of doing this is high among research priorities.

The common regrowth hyphae have been described by Jones (1924), Niell (1944), Magrou (1946) and many other observers. The hyphae emerge mostly from the cut surfaces of an explant: from the end of a severed intramatrical hypha or, as Mosse (1989) described, from a hemispherical mass of cytoplasm extruded from such an intramatrical hypha. Regrowth hyphae also appear to grow from the broken ends of hyphae that are connected with appressoria on the surface of an explant;

exactly where these hyphae have their origin is not clear. There is general agreement that the hyphae are the same whether they arise from a vesicle, an internal spore or an encysted hypha. When the common regrowth hyphae grow out they are initially straight, broad at the base and tapering towards the apex—Jones and Niell both used the description "spear like". As the hyphae increase in length they become flexuous in the liquid medium. The occurrence of branches is very variable. Some hyphae grow to 2 mm or more in length without making a branch, while others branch freely and at short intervals. Anastomoses between adjacent hyphae are common, especially within a short distance (0.1–0.2 mm) of the cut surface of the root. The hyphae grow relatively fast, attaining rates of up to 6 mm per day (Mosse, 1959; Niell, 1944; M. J. Milligan and P. G. Williams, unpubl. res.). Growth of the longer hyphae ceases after 2–4 weeks and the protoplasts withdraw from the apices, progressively laying down retraction (adventitious) septa.

Another, very distinctive kind of regrowth hypha has been recorded (Williams, 1990), but only once, and its origin is a subject of conjecture. Hyphae of this kind were formed by four out of 107 explants incubated in a medium containing gelatin, casein amino acids and sucrose (see below). It is proposed here that the hyphae were active distributive hyphae which were growing partly or wholly under saprotrophic nutrition. The hyphae grew very slowly (0.03–0.10 mm per day) for an unprecedented time, two for 20 weeks and the other two for 27 weeks. They grew among the common type of regrowth hyphae, which they resembled closely except that they periodically formed bulbous structures with numerous projections. This recalled Niell's description of arbuscule primordia (1944). Lateral and intercalary complexes of fine, intensely branched, "coralloid" hyphae, possible arbuscule homologues, were also formed, (see Williams, 1990, Figs 3–6).

## B.  Practice

In the published method (Williams, 1990), the incubation begins when each section of mycorrhiza is placed in a drop of antibiotic solution that also contains bovine albumin. As mentioned above, the antibiotics and albumin treatments are now carried out separately. The incubation therefore can be said to begin when the explants are placed in a solution of albumin (20 g litre$^{-1}$) pH 6.2. The albumin has two functions. One is to promote the contaminants and the other, which is discussed below, is to promote the endophyte. In the first 4–5 days the vigorous moulds appear and are dealt with promptly before the contamination spreads to other incubation drops: either the contaminated root piece is removed

or a crystal of copper sulphate is placed in the incubation drop. In
non-acidic media ($>$ pH 5.5) most bacterial contamination becomes
evident after a few days. Slow-growing fungi, which include endophytes
such as the orchidaceous companion fungi of arbuscular endophytes
(Williams, 1985), may not appear for 4 weeks or more. Actinomycetes,
unicellular algae and protozoa which occur rarely are also slow to
appear.

Root pieces in which an endophyte is present and has survived as a
resting structure are identified by the appearance of the common
regrowth hyphae as already described. In some cases the hyphae appear
after incubation overnight; in others the regrowth hyphae are often not
visible for 2–3 days. Omitting bovine albumin from the incubation
medium delays the appearance of the hyphae and reduces the total
extramatrical growth. Albumin has beneficial effects in other systems
(Kuhl *et al.*, 1971) but a mechanism for its effects is unknown.
Observations of Williams (unpubl. res.) indicate that regrowth hyphae
are also stimulated by horse and foetal calf sera, casein amino acids and
gelatin.

It was reported by Williams (1990) that in 26 experiments an average
of 22% (range 4–64%) of the 4350 root pieces treated were uncontaminated and formed the common type of regrowth hyphae of *G. fasciculatum*, which was the endophyte used to investigate the method. By and
large the same rate of success is obtained today.

The experiment in which distributive hyphae were reported to have
grown in a nutrient medium for about 6 months and to have produced
arbuscule-like structure (Williams, 1990) has been repeated in one form
or another more than 30 times without an unequivocally successful
result. The fact that the medium in which the unique mycelia appeared
to be growing contained gelatin, casein amino acids and sucrose is not
germane to the question of how to obtain those mycelia again. What is
important is the quality of the mycorrhiza and the criteria used to select
them. By comparison, questions about what nutrients to put in the
medium are of minor significance. To make the point in a different way,
no amount of experimentation with different nutrient media will advance the problem of axenic culture until roots can be obtained in which
the endophyte is present in a suitable form to respond to external
nutrition.

## V.   Identity

Claims to have cultivated arbuscular mycorrhizal fungi have been made several times but never substantiated (Barrett, 1947; Janardhanan *et al.*, 1990). Thus, if the claimants achieved nothing else, they ensured that a sceptical climate will exist for the first person(s) to make a sustainable claim to have cultivated such a fungus. The successful investigator(s) will have two pressing tasks: to provide unequivocal evidence of the identity of the fungus in culture and to arrange swift independent confirmation of the discovery. The latter should be easily provided by sympathetic colleagues. The former may not be so simply dealt with.

There may be no problem of identification if the cultured fungus forms arbuscule-like networks of fine, intensely branched hyphae and curious bulbous structures with projecting hyphae (see Figs 3–6 in Williams, 1990). Remembering that "arbuscule-like" is something that is in the eye of the beholder, the diagnosis would be more convincing if an accepted test of function were available. Lacking such a test, a positive cytochemical test for distinctive polysaccharides and absence of chitin in the hyphal walls (Bonfante-Fasolo *et al.*, 1990) would be valuable supporting evidence for suspected *in vitro* arbuscules.

The development of arbuscular endophytes in roots is characteristically determinate, i.e. the infections are discrete units and the fungi do not grow through the cortex indefinitely in the manner of some pathogens. The formation of circumscribed mycelial colonies in axenic culture would therefore be reassuring. If instead the axenic mycelia grew without restraint, how seriously would that count against them in assessing their likely identity as arbuscular endophytes? By the same token, what weight should be placed on the presence or absence in cultured mycelia of the projections which are a feature of intercellular runner hyphae (Brundrett *et al.*, 1985).

Of course it cannot be assumed that in axenic culture an arbuscular mycorrhizal fungus will offer distinctive visible clues to its identity. Indeed, it is arguable that in the artificial conditions of saprotrophic culture it will have no physical resemblance to its symbiotic form. According to this line of argument there will be no structures bearing a credible likeness either to arbuscules, vesicles or chlamydospores. In that event establishing identity will have to rely on biological and biochemical evidence.

The satisfaction of Koch's postulates is a traditionally accepted protocol for establishing identity between a micro-organism in culture and in nature. Having isolated the organism and shown it to multiply in sterile culture, it is necessary then to introduce it once again into the host, establish the original relationship and re-isolate it once again into culture. In the present case, however, as has been emphasized throughout this chapter, the mycelia growing in an axenic culture of one of these endophytes represent the assimilation phase of the symbiont. The mycelia may therefore be phenotypically unable to perform the sequence of steps involved in establishing an infection. This deficiency can probably be overcome, as it was with rust mycelia (Williams *et al.*, 1967), by removing the outer layers of plant cells with a scalpel and placing the cultured mycelia in contact with the exposed host tissues.

In the event that these measures are repeatedly unsuccessful it may be the case that the cultured mycelia, again following the precedent of wheat stem rust mycelia (Maclean, 1982), are an aberrant genotype well adapted to growth in culture but no longer able to behave as a biotrophic symbiont. The identity of the cultured fungus in this circumstance may be resolved by such techniques as DNA:DNA hybridization, restriction fragment length polymorphism analysis or protein or isoenzyme gel electrophoresis.

## VI. Conclusions

There are signs that experimenters wishing to culture arbuscular mycorrhizal fungi are (re)turning to roots and away from chlamydospores as sources of inoculum. This trend toward the use of techniques that are based on those that have proved so effective with rust fungi is likely to bring success well before the millenium. How long it takes before the first breakthrough occurs will be determined by how quickly the appreciation spreads that the test roots must contain "feedable" fungal structures and that the condition of "feedability" is transient. Arbuscular endophytes are probably no more nutritionally fastidious than rust fungi. Therefore it is a safe prediction that the mutualistic symbionts will prove to be only slightly more difficult to culture than the parasitic kind.

## Acknowledgements

I am grateful to R. E. Koske, Botany Department, University of Rhode Island, Kingston, RI, USA for hospitality during preparation of the manuscript. The visit to Rhode Island was funded by a grant, which is gratefully acknowledged, from the Department of Industry, Technology and Commerce, Canberra under the US-Australia Science and Technology Agreement. Thanks are also due to M. J. Milligan for valuable comments and suggestions on the text.

## Appendix I

**Terminology**

The most common terminological blunder is to use axenic when what is meant is aseptic. This is a trifling error compared with such bizarre couplings as "mono-axenic", which is a tautology, and "axenic dual culture", which is an oxymoron (Strullu and Romand, 1986; Mungier and Mosse, 1987; Bécard and Fortin, 1988; Millner, 1988; Bécard and Piché, 1989; Burggraaf and Berringer, 1989). The word "axenic" and its relatives were coined by animal parasitologists (Baker and Ferguson, 1942; Dougherty, 1953), for whom the terms "aseptic", "sterile", "artificial", "pure", "bacteria-free", and so on were ambiguous or imprecise.

The term axenic "pertains to growth of a single species in the absence of living organisms or living cells of any other species". Literally, axenic means "without foreigners", as in xenophobia = 'fear of foreigners". An example of an axenic culture would be: mycelia of *Gigaspora margarita* Becker & Hall or a seedling of *Allium porrum* L. growing on a nutrient medium in a culture tube.

Monoxenic (literally "with one foreigner") describes a culture containing organisms or cells of two species, e.g. mycorrhiza of *Gi. margarita* in the roots of *A. porrum* in an agar slant culture.

Dixenic (literally "with two foreigners") describes a culture containing organisms or cells of three species, e.g. *Gi. margarita*–leek mycorrhiza contaminated by one species of bacterium.

A polyxenic culture needs no explanation.

## References

Alexander, T., Toth, R., Meier, R. and Weber, H. C. (1989). *Can. J. Bot.* **67**, 2505–2513.

Allen, E. A., Blenis, P. V and Hiratsuka, Y. (1988). *Mycologia* **80**, 120–123.

Ames, R. N., Ingham, E. R. and Reid, C. C. P. (1982). *Can. J. Bot.* **28**, 351–355.

Baker, J. A. and Ferguson, M. S. (1942). *Proc. Soc. Exp. Biol. (N.Y.)* **51**, 116–119.

Barrett, J. T. (1947). *Phytopathology* **37**, 359–360.

Bécard, G. and Fortin, J. A. (1988). *New Phytol.* **108**, 211–218.

Bécard, G. and Piché, Y. (1989). *New Phytol.* **112**, 77–83.

Bonde, M. R., Melching, J. S. and Bromfield, K. R. (1976). *Phytopathology* **66**, 1290–1294.

Bonfante-Fasolo, P. (1984). In *VA Mycorrhiza* (C. L. Powell and D. J. Bagyaraj, eds), pp. 5–33. CRC Press, Boca Raton, FL.

Bonfante-Fasolo, P., Faccio, A., Perotto, S. and Schubert, A. (1990). *Mycol. Res.* **94**, 157–165.

Brundett, M. C. and Kendrick, B. (1990). *New Phytol.* **114**, 457–468.

Brundett, M. C., Piché, Y. and Peterson, R. L. (1985). *Can. J. Bot.* **63**, 184–194.

Burggraaf, A. J. P. and Berringer, J. E. (1989). *New Phytol.* **111**, 25–33.

Dougherty, E. C. (1953). *Parasitology* **42**, 259–261.

Gerdemann, J. W. (1955a). *Mycologia* **47**, 916–918.

Gerdemann, J. W. (1955b). *Mycologia* **47**, 619–632.

Hamel, C., Fyles, H. and Smith, D. L. (1990). *New Phytol.* **114**, 297–302.

Hepper, C. M. (1984). In *VA Mycorrhiza* (C. L. Powell and D. J. Bagyaraj, eds), pp. 95–112. CRC Press, Boca Raton, FL.

Jabaji-Hare, S. H., Perumalla, C. J. and Kendrick, W. B. (1984). *Can. J. Bot.* **62**, 2665–2669.

Janardhanan, K. K., Gupta, M. L. and Husain, A. (1990). *Curr. Sci.* **59**, 509–513.

Jones, F. R. (1924). *J. Agric. Res.* **29**, 459–470.

Kuhl, J. L., Maclean, D. J., Scott, K. J. and Williams, P. G. (1971). *Can. J. Bot.* **49**, 201–209.

Lane, W. D. and Shaw, M. (1974). *Can. J. Bot.* **52**, 2228–2229.

Maclean, D. J. (1982). In *The Rust Fungi* (K. J. Scott and A. K. Chakravorty, eds), pp. 37–149. Academic Press, London and New York.

Magrou, J. (1946). *Rev. Gén. Bot.* **53**, 49–77.

Millner, P. D. (1988). In *Developments in Industrial Microbiology* (G. Pierce, ed.), Vol. 29, pp. 149–158. (*J. Ind,. Microbiol.* Suppl. 3).

Mosse, B. (1953). *Nature (Lond.)* **171**, 974.

Mosse, B. (1959). *Trans. Br. Mycol. Soc.* **42**, 273–286.

Mosse, B. (1989). *Can. J. Bot.* **66**, 2533–2540.

Mosse, B. and Phillips, J. M. (1971). *J. Gen. Microbiol.* **69**, 157–166.

Mungier, J. and Mosse, B. (1987). *Phytopathology* **77**, 1045–1050.

Niell, J. C. (1944). *N. Z. J. Sci. Technol.* **25**, 191–201.

Pearson, V. and Read, D. J. (1973). *New Phytol.* **72**, 371–379.

Peyronel, B. (1923). *Bull. Soc. Mycol. France* **39**, 119–126.

Reed, M. L. (1989). *Austral. J. Plant Physiol.* **16**, 155–160.

St. John, T. V., Hays, R. I. and Reid, C. C. P. (1981). *New Phytol.* **89**, 81–86.

Strullu, D.-G. and Romand, C. (1986). *C. R. Acad. Sci., Paris* **303**, 245–250.

Tolle, R. (1958). *Arch. Mikrobiol.* **30**, 285–291.

Tommerup, I. C. and Abbott, L. K. (1981). *Soil Biol. Biochem.* **13**, 431–433.

Tommerup, I. C. and Kidby, D. K. (1980). *Appl. Environ. Microbiol.* **39**, 1111–1119.

Trappe, J. M., Molina, R. and Castellano, M. (1984). *Ann. Rev. Phytopathol.* **22**, 331–359.

Trinci, A. P. J. (1971). *J. Gen. Microbiol.* **67**, 325–344.

Tsao, P. H. (1970). *Ann. Rev. Phytopathol.* **8**, 157–186.

Waterhouse, W. L. (1921). *Ann. Bot.* **35**, 557–564.

Williams, P. G. (1985). *Can. J. Bot.* **63**, 1329–1333.

Williams, P. G. (1990). *Mycol. Res.* **94**, 995–997.

Williams, P. G., Scott, K. J., Kuhl, J. L. and Maclean, D. J. (1967). *Phytopathology* **57**, 326–327.
Yamazaki, S. and Katsuya, K. (1987). *Ann. Phytopathol. Soc. Japan* **53**, 643–646.

# 12

# Use of Monoclonal Antibodies to Study Mycorrhiza: Present Applications and Perspectives

SILVIA PEROTTO

*John Innes Institute, John Innes Centre for Plant Research, Colney Lane, Norwich NR4 7UH, UK*

FABIO MALAVASI

*Department of Medical Genetics, Biology and Medical Chemistry, University of Turin, Centre of Immunogenetics and Histocompatibility, via Santena 19, 10126 Torino, Italy*

GEOFFREY W. BUTCHER

*Monoclonal Antibody Centre, AFRC Institute of Animal Physiology and Genetics Research, Babraham, Cambridge CB2 4AT, UK*

METHODS IN MICROBIOLOGY
VOLUME 24   ISBN 0-12-521524-X

Copyright © 1992 by Academic Press Limited
All rights of reproduction in any form reserved

## I. Introduction

Antibodies have received considerable attention in biological sciences for their ability to identify specific molecules in complex preparations. In particular, the development of hybridoma technology has allowed the generation of specific immunological probes, namely the monoclonal antibodies (McAbs), whose main feature is not only to identify single molecules, but also to discriminate different regions within the same molecule. Another advantage of monoclonal antibodies is that they can be generated from crude preparations of complex and uncharacterized antigens. These features have already made monoclonal antibodies a powerful tool with which to study complex and dynamic structures such as cell surfaces, which are involved in many aspects of development and cell–cell interactions both in animals (Shur, 1989) and in plants (Roberts, 1990). In plants, for example, the use of McAbs has led to the identification of components of the cell surface which are expressed at discrete stages of plant development and with distinct patterns of tissue distribution (Knox *et al.*, 1989, 1990; Knox; 1990).

In plant–microbe interactions, the contact between the cell surfaces of both plant and micro-organisms leads to the formation of an interface whose role can differ depending on the type of interaction. Plant cells and fungal hyphae are in intimate contact during the colonization of host roots by mycorrhizal fungi, and the interface which is formed plays a role of primary importance in the regulation of the fungal infection and in the exchange of information between the partners (Smith and Smith, 1990; Bonfante-Fasolo and Perotto, 1991a,b).

The application of hybridoma technology to the study of plant–microbe interactions, and in particular of the interfaces which are formed during the interaction, allows us to develop unique tools which can dissect immunologically both plant and microbial structures into single molecular components. The characteristics of monoclonal antibodies also allow detection of changes in the expression of these components during the establishment of different types of symbiotic or pathogenic interactions. This information can provide clues to understanding the molecular mechanisms involved in the dialogue occurring between the plant and its inhabitants.

Further applications include the production of McAbs capable of detecting and discriminating between different species of pathogenic micro-organisms in the soil or in infected tissues (Halk and De Boer, 1985; Torrance *et al.*, 1986; Dewey *et al.*, 1989). Monoclonal antibodies can also be exploited as probes in particular immunotechniques where unique specificity is required, or when more conventional probes such as polyclonal antibodies are difficult to obtain. The use of the hybridoma technology as a simple means to generate immunoreagents for already characterized molecular components will not be the main subject of this chapter, and we would refer the reader to alternative publications (Harlow and Lane, 1988).

The achievement of a fusion and the subsequent culturing of the hybridomas nowadays involves well standardized procedures, which are detailed in several excellent books and publications (in the Reference section, e.g. Harlow and Lane, 1988). Therefore, the aim of this chapter is not to describe technical steps involved in the production of hybridomas, but rather to discuss the key principles of the technology and their significance to research on mycorrhiza. We will outline those most relevant characteristics of monoclonal antibodies which are important for an understanding of their potential applications to the study of plant–microbe interactions. The main steps in the production of hybridomas and in the selection of clones will then be discussed and some practical examples of screening strategies provided.

The application of hybridoma technology to the study of mycorrhiza is still very infrequent and the few McAbs so far derived to plant and fungal components will be described. Particularly useful, however, is the experience collected in the production of monoclonal antibodies to other types of plant–microbe interactions, and these are highlighted to reveal problems and strategies that may be of a general application.

## A.   What is a monoclonal antibody?

In 1975 Milstein and Kohler described for the first time a technique to produce unlimited quantities of antibodies of defined specificity from cell lines growing in monoculture. This technique was based on the fusion between murine myeloma cells capable of growing in culture and antibody-secreting cells from the spleen of immunized animals. Some of the hybrid cells derived from the somatic fusion inherit from the parent cells both the ability to secrete antibodies and to grow in culture. The possibility of subsequent cloning of these hybrid cells leads to cell lines, each producing an inexhaustible supply of a single type of antibody with unique specificity.

The antibodies are part of the soluble effectors of the immune response of vertebrates, the main functions of which are to block infectious organisms and their toxic products. Such a role requires the availability of an extensive array of antibodies which can specifically recognise a large variety of biochemically different antigens. Thus, antibodies can provide specific probes to a large range of molecular components. This feature lends antibodies a wider applicability compared to other affinity probes such as lectins, which only bind to a limited range of monomeric or oligomeric sugars.

## 1. Antibody–antigen interaction

Antibodies are proteins which belong to the large family of immunoglobulins. Though distinct groups of antibodies are recognized inside this family (IgG, IgA, IgE and IgM), the structure of the basic unit is maintained in the different groups (Fig. 1) and consists of two main parts, the Fab and the Fc. The Fab region encompasses the region that binds specifically to the antigen, which is highly variable in amino acid composition and forms the *binding site*. In contrast to the variability observed in the Fab region of different antibodies (probably reflecting their power to adapt to antigens of different structure), the Fc region includes very conserved sequences. These sequences define the class of immunoglobulin and mediate the effector functions of antibodies in animals. The Fab region also contains protein sequences which are typical of the animal species in which they are produced.

The basis of the antigen–antibody binding is the fitting together of two molecules of complementary shape (Fig. 2). Non-covalent bonds involving surface charges, polar and hydrophobic interactions, are established between the facing regions of the two molecules involved, and the strength of this interaction depends on the degree of stereochemical

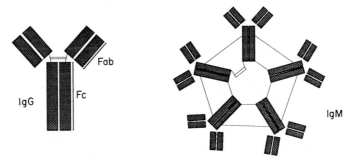

**Fig. 1.**   The IgG molecule is the simplest type of immunoglobulin. Other classes of immunoglobulin (e.g. IgM) are multimers of this subunit.

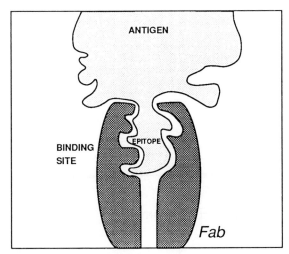

**Fig. 2.** The affinity of an antibody depends upon the stereochemical complementarity between the binding site and the epitope. The strength of binding is dependent upon the number of non-covalent bonds between the facing regions of the two molecules.

complementary of the two molecules and the energetic value of non-covalent bonds. These parameters define the affinity of an antibody for the antigen. The absence of covalent bonds means that the antigen–antibody complex is in equilibrium with the dissociated components, and the affinity of the antibody for the antigen defines the equilibrium of the free and bound components in solution. The degree of affinity of an antibody for its antigen can be crucial in the application of certain immunotechniques.

The particular region of the antigen which is recognized by the binding site of the immunoglobulin is termed the *epitope*. Large macromolecule antigens usually contain several epitopes that can each be recognized by a different antibody (Fig. 3).

## B.   Why choose monoclonal antibodies?

When immunoprobes are required, monoclonal antibodies are not necessarily the best choice in all cases. Polyclonal antibodies (i.e. antisera) may offer a good and cheaper alternative, and the choice between these two different types of immunoreagents depends upon the specific needs. Due to the high cost and man hours required, it is important to define in which cases the use of McAbs can or cannot be substituted by other probes.

a. POLYCLONAL ANTIBODIES

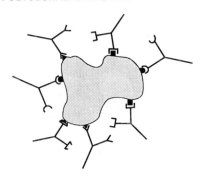

b. MONOCLONAL ANTIBODIES

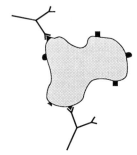

**Fig. 3.** Antigens are usually multi-epitope structures. Monospecific antisera contain antibodies that recognise many of the epitopes (a), while monoclonal antibodies only recognize a single epitope (b). Epitopes can be present on the molecule in multiple copies (e.g. glycosidic chains).

Polyclonal antibodies consist of the total population of antibodies circulating in the blood of an immunized animal and available from the serum. This type of immunoreagent is preferable for the application of some immunotechniques that involve the formation of an insoluble antigen–antibody complex, such as agglutination or immune precipitation. The main practical limitation is that it is usually necessary to elicit them by immunization with a highly purified antigenic preparation.

In the study of symbiosis, however, polyclonal antibodies have still proved to be useful reagents. For example, antisera have been used to isolate nodule-specific components from soybean (Lagocki and Verma, 1980). The same kind of probe has also allowed the identification of some symbiosis-specific proteins in mycorrhizal plants (Wyss *et al.*, 1990).

A polyclonal antiserum contains a mixture of antibodies produced by different clones of B lymphocytes. It is therefore a reagent which generally displays a high heterogeneity both in the classes of antibodies and in their specificities. Even if a polyclonal antiserum only contains antibodies to a single macromolecule, still the multi-epitope structure of the antigen induces the production of different antibodies, each recognizing a discrete epitope (Fig. 3a). It is not simple to further increase the specificity of a polyclonal antiserum, which remains, therefore, a heterogeneous reagent recognizing a molecule as a complex of epitopes. A monoclonal antibody, however, is the product of a homogeneous cell line which secretes only one type of antibody directed to a single epitope (Fig. 3b). A set of monoclonal antibodies that react with different epitopes on the same macromolecule can be used to learn about the intramolecular structure of that molecule (epitope mapping). This feature makes McAbs unique compared to polyclonal antibodies and makes it possible to follow subtle modification in the intramolecular structure of cell components during biological processes.

## C.   Use of monoclonal antibodies in different techniques

The ability of antibodies to bind specifically to an epitope forms the basis of all immunochemical techniques. A wide range of techniques have been developed for the identification, isolation and cellular localization of the antigen, and they all rely on two main strategies: the capture of antibodies by immobilized antigens and the capture of antigens by immobilized antibodies.

Only a few of the available techniques derived from each strategy are described here and the reader is referred to more specific texts (e.g. Harlow and Lane, 1988) for practical aspects.

### 1.   Capture of antibodies by immobilized antigen (Fig. 4)

*Dot blot and ELISA.* Nitrocellulose membranes or polyvinyl chloride (PVC) wells, which bind non-specifically various types of molecules, are the most common solid substrates used to absorb the antigens. Dilutions of culture supernatant containing antibodies can be applied on to the immobilized antigen and specific binding detected and visualized by using a secondary antibody conjugated to a variety of indicators. Enzymes such as horseradish peroxidase and alkaline phosphatase are the most common indicators for both techniques, but radiolabelled antibodies can be a more sensitive alternative. Both techniques are very

**ANTIBODY CAPTURE**

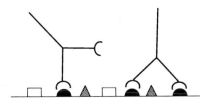

**Fig. 4.** Antibody capture assays. The antigen is immobilized on a solid phase, and the antibody is allowed to bind.

powerful, but the adsorption of the antigen to the substrate must be carefully checked to avoid pitfalls during the screening of the fusion.

*Western blot.* The main difference between this technique and dot blot is the fact that the antigen is separated by polyacrylamide gel electrophoresis before being transferred on to the membranes that act as support. This technique is extremely powerful because it provides information on the biochemical characteristics of the antigen. However, it permits the detection only of McAbs reactive with the denatured antigen, which may be an infrequent feature for reagents to some cellular components.

*Immunolocalization.* In this technique the antigens are immobilized in their cellular compartment by chemical fixation. Whole cells or tissue sections are incubated with the antibody solution, and specific binding can be visualized in bright and fluorescence microscopy, depending on the type of label. The intracellular distribution of antigens can be localized on tissue embedded in resin by using a secondary antibody conjugated to electron-dense particles, such as colloidal gold. Surface antigens such as cell wall and membrane components can be incubated with antibodies prior to chemical fixation. As the antigen–antibody complex is often more stable during fixation than the antigen alone, pre-labelling techniques are advisable with some surface components.

## 2. *Capture of antigen by immobilized antibody* (Fig. 5)

*Two antibodies-sandwich technique.* This technique can be very useful when antisera against defined or crude antigens (for example rabbit antisera) are already available. The polyclonal antibodies present are first immobilized on a solid substrate, such as PVC wells, and the

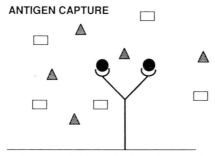

**Fig. 5.**   Antigen capture assays. The antibody is immobilized on a solid phase, and the antigen is allowed to bind.

antigen is then added to the wells. The wells containing the antibody–antigen complex can then be used to assay culture supernatants from hybridomas. One of the advantages of this technique is that it provides immobilization of those cell components which adsorb poorly to the solid substrate.

*Immunoaffinity chromatography.* Purification of specific antigens from a complex mixture of macromolecules can be achieved by using antibodies immobilized onto an insoluble matrix, commonly on a chromatographic column. The mixed antigen solution is incubated with the antibody immobilized on the column and the antigen specifically retained on the matrix can be recovered after washing the column with various buffers.

## 3.   Applications to genetic analysis and molecular biology

Some progress has been made recently in the use of molecular techniques to study mycorrhiza (described in other chapters in this volume, and reviewed in Bonfante-Fasolo and Perotto, 1991b). Though these applications remain still infrequent their development opens up a new field of research where McAbs can be used as immunoprobes.

Some examples of the possible use of McAbs are illustrated by the application of a range of molecular biology techniques to the study of other plant–microbe interactions. McAbs can be used to identify and isolate mutants for particular epitopes. If the epitope is relevant to the establishment of the plant–microbe interaction, such mutants can provide useful information both on the role of these epitopes and on the stage at which they are required during the establishment of the association. A good example of this use of McAbs was the identification of bacterial mutants of *Rhizobium leguminosarum* (see Section IV.A).

Another possible application of the ability of McAbs to recognize protein components is their use as immunoprobes to screen expression libraries of cDNA and isolate the gene encoding a specific polypeptide.

## II. Production of monoclonal antibodies

### A. Principles

An essential principle of immunology is that each antibody-secreting cell in the lymphoid tissue (plasma cells of the B lineage) only produces one type of antibody. If B lymphocytes could be isolated and grown in culture after immunization, specific cell lines producing antibodies to a defined antigen could be selected simply by cloning. However, many attempts to grow such cells in culture have failed. The alternative strategy explored by Kohler and Milstein (1975) was to immortalize B cells through somatic fusion with tumour cells which are able to grow in culture. The success achieved by using myeloma (plasmacytoma) cells, a tumoural form of B cells, allowed growth of the hybrids in culture and their cloning in order to get cell lines producing homogeneous culture supernatants each containing a single type of immunoglobulin.

Parent myeloma cells in culture would normally outgrow their fused products, the hybridomas. To overcome this problem, myeloma cell mutants were used which are sensitive to metabolic selectors that do not affect the hybrids. The myeloma cells used for the fusion are deficient for the production of the enzyme hypoxanthine phosphoribosyl transferase (HPRT). This enzyme is involved in DNA synthesis, and the myeloma cells cannot grow in a culture medium containing a mixture of aminopterin, hypoxanthine and thymidine (HAT medium) or hypoxanthine and azaserine (HAZA medium). The hybrid cells acquire an HPRT gene set from the parent B lymphocytes during the fusion process and can grow in either HAT or HAZA selection medium (Fig. 6).

### B. Immunization

The success of a fusion experiment depends upon the immunization conditions of the mouse or rat, as a good immunization increases both the number and the variety of cells producing specific antibodies. One advantage of hybridoma technology is that it allows the use of small amounts of crude, uncharacterized antigens to generate antibodies that

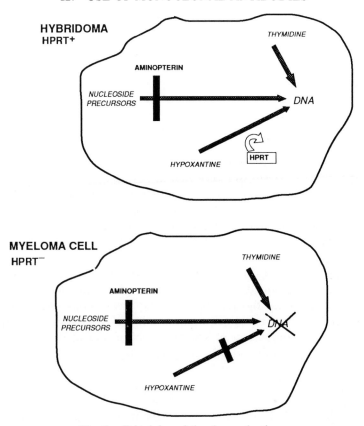

**Fig. 6.** Principles of the drug selection.

identify single molecular components. Particulate antigens, such as whole-cells preparations, have been found to induce a better immunogenic response than soluble antigens. One way to induce a stronger response to a weak immunogen is to use adjuvants, which can contain inactivated cells of pathogenic mycobacteria (complete adjuvants) that stimulate the immune response. Even without bacterial cells (incomplete adjuvants), they increase the immunogenicity, probably by providing a longer exposure of the antigen to the immune system by slowing its catabolism in the organism. Adjuvants can be extremely useful, especially with weak immunogens. However, they must be used with care as they may have serious side effects on the animal.

The length of the immunization protocol and the number of boosts can vary widely. Due to competition for antigen in the animal, successive boosts will select those B cells producing antibodies with higher affinity for the epitope by inducing their clonal expansion. Also, a concomitant switch from the production of immunoglobulin of class M to class G, which is preferable for some immunotechniques, is usually observed in immunization studies with protein antigens. However, IgM seems to be a very common class of monoclonal antibody isolated against plant or fungal components (Hahn et al., 1987; Bradley et al., 1988; Dewey et al., 1989), even after long immunization protocols where IgM are expected to switch naturally to IgG. The majority of these antibodies recognizes carbohydrate epitopes, which is not surprising as plant and fungal surfaces are very rich in polysaccharides and glycoconjugates. Carbohydrate antigens are often strong immunogens but do not seem to induce the switch from IgM to IgG production as well as protein antigens, and shorter immunization protocols may therefore give results similar to longer ones. If IgM are unsuitable for a specific use, an alternative is to induce this switch in vitro (Aguila et al., 1986).

The assays planned for screening should be validated before the fusion. Antisera obtained from a test bleed of the immunized animals can be used to assess the feasibility of the assay. Once the assay has been devised, the last boost can be injected, 3–4 days before the fusion to maximize the stimulation of specific B cell clones at the moment of the fusion (de St. Groth and Scheidegger, 1980).

## C.  Fusion

Polyethylene glycol (PEG) is the most common reagent used to induce the fusion of spleen cells with myeloma cells. The role of PEG is to reduce the surface tension of the cell surface and facilitate cytoplasmic fusion.

Somatic fusion (Fig. 7) is not difficult but requires careful laboratory organization. The original protocol described by Kohler and Milstein (1975) has been successively modified: one of the most popular protocols is that described by Galfre and Milstein (1981) or, more recently, by French et al. (1986).

Somatic fusions between spleen cells and myeloma cells from different animal species are possible, especially if the parent cells derive from animals which are phylogenetically related. However, hybrid cells are generally more stable when derived from myeloma and spleen cells from the same animal strain. A list of the most common cell lines with their derivation and main characteristics is shown in Table I.

Some critical issues to consider when performing a fusion experiments are:

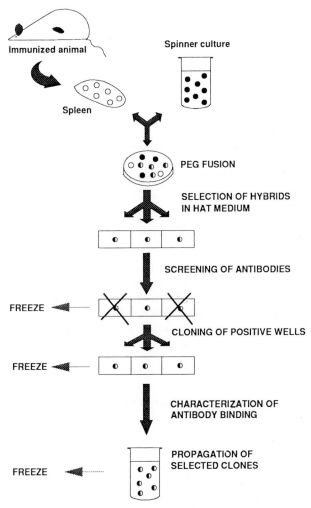

**Fig. 7.** Schematic protocol for the derivation of hybrid myelomas (modified from Galfre and Butcher, 1986).

## 1. During the fusion

To provide protection from the impact of gross microbial contaminations, it is advisable not to use the whole spleen in one experiment, especially if larger animals (like rats) are used and they have been subjected to a long and complex immunization protocol. Half the spleen can be frozen, or it can be fused with myeloma cells derived from a different culture.

## 2. *After the fusion*

*Plating.* The fusion can be plated in 24- or 96-well plates depending on various factors such as the number of positive clones expected, the labour involved in the screening procedure and the eventual presence of immunodominant molecules (see Section III.C). The use of 24-well plates is recommended to reduce the labour of screening when only a few positives are expected, as several clones are usually growing in the same well and are screened together. However, 96-well plates should be preferred when strong immunodominant epitopes are present in the antigen preparation, as they increase the chances of segregating different clones in separate wells.

*Feeding.* Feeding the plates with fresh HAT (or HAZA) medium between the fusion and the beginning of the screening is a step which is highly recommended to avoid false positives during the search for positive clones. Unfused B cells can in fact survive in culture for a few days after the fusion, and during this time they secrete antibodies which accumulate in the culture medium. Feeding the plates before the screening will therefore dilute out these antibodies.

If fusion is successful, the first clones become visible around day 4–5 after fusion, and will be ready to screen starting from day 7–8.

## III.   Screening and cloning of monoclonal antibodies

Even if the proportion of specific B cells has been enhanced during immunization, this component still represents a very small percentage of the total number of cells present in the spleen at the time of the fusion. An even smaller number of cells producing antibodies with the required specificities will fuse successfully with the myeloma cells and survive in culture. It must be established as soon as possible which clones are of interest and which ones can be discarded, as the labour of cell culturing is very intensive and only a limited number of clones can be managed efficiently by a single person.

## A.   Screening

The assay of clones is crucial to success in isolating antibodies with desired specificities. The assay should be quick, reliable, inexpensive and applicable to small samples. It is often useful to combine a simple

**TABLE I**

Myeloma cells most commonly used for hybridoma derivation

| Cell line | Strain | References | Secreting |
|---|---|---|---|
| *Mouse lines* | | | |
| P3-X63Ag8 | BALB/c | Kohler and Milstein (1975) | IgG$_1$ |
| X63Ag8.653 | BALB/c | Kearney *et al.* (1979) | No |
| NSI/Ag4.1 | BALB/c | Kohler *et al.*(1976) | No |
| NSO/1 | BALB/c | Galfre and Milstein (1981) | No |
| Sp2/0-Ag14 | BALB/c | Schulman *et al.* (1978) | No |
| S194/5.XX0.Bul | BALB/ | Trowbridge (1978) | No |
| | | | |
| *Rat lines* | | | |
| IR983F | LOU | Bazin (1982) | No |
| Y3Ag1.2.3 | LOU | Galfre *et al.* (1979) | No |
| YB2/0 | (LOU × AO)F1 | Kilmartin *et al.* (1982) | No |

primary screening with a more specific one. The primary screening will identify the clones which produce antibodies to the antigen, while the secondary screening will select antibodies with the required binding specificities. This is particularly important when a high number of clones grow after the fusion, but only epitopes with defined characteristics are being sought. Another strategy, where rapid assay is not possible, is to freeze plates *prior to* screening. The cell culture supernatants are collected from the plates and can be tested in some slower assay or a set of assays. The selected cell lines can then be thawed and expanded for further freezing and cloning.

Most of the techniques described (see Section I.C) can be used, but the type of screening should reflect the future use of the antibody in particular immunotechniques.

Screening can be relatively simple if purified antigens were used for immunization. In this case, the reactivity of culture supernatants on the same antigen used as immunogen immobilized on a solid substrate (dot blot or ELISA technique) will be sufficient to select the clones. In general, more complex antigens such as whole cells are used as immunogen. To identify epitopes with particular characteristics the secondary screening should provide additional information on the particular epitope recognized. For example, Wright *et al.* (1987) described a protocol to select by ELISA antibodies specific to spores of the vesicular-arbuscular mycorrhizal fungus *Glomus occultum*. In this case, a secondary screening involved testing the same McAbs for their cross-reactivity with other fungal species.

As cell surface components are of particular interest in the study of symbiotic interactions, an additional screening to detect antibodies recognizing surface epitopes should involve the localization of the antigen *in situ*. Some immunolocalization techniques such as immuno-gold labelling for electron and light microscopy are too time-consuming and expensive to be suitable for screening of a large number of samples. Immunofluorescence techniques (IIF) can be used, but the samples should be easy to handle. For example, Mackie *et al.* (1991) used haustoria purified from infected leaves attached to poly-L-lysine coated slides for the screening of hybridoma supernatants. Also, a quick screening was carried out on ericoid fungi by using fungal mycelium grown in liquid culture and ruptured in a Dounce homogenizer (Perotto *et al.*, 1987). Labelling and rinsing were then carried out on samples handled in small plastic tubes. Alternatively, fungal mycelium was grown adhering to glass cover-slips for two weeks and the glass cover-slips used for assay of cultures which were positive after dot blot assay on the original immunogen. Modifications of the ELISA technique to detect surface antigens on plant protoplasts are also described by Hahn *et al.* (1987), and may be adapted to other samples.

Finally, components which are newly expressed, enhanced or down-regulated during pathogenic and symbiotic interactions would be of particular interest. Some of these components have been detected both in plants and micro-organisms (see Section IV) by using a differential screening of the same culture supernatant on fractions derived from plant and micro-organisms grown in association or in the free-living phase.

Whatever technique is going to be used on the basis of the specific needs, it is important that the screening procedure is planned before the fusion, and the material necessary for the assay prepared in advance and in excess.

## B.   Cloning

Multiple clones are often growing in the same well after fusion. Hybrid cells will therefore compete for growth and this, together with the occurrence of chromosome segregation during the early stages of hybridoma culturing, will lead to dilution of those cells producing antibodies of interest. It is therefore good practice to clone quite early after fusion.

Immediately after screening, samples of cells from each clone should be frozen and stored in liquid nitrogen. This step is very important, especially during the early stages of culture, as it will preserve the clone in case of microbial contamination.

## C. Alternative strategies

It is certainly frustrating when the screening assay does not indicate the presence of any antibodies with the required specificities. Immunization and screening are the two crucial steps that should be reconsidered carefully before starting a new fusion experiment.

The components of a complex antigen are often able to induce the immune response at a different rate. Particular components present in the antigen preparation used for the immunization can be very weak immunogens, while others can be particularly immunogenic (immunodominant). It is not fully understood which characteristics make some molecules more immunogenic than others, but a completely negative screening indicates that the particular antigen of interest stimulates only poorly the proliferation of B cells. Immunized animals which do not show a high titre of antibodies to the antigen are not suitable for fusion, and test bleed samples should be checked during the immunization protocol. If not adopted already, adjuvants could be used to increase stimulation of the immune response during immunization.

Data accumulating from fusion experiments to generate monoclonal antibodies to micro-organisms and plant material indicate that immunodominance can be a common phenomenon with these antigens. One possible explanation for the high immunogenicity of bacterial molecules such as those present on the surface of symbiotic rhizobia may be the similarity of these molecules to those present on the surface of many pathogenic bacteria. The efficiency of the immune system in defending the organism against microbial invasion depends on the particular ability to recognize any potential pathogen and to produce antibodies to it.

Immunodominance seems also to be a common phenomenon in responses against plants. Anderson *et al.* (1984) first reported the apparent immunodominance of carbohydrate epitopes on soluble arabinogalactan protein (AGPs). This particular class of molecules is unique to plants, and subsequent studies of plant membrane antigens have identified immunodominance of some membrane-associated forms of these AGPs (Hahn *et al.*, 1987; Norman *et al.*, 1990). The presence of immunodominant antigens has also been observed in our experience during the production of antibodies to the peribacteroid membrane of pea nodules (Brewin *et al.*, 1987; Perotto *et al.*, 1991b). Some of these antigens were subsequently identified as membrane-associated AGPs (Pennell *et al.*, 1989). The reason for the immunodominance of AGPs is so far unknown, and different explanations are equally plausible. AGPs belong to a group of proteins which is apparently absent from animals. They are therefore easily recognized as "foreign molecules" by the immune system. Furthermore, the animals used for immunization are usually fed with material of plant origin, and may be pre-sensitized to particular components of it such as AGPs.

Another interesting observation is the fact that the glycosidic chain present on the AGP molecule, which is the part of the molecule usually recognized by the McAbs, contains sugar residues similar to arabinogalactan molecules present on some pathogenic micro-organisms such as *Mycobacterium leprae*. Although there are clear differences between AGPs and bacterial arabinogalactan molecules, the defence role of the immune response may identify on the plant molecule epitopes similar to those present on some pathogenic micro-organisms.

Other possible mechanisms which would explain the isolation of antibodies to a single molecular structure may also depend on a preferential fusion of spleen cells producing antibodies with particular specificities. This possibility, however, has not been explored in detail so far (Goding, 1980).

Alternative strategies can be suggested to reduce the phenomenon of immunodominance. One possibility is to intervene on the immunization protocol, by using antigen preparations where the most immunogenic components have been separated, for example by affinity chromatography (Brewin *et al.*, 1987). Another possibility is to deglycosylate the antigen prior to immunization, as the most immunogenic epitopes are usually carbohydrates (Bradley *et al.*, 1988). The pre-treatment of course is only advisable when the glycosidic chain is not the part of the molecule of interest.


## IV.   Applications in other plant–microbe interactions

### A.   *Rhizobium*–legume interactions

The association between *Rhizobium* and the roots of leguminous plants offers an interesting model to dissect the sequence of events and the molecular mechanisms leading to the establishment of a mutualistic symbiosis. The various plant and bacterial components that are found in a root nodule homogenate can be separated through a series of centrifugations on sucrose gradients and ultracentrifugations (Brewin *et al.*, 1985). These fractions, together with preparations from plant and bacteria in the free-living phase, have been used as antigens in a differential screening of McAbs to plant or bacterial antigens on dot blot or ELISA (Bradley *et al.*, 1988). This strategy allows selection of epitopes that are enhanced during the symbiotic phase. A number of epitopes have been identified with these characteristics on both the bacterial and the plant cell surfaces.


## 1.  Bacterial antigens

On the surface of *Rhizobium*, as in other Gram-negative bacteria, the outer membrane is mostly formed by lipopolysaccharide molecules (LPS). LPS molecules are strongly immunogenic and antibodies to these components are often a majority, even when LPS molecules only represent a contaminant in the antigen sample used for immunization. LPS molecules of *Rhizobium leguminosarum* bv *viciae* seem to be involved in surface interaction with the membranes of their host plant (Bradley *et al.*, 1986), and the expression of some of the epitopes during symbiosis is developmentally regulated (VandenBosch *et al.*, 1989b). A McAb recognizing one of these epitopes (MAC 203) is a particularly good example of the potential of the hybridoma technology in a vast range of fields such as cytology, biochemistry, molecular genetics and physiology. MAC 203 McAb has been used to study *in vitro* the regulation of expression of this specific epitope on the LPS molecule, which has been found to depend both on pH and oxygen level (Kannenberg and Brewin, 1989). The use of MAC 203 McAb as an immunoprobe also allowed the isolation of bacterial mutants which were constitutively unable to produce the particular epitope recognized by this antibody (MAC 203[−]). The inoculation of pea plants with some of these MAC 203[−] mutants induced the formation of nodules which contain intracellular bacteria but are unable to fix nitrogen (Brewin *et al.*, 1991).


## 2.  Plant antigens

Different classes of plant antigens have been identified whose expression seems to be regulated during the development of pea nodules.

A group of McAbs identify in the nodule an extracellular glycoprotein that forms the embedding medium for the bacteria inside the infection threads (VandenBosch *et al.*, 1989a) (Fig. 8). The localization of the same antigen in the material filling intercellular spaces between plant cells may provide some interesting clues to understanding the type of reaction of the plant during the invasion by *Rhizobium* and the formation of the infection thread.

A set of McAbs has also been raised to the plant membrane which enfolds the endosymbiotic *Rhizobium* bacteroids in the infected cells

240 S. PEROTTO *et al.*

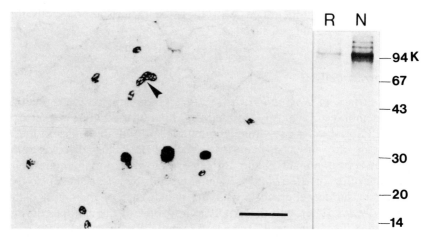

**Fig. 8.** Labelling of MAC 265 McAb on nodule tissues and on Western blot. The antigen is a 95 kDa glycoprotein whose expression is enhanced in the nodule (N) compared with the uninfected root (R). Immunogold/silver enhancement on the infection zone of the nodule shows that MAC 265 antigen is abundantly expressed in the infection threads, where it embeds the *Rhizobium* bacteria (arrowhead). Bar, 50 μm.

(Bradley *et al.*, 1988; Perotto *et al.*, 1991b) (Fig. 9). The expression of some epitopes present on antigens associated with this membrane seems to correlate with different stages of nodule development and maturation of bacteroids, indicating their different roles during the symbiosis (Perotto *et al.*, 1991b). Some of these epitopes are associated with intracellular invasion by rhizobia and may be involved in the endocytosis of bacteria. Antibodies to these particular components would be useful tools to investigate more closely the nature of endocytosis in plants and the particular type of surface interactions occurring between plant and bacterium.

Another class of plant glycoproteins identified with McAbs in the nodule may be involved in the lysis of the bacteroids during senescence of the nodule (Perotto *et al.*, 1991b). The McAbs recognizing these glycoproteins have been used as immunoprobes to purify the corresponding antigens in order to determine their protein sequence (M. F. LeGal, unpubl. res.).

The identification of the functional role that this array of antigens plays in the pea nodule will provide useful information for comparison with mycorrhizal pea plants, where most of these antigens are also expressed (see Section V.B).

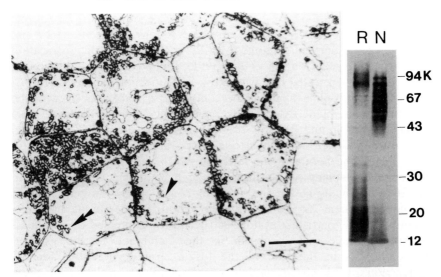

**Fig. 9.** Labelling of MAC 268 McAb on nodule tissue and on Western blot. The epitope is present on a series of molecules, where it represents part of the glycosidic chain. The distribution pattern of these molecules is different in the nodule compared with the uninfected root (R). The immunogold/silver enhancement staining shows that the MAC 268 epitope is present on plant cell membranes (arrowhead), and is strongly expressed at the time of bacterial invasion (double arrowhead). Bar, 20 μm.

## B. Plant–pathogen interactions

The early detection and identification of plant–pathogenic micro-organisms in the soil or in infected tissue is crucial for crop-protection. Polyclonal antisera have been extensively used for this purpose, but hybridoma technology has been recently adopted to develop more specific and homogeneous diagnostic immunoreagents. Many McAbs are now commercially available which identify a large range of micro-organisms such as viruses, pathogenic bacteria, fungi (Halk and De Boer, 1985; Hardham *et al.*, 1986; Torrance *et al.*, 1986; Dewey *et al.*, 1989) and even toxic fungal metabolites such as mycotoxins (Candlish *et al.*, 1989). The strategy followed for the production of these immunoreagents involved a large screening of the McAbs on related and unrelated species and biovars, in order to assess the degree of specificity. These screenings also select those McAbs featuring high affinity and ability to detect the target antigen even at high dilutions in complex mixtures, such as soil or tissues.

Monoclonal antibodies have also been generated to pathogenic systems in order to investigate the interactions between plants and pathogens. In particular, some McAbs generated to the association *Erysiphe pisi*–pea (Mackie *et al.*, 1991) could provide an interesting comparison when tested on mycorrhiza. The infection process of *E. pisi* and other biotrophic fungi can be compared in some aspects with the colonization of plant tissues by vesicular-arbuscular mycorrhizal fungi (Bonfante-Fasolo and Perotto, 1991a). These obligately biotrophic fungi can invade cells, forming haustoria which are surrounded by a plant membrane. The nature of the interface which is formed by the contact of fungal and plant cell surfaces in the haustorial complex has been investigated by using McAbs (Mackie *et al.*, 1991). In particular, a fungal component associated with the haustorial membrane has been identified. This particular epitope is specifically expressed on the haustorial membrane and is absent in the membrane of the external mycelium. As in the case of the fusion strategy described for the production of McAbs to nodule components, the isolation of these pathogenesis-related antigens relies on fractionation techniques that yield a relatively pure sample containing the intact haustorial complex from infected plant tissues. This fraction has been used both for immunization and screening of clones. A differential screening of haustorial material and plant membranes from uninfected tissues has also allowed the identification of plant antigens expressed on the membrane only during the pathogenic interaction (J. R. Green, pers. commun.). These plant membrane antigens may prove to be interesting in comparing the response of the same plant to pathogenic or mycorrhizal fungi, as the physiological and structural role of these membranes appear to be very different.

## V.  Applications in plant–mycorrhizal fungal interactions

Hybridoma technology has recently been applied to mycorrhizal associations, but specific applications to this symbiotic association so far remain few. As a result, all the monoclonal antibodies described that recognize components of the host plant were actually derived from studies on root nodules, pathogenic interactions or plant development.

The same factors that make mycorrhiza difficult to study compared with other symbiotic and pathogenic systems also pose some limitations on the generation of McAbs to specific components. Vesicular-arbuscular mycorrhizal fungi, for example, cannot be cultured *in vitro* in the absence of a host root. This feature limits the availability of fungal material both for immunization and screening of clones. Fusions to generate McAbs to the fungal mycelium, therefore, must be carefully planned.

Another limitation is the absence of satisfactory protocols to separate the various plant and fungal components from mycorrhizal roots without affecting their cell surface components. These fractions, as demonstrated in other fusion experiments, would be extremely useful to detect plant or fungal epitopes specifically enhanced or repressed during symbiosis.

## A.  Antibodies recognizing mycorrhizal fungi

### 1.  Vesicular-arbuscular mycorrhiza

Two monoclonal antibodies directed against spores of *Glomus occultum* have been isolated (Wright *et al.*, 1987). The aim was to isolate McAbs able to discriminate between species of vesicular-arbuscular mycorrhizal fungi with similar morphology and to identify them in the soil or in infected plant tissues. Previous experiments with polyclonal antibodies were limited by the low antiserum titre and strong cross-reactivity among different species (Kough *et al.*, 1983; Wilson *et al.*, 1983). The McAbs produced have high specificity for *G. occultum*, and their use in ELISA can discriminate between geographically distinct strains. The protocol used involved immunization with a suspension of crushed spores and the same antigen was used during the screening using ELISA. The two antibodies isolated are both immunoglobulin of class G, and the sensitivity of the epitope to high temperature and formalin suggests it is a protein. The exact location of these epitopes on the fungal structures is yet to be defined.

### 2.  Ericoid fungi

Mycorrhizal fungi that can be grown in pure axenic culture, such as those forming ericoid associations, provide large quantities of material for immunization and screening. McAbs have been generated to identify components of the cell surface of ericoid strains, and part of the screening procedure that has been adopted is described in Section III.A. A McAb was isolated as being able to discriminate between infective and non-infective ericoid strains (Perotto *et al.*, 1987). The antigen was localized on the thick coat of extracellular fibrillae present in these fungi (Fig. 10), which is probably involved in the adhesion process of the fungus to the host root (Gianinazzi-Pearson *et al.*, 1986). An antigen with a similar location (Perotto *et al.*, 1991a) was also identified by using a McAb (GMP-1) which was originally raised against the pathogenic fungus *Candida albicans* (Cassone *et al.*, 1988). Biochemical analysis of

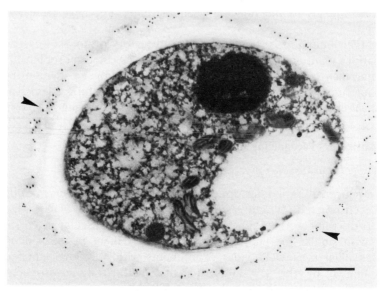

**Fig. 10.** Immunogold of CB25 McAb on the surface of the ericoid fungus *Hymenoscyphus ericae*. Gold particles are distributed on the fibrillar sheath typical of this fungus. Bar, 200 μm.

the epitope recognized on ericoid strains by GMP-1, which is an IgM, indicates its carbohydrate nature, and the difficulty of separating it on polyacrylamide gel may suggest that it is part of a polysaccharide (S. Perotto *et al.*, unpubl. res.). In *C. albicans*, this antibody binds to fibrillae which are produced around the hyphae and have been correlated with the adhesion process to the human host. The presence of a common epitope on the fibrillae surrounding the hyphae of both *C. albicans* and ericoid fungi suggests that there may be some simlarities in the molecular mechanisms of the adhesion process in these two organisms.

## B.   Antibodies recognizing the host

A legume plant can harbour both nodules and vesicular-arbuscular mycorrhiza at the same time, and some interesting comparisons can be made on particular aspects of the interactions by using the same immunoprobes. For example, the plant responses during the infection process by rhizobia (VandeBosch *et al.*, 1989a) and by vesicular-arbuscular mycorrhizal fungi (Perotto *et al.*, 1989; Bonfante-Fasolo *et al.*, 1990) have been investigated using McAbs and other affinity probes to plant cell wall components. The results indicate a similar array of cell wall components in the early stages of colonization by rhizobia and vesicular-arbuscular mycorrhizal fungi (Perotto *et al.*, 1989), but also reveal the formation of a different type of interface during the intracellular stages (Bonfante-Fasolo *et al.*, 1990).

Antibodies recognizing the peribacteroid membrane in pea nodules have also been tested on the periarbuscular membrane in mycorrhizal pea roots, as both membranes are actively involved in surface interaction and metabolite exchange with the endosymbiont. As mentioned already, techniques to separate efficiently fractions of the periarbuscular membranes from vesicular-arbuscular mycorrhizal roots for immunization and screening are still missing, and the study of this membrane relies so far on the availability of antibodies derived from other systems. Reports indicate that an epitope which is present on the peribacteroid membrane is also expressed on the periarbuscular membrane (Gianinazzi *et al.*, 1989). This analysis of the periarbuscular membrane has been recently extended by using a wide range of antibodies reacting with the peribacteroid membrane, whose epitopes has been partially characterized in the nodule (Fig. 11). Some of these antibodies identify

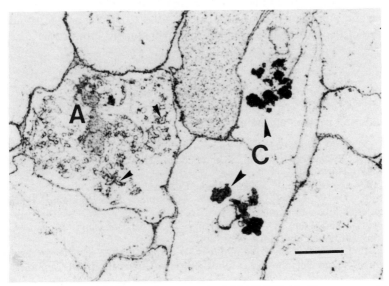

**Fig. 11.** Immunogold/silver enhancement of MAC 266 McAb on pea mycorrhizal roots. The epitope recognized by this McAb is expressed on the plant cell membranes of uninfected cells and on the periarbuscular membrane (small arrowheads) of infected cells (A). Its expression is strongly enhanced around collapsed arbuscules (C) (arrowheads). Bar, 15 μm.

epitopes which are differentially expressed on the membrane surrounding the vesicular-arbuscular mycorrhizal fungus during particular stages of the fungal colonization and arbuscule development (Bonfante-Fasalo *et al.*, 1991c).

## VI. Perspectives and conclusions

Hybridoma technology has been adopted to investigate various aspects of plant differentiation and plant–microbe interactions. The use of monoclonal antibodies has revealed the presence of particular epitopes on the cell surfaces of both plants and micro-organisms whose expression is modulated during development and during the establishment of symbiotic or pathogenic interactions.

The use of monoclonal antibodies for the study of mycorrhizal symbiosis is still restricted, and mostly relies on immunoprobes derived from other symbiotic or pathogenic systems. The fusion strategies followed in these systems may provide some guidelines for a wider application of the hybridoma technology to mycorrhiza. The generation of more McAbs to plant and fungal components expressed in mycorrhiza will certainly provide probes that can be used to increase understanding of some of the molecular aspects of the plant–fungus interaction.

## References

**Manuals and reviews on the production of monoclonal antibodies:**

Catty, D. (1988) *Antibodies: a Practical Approach*. IRL Press, Oxford.
French, D., Fischberg, E., Buhl, S. and Scharff, M. D. (1986). *Immunol. Today* **11**, 344–346.
Galfre, G. and Milstein, C. (1981). *Meth. Enzymol.* **73**, 3–46.
Godwing, J. W. (1986). *Monoclonal Antibodies: Principles and Practice*. Academic Press, London.
Harlow, E. and Lane, D. (1988). *Antibodies: A Laboratory Manual*. Cold Spring Harbour Laboratory, New York.
Wang, T. L. (ed.) (1986). *Immunology in Plant Science*, Society for Experimental Biology Seminar Series 29. Cambridge University Press, Cambridge.

## References

Aguila, H. L., Pollock, R. P., Spira, G. and Sharff, M. D. (1986). *Immunol. Today* **12**, 380–383.

Anderson, M. A., Sandrin, M. S. and Clarke, A. E. (1984). *Plant Physiol.* **75**, 1013–1016.
Bazin, H. (1982). In *Protides of the Biological Fluids* (H. Peters, ed.). Pergamon Press, New York.
Bonfante-Fasolo, P., Vian, B., Perotto, S., Faccio, A. and Knox, J. P. (1990). *Planta* **180**, 537–547.
Bonfante-Fasolo, P. and Perotto, S. (1991a). In *Electron Microscopy Applied in Plant Pathology* (K. Mendgen, ed.), pp. 265–275. Springer-Verlag, Konstanz.
Bonfante-Fasolo, P. and Perotto, S. (1991b). In *Molecular Signals in Plant–Microbe Communication* (D. P. S. Verma, ed.). CRC Press, Boca Raton, FL (in press).
Bonfante-Fasolo, P., Peretto, R. and Perotto, S. (1991c). In *SEB Series, Topics in Environmental Physiology* (W. J. Davies, ed.). BIOS Press (in press).
Bradley, D. J., Butcher, G. W., Galfre, G., Wood, E. A. and Brewin, N. J. (1986). *J. Cell Sci.* **85**, 47–61.
Bradley, D. J., Wood, E. A., Larkins, A. P., Galfre, G., Butcher, G.W. and Brewin, N. J. (1988). *Planta* **173**, 149–160.
Brewin, N. J., Robertson, J. G., Wood, E. A., Wells, B., Larkins, A. P., Galfre, G. and Butcher, G. W. (1985). *EMBO J.* **4**, 605–611.
Brewin, N. J., Davies, D. D. and Robins, R. J. (1987). In *The Biochemistry of Plants* (D. D. Davies, ed.), Vol. 13, pp. 1–31. Academic Press, London.
Brewin, N. J., Rathbun, E. A., Perotto, S., Gunder, A., Rae, A. L. and Kannenberg, E. L. (1991). In *SEB Series, Topics in Environmental Physiology* (W. J. Davies, ed.). BIOS Press (in press).
Candlish, A. A. G., Smith, J. E. and Stimson, W. H. (1989). *Biotechnol. Adv.* **7**, 401–418.
Cassone, A., Torosantucci, A., Boccanera, M., Pellegrini, G., Palma, G. and Malavasi, F. (1988). *J. Med. Microbiol.* **27**, 233–242.
Dewey, F. M., MacDonald, M. M. and Phillips, S. I. (1989). *J. Gen. Microbiol.* **135**, 361–374.
De St. Groth, F. and Scheidegger, D. (1980). *J. Immunol. Meth.* **35**, 1–21.
Galfre, G. and Butcher, G. W. (1986). In *Immunology in Plant Science*, Society for Experimental Biology, Seminar Series 29 (T. L. Wang, ed.), pp. 1–26. Cambridge University Press, Cambridge.
Galfre, G. and Milstein, C. (1981). *Meth. Enzymol.* **73**, 3–46.
Galfre, G., Milstein, C. and Wright, B. (1979). *Nature* **277**, 131–133.
Gianinazzi-Pearson, V., Bonfante-Fasolo, P. and Dexheimer, J. (1986). In *Recognition in Microbe–Plant Symbiotic and Pathogenic Interactions* (B. Lugtenberg, ed.), pp. 273–281. Springer-Verlag, New York.
Gianinazzi-Pearson, V., Gianinazzi, S. and Brewin, N. J. (1989). In *Endocytobiology IV* (P. Nardon, V. Gianinazzi-Pearson, A. M. Grenier, L. Margulis and D. C. Smith, eds), pp. 127–131. INRA, Paris.
Goding, J. (1980). *J. Immunol. Meth.* **39**, 285–308.
Hahn, M. G., Lerner, D. R., Fitter, M. S., Norman, P. M. and Lamb, C. J. (1987). *Planta* **171**, 453.
Halk, E. L. and De Boer, S. H. (1985). *Ann. Rev. Phytopatol.* **23**, 321–350.
Hardham, A. R., Suzaki, E. and Perkin, J. L. (1986). *Can. J. Bot.* **64**, 311–321.
Kannenberg, E. L. and Brewin, N. J. (1989). *J. Bacteriol.* **171**, 4543–4548.
Kearney, J. F., Radbruch, A., Liesegang, B. and Rajewsky, K. (1979). *J. Immunol.* **123**, 1548–1550.

248    S. PEROTTO et al.

Kilmartin, J. V., Wright, B. and Milstein, C. (1982). *J. Cell Biol.* **93**, 576–582.
Knox, J. P. (1990). *J. Cell Sci.* **96**, 557–561.
Knox, J. P., Day, S. and Roberts, K. (1989). *Development* **106**, 47–56.
Knox, J. P., Linstead, P. J., King, J., Cooper, C. and Roberts, K. (1990). *Planta* **181**, 512–521.
Kohler, G. and Milstein, C. (1975). *Nature* **256**, 495–497.
Kohler, G., Howe, S. C. and Milstein, C. (1976). *Eur. J. Immunol.* **6**, 292–295.
Kough, J., Malajczuk, N. and Linderman, R. G. (1983). *New Phytol.* **94**, 57–62.
Legocki, R. P. and Verma, D. P. S. (1980). *Cell* **20**, 153–163.
Mackie, A. J., Roberts, A. M., Callow, J. A. and Green, J. R. (1991). *Planta* **183**, 399–408.
Norman, P. M., Kjellbom, P., Bradley, D. J., Hahn, G. C. and Lamb, C. J. (1990). *Planta* **181**, 365–373.
Pennell, R. I., Knox, J. P., Scofield, G. N., Selvendran, R. R. and Roberts, K. (1989). *J. Cell Biol.* **108**, 1967–1977.
Perotto, S., Faccio, A., Malavasi, F. and Bonfante-Fasolo, P. (1987). *Giorn. Bot. It.* **120**, 78–79.
Perotto, S., VandenBosch, K. A., Brewin, N. J., Faccio, A., Knox, J. P. and Bonfante-Fasolo, P. (1989). In *Endocytobiology IV* (P. Nardon, V. Gianinazzi-Pearson, A. M. Grenier, L. Margulis and D. C. Smith, eds), pp. 115–117. INRA, Paris.
Perotto, S., Peretto, R., More, D. and Bonfante-Fasolo, P. (1991a). *Symbiosis* **9**, 167–172.
Perotto, S., VandenBosch, K. A., Butcher, G.W. and Brewin, N. J. (1991b). *Development* **112**, 763–774.
Roberts, K. (1990). *Curr. Opinion Cell Biol.* **1**, 1020–1027.
Schulman, M., Wilde, C. D. and Kohler, G. (1978). *Nature* **276**, 269–270.
Shur, B. D. (1989). *Curr. Opinion Cell Biol.* **1**, 905–912.
Smith, S. E. and Smith, F. A. (1990). *New Phytol.* **114**, 1–38.
Torrance, L., Larkins, A. P., Pead, M. T. and Butcher, G. W. (1986). *J. Gen. Virol.* **67**, 549–556.
Trowbridge, I. (1978). *J. Exp. Med.* **148**, 313–325.
VandenBosch, K. A., Bradley, D. J., Knox, J. P., Perotto, S., Butcher, G. W. and Brewin, N. J. (1989a). *EMBO J.* **8**, 335–342.
VandenBosch, K. A., Brewin, N. J. and Kannenberg, E. L. (1989b). *J. Bacteriol.* **171**, 4537–4542.
Wilson, J. M., Trinick, M. J. and Parker, C. A. (1983). *Soil Biol. Biochem.* **15**, 119–123.
Wright, S. F., Morton, J. B. and Sworobuk, J. E. (1987). *Appl. Environ. Microbiol.* **53**, 2222–2225.
Wyss, P., Mellor, R. B. and Wimken, A. (1990). *Planta* **182**, 22–26.

# 13

# Phytohormone Analysis by Enzyme Immunoassays

B. HOCK, S. LIEBMANN, H. BEYRLE and K. DRESSEL

*Department of Botany, Techische Universität of München at Weihenstephan,*
*W-8050 Freising 12, Germany*

## I. Introduction

Phytohormones serve in higher plants as signal substances which are produced in very small amounts and trigger a variety of developmental responses. Evidence has been obtained during the last few decades that changes in phytohormonal levels are caused by fungi in symbiotic as well as in parasitic interactions. Infection induces reactions in host tissues which enable, for example, the maintenance of source functions. An extreme situation is experienced in those cases where higher plants are externally provided with excess levels of phytohormones. For instance,

METHODS IN MICROBIOLOGY
VOLUME 24   ISBN 0-12-521524-X

Copyright © 1992 by Academic Press Limited
All rights of reproduction in any form reserved

morphogenetic changes in pine roots during mycorrhiza formation have
been related to the production of auxin by the fungal partner (Slankis,
1951, 1958). More recently, the growth of orchid protocorms was found
to be dependent on auxin and cytokinin supply by the endosymbiotic
fungus (Beyrle *et al.*, 1991).

Research on the function and importance of the phytohormonal status
in mycorrhizal relations depends upon the availability of highly sensitive
and specific methods for the quantification of individual phytohormones.
The rapid development of sophisticated analytical tools has provided a
respectable choice for the performance of complicated phytohormone
determinations. Gas chromatography (GC), gas chromatography/mass
spectroscopy (GC/MS), gas chromatography-mass spectroscopy
(GC-MS), high pressure liquid chromatography (HPLC) and proton
nuclear magnetic resonance ($^1$H-NMR) spectrometry are at present the
most important examples of physico-chemical tools which have been
applied for phytohormone research.

However, in addition to high costs and the restricted availability,
there are other drawbacks, the most important one being the need for
rather large amounts of sample material which may not be available in
mycorrhizal studies, which are often handling tissues in only milligram
quantities (e.g. Liebmann and Hock, 1989; Münzenberger *et al.*, 1990).

As has been shown in other chapters of this volume, serological
techniques are best suited for the analyses of trace quantities in small
samples. They have been successfully employed in the past for the
quantitation of phytohormones such as auxins, cytokinins, gibberellins
and abscisic acid (ABA). The subject matter has been recently reviewed
by Weiler (1990).

This chapter deals with the technique of enzyme immunoassays
(EIAs), their advantages and limitations. With proper care these can be
applied successfully to mycorrhizal research.


## II.  Immunoassays

### A.  General principles

Immunoassays for the determination of phytohormones use the high
affinity and specificity of suitable antibodies towards individual phytohormones. In immunological terms, phytohormones are haptens: they
can be bound by appropriate antibodies but they elicit the antibody
production only after their conjugation to a higher molecular mass
immunogenic carrier in order to produce a suitable antigen.

The common principle of all immunoassays is based upon the law of mass action which is formulated here for haptens such as phytohormones. In the equilibrium reaction between the free hapten H and the antibody Ab towards the hapten–antibody complex HAb (= bound hapten)

$$H + Ab = HAb \qquad (1)$$

the affinity constant $K$ determines the concentration ratio between the bound hapten and the free reaction partners:

$$K = \frac{[HAb]}{[H] \times [Ab]} \ (1 \ mol^{-1}) \qquad (2)$$

A low detection limit in immunoassays requires a high affinity of the antibodies towards the analyte, which is expressed by a high affinity constant.

In order to explain the rationale behind the immunoassays being used for hapten determination, i.e. competitive immunoassays, let us assume a set of assays with constant antibody concentrations. In this case, the total amount of hapten (i.e. the phytohormone concentration in a sample) determines the ratio between bound and free hapten (b/f) in the equilibrium. Therefore, unknown hapten concentrations can be determined from the ratio b/f.

Figure 1 shows b/f as a function of the logarithm of the total hapten

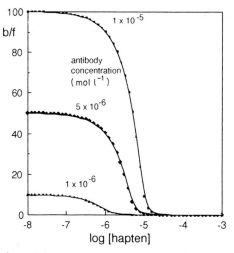

**Fig. 1**   Influence of total hapten concentration on the ratio of bound to free hapten (b/f) at an affinity constant of $K = 10^7 \ 1 \ mol^{-1}$ and at three different antibody concentrations.

concentration $[H_{tot}]$ for three different antibody concentrations $[Ab_{tot}]$ with an affinity constant of $10^7 \, l \, mol^{-1}$. b/f has been calculated by the formula (cf. Chan, 1987)

$$(b/f)^2 - (K[H_{tot}] - K[Ab_{tot}] - 1) \times (b/f) - K[Ab_{tot}] = 0 \qquad (3)$$

It is obvious from these curves that with increasing antibody concentrations the sensitivity of the assay increases and the detectability decreases. The detectability is given by the detection limit, i.e. the lowest concentration, which significantly differs from the zero sample. Sensitivity* is usually defined as the slope of the curve at the inflection point (= middle of the test). Therefore, an improvement of the sensitivity is paid for by a deterioration of the detection limit, and vice versa.

Under real conditions, the ratio b/f cannot be determined directly. Therefore, a helper reagent is added which produces the measuring signal. In the case of competitive immunoassays†, which must be applied for hapten determinations such as phytohormone analyses, a labelled hapten (= tracer) is added to the system. This tracer is a radioactively labelled hapten in the radioimmunoassay (RIA) or an enzyme–hapten conjugate in the enzyme immunoassay (EIA).

Enzyme tracers produce a signal amplification because of the catalytic activity of their enzymes—a single enzyme tracer molecule can convert many substrate molecules. General reviews on EIAs are found in Maggio (1980) and Tijssen (1985); reviews on RIAs are found in Chard (1982).

The basic principle of a competitive immunoassay is very simple. If the antibody and the tracer concentrations are kept constant in an assay and if there are a limited number of antibody binding sites, the proportion of bound tracer decreases with increasing hapten concentration; this is illustrated in Fig. 2. Here, three standard situations are compared: to the left an assay in the absence of the hapten, in the middle in the presence of hapten (a hapten:tracer ratio of 1:2 is assumed) and to the right in an excess of hapten. The upper row shows the reaction partners, the rows below the results after completion of the reaction. The middle row represents an assay without phase separation, as is used in the homogeneous EIA, i.e. bound and unbound haptens remain in the same system. The lower row represents an assay with phase separation, as is used in the RIA or in the heterogeneous EIA. Here, the antibody-bound and the free haptens are separated from one

*However, some authors use this term for the detectability (cf. Ekins, 1970).
†Non-competitive immunoassays (= immunometric assays) are more useful for the determination of certain antigens such as immunoglobulins or viruses and bacteria. Usually labelled antibodies are used as tracers.

another. In both cases, the number of antibody binding sites which are available for the tracer depends upon the hapten concentration. In the absence of hapten (left columns) only tracer binds to the antibody, the tracer is almost completely displaced from the antibody at a large excess of hapten (right columns).

## B.    The heterogeneous enzyme immunoassay

The separation of the antibody-bound and the free ligand provides the basis of the heterogeneous EIA. In the heterogeneous EIA either the bound or (more rarely) the free fraction is used for the measurement. As the enzyme tracer produces the signal, in the first case it is inversely proportional to the hapten concentration (cf. Fig. 2, upper part of the lower row); in the second case (cf. Fig. 2, lower part of the lower row) it is proportional to the hapten concentration.

The procedure which is most often applied for phase separation uses the binding of antibodies to a solid phase. Figure 3 illustrates the most commonly used variant, the enzyme-linked immunosorbent assay (ELISA). In the first step, the antibodies are adsorbed to a solid surface, e.g. the cavities of a microtitre plate, this procedure being known as coating (Fig. 3a, upper row). Immobilization in this case is based on the passive adsorption of the antibodies to a plastic surface, e.g. polystyrene. After washing in order to eliminate unbound molecules, hapten (sample) and enzyme tracer are added (Fig. 3a, middle) and incubated for the immunoreaction (Fig. 3a, lower part). At this stage it is very important to apply antibody and tracer in constant amounts and to limit the amount of antibodies. Note that the arrangement corresponds to the general scheme of Fig. 2. Again the three standard situations are used: (1) the absence of hapten (left), where only enzyme tracer can be bound to the antibody; (2) the presence of hapten and enzyme tracer, which both compete for the available antibody-binding sites (middle); and (3) excess of hapten, which then displaces the tracer from the antibody-binding sites. Therefore, the less hapten that is available in the assay, the more enzyme tracer is bound by the antibodies. This competing situation becomes apparent after the washing step that follows (Fig. 3b, upper part). After the subsequent addition of enzyme substrate the enzyme activity is determined (Fig. 3b, middle), which is proportional to the concentration of the bound tracer and therefore inversely proportional to the applied hapten concentration (Fig. 3b, lower part). The concentration of unknown samples therefore can be determined by means of calibration curves.

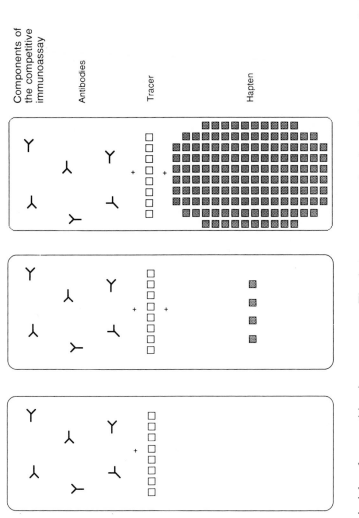

**Fig. 2** The principle of competitive immunoassays. The hapten concentration is varied at constant antibody and tracer concentrations. The upper row shows the reaction partners, the hapten concentration increases from left to right. Below, the results are presented after the completion of the reaction in an assay without phase separation (middle) and in an assay with phase separation (lower row). Each antibody molecule can bind two ligands which is omitted here for the sake of simplicity.

Components of the competitive immunoassay

Antibodies

Tracer

Hapten

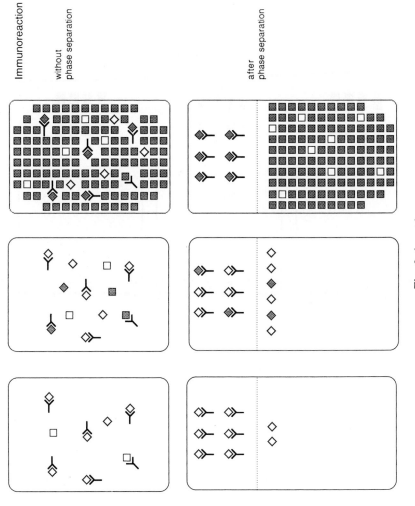

**Fig. 2** (*cont.*)

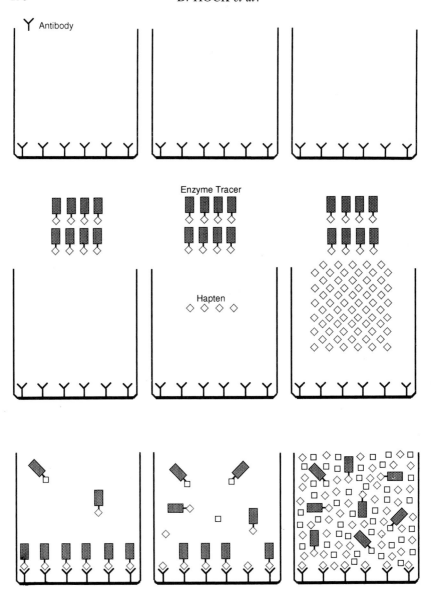

**Fig. 3** The heterogeneous enzyme immunoassay using ELISA as an example. (a) After the immobilisation of the antibody at a solid phase (e.g. the wall of a microtitre plate) and a washing step (upper row), the enzyme tracer and the hapten are added (middle row). The hapten (e.g. auxin) concentration increases from left to right. The lower row shows the immunoreaction.

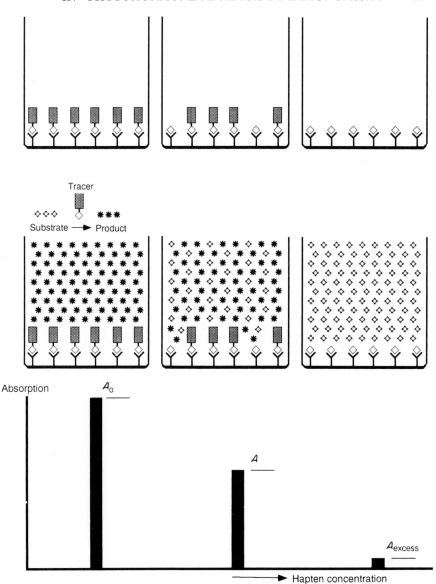

**Fig. 3** (b) After a further washing step for the removal of the unbound component, substrate is added, which is converted to a coloured product according to the concentration of the bound enzyme tracer (middle row). Photometric determination of the enzyme activity by the absorption $A$ therefore yields decreasing enzyme activities with increasing hapten concentrations. $A_0$, absorption in the absence of hapten.

Figure 4a shows a calibration curve for the determination of the auxin indole-3-acetic acid (IAA) on the basis of polyclonal antibodies and an alkaline phosphatase tracer. The middle of the test lies at sample concentrations of *c*. 12 nmol litre$^{-1}$ IAA (*c*. 2.1 ng ml$^{-1}$), the detection limit at *c*. 0.57 nmol litre$^{-1}$ IAA (*c*. 0.1 ng ml$^{-1}$). Sample volumes of 100 μl per cavity are used for the microtitre plate test.

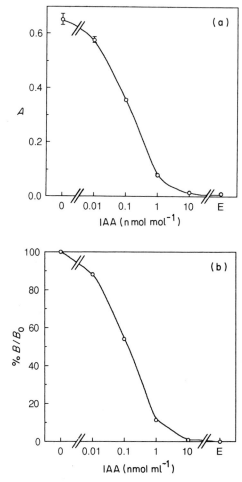

**Fig. 4**  ELISA for the determination of auxin with polyclonal antibodies from rabbit and an alkaline phosphatase enzyme tracer. (a) Absorption curve (means of four determinations ± standard deviations), (b) % $B/B_0$ curve, (c) logit–log transformation by the two-parameter fit.

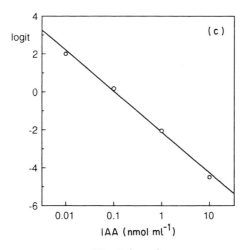

**Fig. 4** (*cont.*)

## C.  Data processing and statistics

A linearization of the calibration curves is useful for many purposes, for instance for their direct comparison or for the evaluation of unknown concentrations. The most sophisticated procedures are based upon the logistic model (cf. Rodgers, 1984; Dudley *et al.*, 1985). A reasonable compromise between the quality of the data fit and the computing effort is met by the four-parameter model which relates the expected response $y$, i.e. the absorption, to the concentration of the analyte, $x$, by

$$y = \frac{a - d}{1 + (x/c)^b} + d \tag{4}$$

where $a$, $b$, $c$ and $d$ are constants.

In formula (4), $a$ and $d$ correspond to the upper and lower asymptotes of the curve, respectively, $c$ to the concentration at the inflection point of the curve, and $b$ to the slope at the inflection point.

A linear relation is obtained after reformulating equation (4):

$$\log\left(\frac{y - d}{a - y}\right) = (-b) \times \log(x) + b \times \log(c) \tag{5}$$

If $\log(y - d/a - y)$ is set equal to $Y$, which is the logit transformation, equation (5) becomes

$$Y = (-b) \times \log(x) + [b \times \log(c)] \tag{6}$$

However, (6) is the equation for a straight line for $Y$ plotted against $\log(x)$ with slope $(-b)$ and $Y$ intercept $[b \times \log(c)]$.

Computer programs are available to determine the constants $a$, $b$, $c$, and $d$ for the four-parameter fit of a given calibration curve using an iterative procedure (e.g. AssayZap by Biosoft, Milltown, NJ, USA).

A simplified model chooses $a = 1$ (or 100%) and $d = 0$ (two-parameter model). Therefore

$$y = \frac{1}{1 + (x/c)^b} \tag{7}$$

The logit relationship is given by

$$Y = \log\left(\frac{y}{1 - y}\right) = (-b) \times \log(x) + b \times \log(c) \tag{8}$$

which again produces a straight line when $Y$ is plotted against $\log(x)$. The logit–log transformation can easily be carried out with a pocket calculator. For normal data sets in well functioning assays, this two-parameter fit is not inferior to the four-parameter fit explained above.

The two-parameter fit is carried out in two steps. First, the absorptions are converted to % $B/B_0$ values, which represent the ratio

$$\left(\frac{\text{Bound tracer in the presence of hapten}}{\text{Bound tracer in the absence of hapten}}\right) \times 100 \tag{9}$$

and lie between 100% ($= A_0$, the upper asymptote of the curve) and 0% ($= A_{\text{excess}}$, the lower asymptote) by the formula

$$\% \, B/B_0 = \frac{A - A_{\text{excess}}}{A_0 - A_{\text{excess}}} \times 100 \tag{10}$$

The linearization is obtained after the logit–log transformation

$$\text{logit}(\% \, B/B_0) = \ln\left(\frac{\% \, B/B_0}{100 - \% \, B/B_0}\right) \tag{11}$$

by plotting the logits against the log of the concentrations. An example for this type of linearization is shown in Fig. 4.

The measuring range of the calibration curves is delimited by the lower and the upper detection limits. They are defined as those concentrations, which differ significantly from the zero and the excess concentrations, respectively. The experimental errors increase towards these limits (cf. Fig. 5). Consequently, the most precise measurements are obtained in the region of the middle of the test (at 50% $B/B_0$).

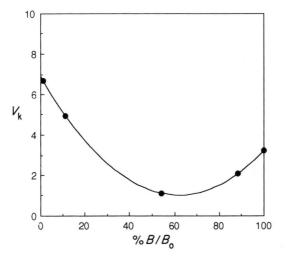

**Fig. 5** Distribution of the statistical errors (variation coefficient, $V_k$ in dependence of % $B/B_0$). The data of Fig. 4a are used.

## D.  Error sources

In addition to errors due to insufficient precision, which reflect the scattering of the data as measured by the standard deviation or the variation coefficient, the most commonly encountered error in immunoassays is insufficient accuracy. This term expresses the ability of the test to measure the true value of the analyte. Losses of accuracy stem either from cross-reactivities or from matrix effects.

The cross-reactivity of an antibody or an antibody mixture designates in our context the extent to which an antibody reacts with related phytohormones—or even worse—with other compounds. Cross-reactivity can be due to the existence of a heterogeneous antibody mixture, as is usually the case with polyclonal antibodies where individual antibody molecules may react with different analytes. But even with a homogeneous population of antibodies, such as monoclonal antibodies, cross-reactivity cannot be avoided because it is an intrinsic property of practically any antibody, especially those which are directed against small haptens, to bind related compounds, though usually with different affinities.

Cross-reactivity is calculated according to the formula.

% Cross-reactivity =

$$\frac{\text{Hapten concentration at 50\% } B/B_0}{\text{Concentration of the cross-reacting hapten at 50\% } B/B_0} \times 100 \quad (12)$$

For this purpose, calibration curves are constructed for the phytohormone to be measured and those compounds to be tested for their cross-reactivities, using the same antibody. Then, the concentrations at the middle of the tests are determined, e.g. by calculation of the 50% $B/B_0$ values.

Figure 6 shows the cross-reactivities of an auxin antibody raised in the laboratory of the authors. It is evident that as well as reacting to IAA (100%) it also reacts to a minor extent with indole-3-acetone and indole-3-acetonitrile. However, this can be neglected in most cases because these compounds are not known to be major constituents of mycorrhizal roots.

Another serious source of error is the occurrence of matrix effects, which may be due to substances in the sample interfering with the assay. A very important goal of validation (see below) is the detection of matrix effects, which then have to be eliminated by a more rigorous clean-up of plant extracts.

Finally, it should be pointed out that deterioration of the immunoassay may be due to the limited stability of the antibody and of the enzyme tracer. Proper controls, for instance the performance of calibration curves *before* the measurement of valuable samples, should be carried out to detect this common type of error.

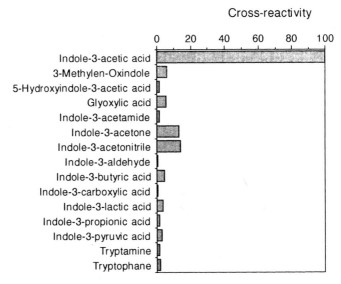

**Fig. 6** Cross-reactivities of a polyclonal antibody for auxin determination. Test conditions as in Fig.4.

### III. Validation of enzyme immunoassays

A significant step during the establishment of immunoassays for routine analyses is the validation by independent methods. Comparison of results with those obtained by physico-chemical methods, which have been recently reviewed with respect to phytohormone analyses by Sembdner *et al.* (1988), is necessary. However, difficulties may arise not only due to the limited availability of proper analytical procedures, which also have to be adapted to the specific conditions of plant extracts, but also due to the limited supply of plant material. Again this is especially true for mycorrhizal samples.

Therefore, it is very important to detect potential interferences by constituents of cell extracts with the immunoassays themselves. Such compounds include non-specific and competitive inhibitors which can produce false positives (for a theoretical treatment, see Pengelly, 1986). The extent of such inhibition must be determined in order to correct the results for this effect. The usual method of validation is dilution analysis with internal standardization.

Figure 7 shows the internal standardization of plant extracts. In this example by Wang *et al.* (1986) two dilutions of immunoreactive material were assayed after the addition of different amounts of ABA. The lines

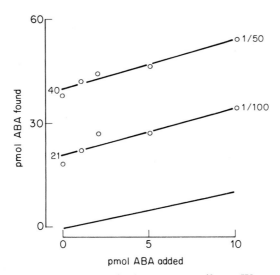

**Fig. 7** Internal standardization of plant extract (from Wang *et al.*, 1986). Immunoreactive material from a pea embryo extract in two dilutions was determined by an (+)-ABA enzyme immunoassay after the addition of different amounts of the phytohormone.

are parallel and show an additive effect. The quantity added always produces the value of the endogenous ABA plus the amount added. If interference influences the binding of the antigen to the antibody, then the lines obtained are not parallel to the standard line, have different slopes and do not show an additive effect. Successive approximation is then essential, with analyses at various levels of purification to remove selectively interfering compounds such as lipids, pigments and phenols to concentrate the phytohormones of interest.

### IV. Application of enzyme immunoassays for phytohormone analyses

Serological techniques such as the RIA or the ELISA have gained a growing importance in phytohormone analysis because of their sensitivity and their simplicity in performance. The presently available tests, which have been extensively reviewed by Weiler (1990), cover a significant part of phytohormonal analysis.

For details the following references are useful: auxins, Weiler (1981), Weiler *et al.* (1981); cytokinins, Ernst (1986); gibberellins, Crozier *et al.* (1986), Schneider *et al.* (1990); abscisic acid, Mertens *et al.* (1983). IAA is taken here for the step-by-step description of a phytohormone ELISA (cf. Protocol A), which in principle can be applied for other assays after necessary modifications. Although it is possible to use commercially available tests, those procedures are added which are necessary for the synthesis of the required enzyme tracer (cf. Protocol B) and finally for raising polyclonal antibodies in rabbits (Protocol C).

### V. Use of monoclonal antibodies

There are several advantages of monoclonal antibodies for application in enzyme immunoassays. They can be obtained in an unlimited quantity with the same characteristics, although one should be aware of the fact that even immortal hybridomas sometimes become unstable, e.g. by chromosome losses. Since monoclonal antibodies represent a homogeneous population of identical antibody molecules, the calibration of dose–response curves is facilitated. The need of purification steps to remove any interfering serum components is eliminated.

However, it is important to realize that cross-reactivities with related haptens are an inherent property of antibodies because the binding sites usually recognize similar compounds in addition to the preferred analyte. Even cases in hapten research (pesticides) have been observed where polyclonal antibodies outperform monoclonal antibodies with respect to specificity in spite of extensive screening (Wittmann and Hock, 1990). In addition, the affinity of the antibodies towards the hapten analyte has to be considered. This is usually lower in the case of monoclonal compared with polyclonal antibodies.

At present, it is not possible to generalize with respect to the situation. If the facilities are available, the production of monoclonal antibodies is certainly recommended. However, one has to be aware of the high costs of production, selection, and maintenance of hybridomas. In any case, it is highly recommended to begin with polyclonal antibodies in order to guarantee functioning enzyme tracers before starting work with monoclonal antibodies.

It is beyond the scope of this chapter to review the procedures which are required for the production of monoclonal antibodies. Several laboratory manuals are available describing all the steps in raising monoclonal antibodies, e.g. Peters *et al.* (1985), Bartal and Hitshaut (1987) or Zola (1987).

## VI.    Clean-up of mycorrhizal extracts

The aim of the extraction procedure is to quantitatively and selectively remove phytohormones from the matrix. It is the most critical step of the hormone analysis which cannot readily be checked by other techniques. The choice of the extraction procedure has to take account of the following factors (cf. Sembdner *et al.*, 1988): (1) representative sampling; (2) immediate stopping of any enzyme activities at the moment of decompartmentation; and (3) optimal removal of the phytohormones under strictly standardized and reproducible conditions.

Extraction rarely provides a sufficient concentration step for the subsequent analytical procedure. Usually, further separation and clean-up steps are required which depend upon the particular analytical method in use.

There is no general extraction procedure to be applied for all phytohormones. The specific method has to be independently worked out for each case and adjusted to each type of sample. However, there is a wealth of information in the literature on this topic which has been reviewed by Sembdner *et al.* (1988) and can be used as a starting point for the optimization of the extraction and further clean-up steps.

In Protocol D, a detailed description is given for the clean-up of mycorrhizal extracts for the extraction of auxins as it has been applied to orchid protocorms (Beyrle *et al.*, 1991) and ectomycorrhizal roots of spruce (Liebmann and Hock, 1989).


## VII.  Conclusions

The opportunity to accelerate physiological studies of mycorrhiza depends on plant hormone analysis. If the problems of validation are overcome then the ELISA technique allows for the convenient analysis of rather impure samples for the presence of specific plant growth regulators with high specificity, sensitivity and ease of operation. No extensive work-up of samples is necessary and a great number of samples can be processed in a short time. This, and the fact that highly specialized laboratory facilities are not required, makes the ELISA very economical.

It is to be expected that the ELISA technique with the inclusion of methodological modifications will gain growing importance. This especially concerns combinations of serological and physico-chemical methods such as HPLC.


## VIII.  Protocols

### A.  ELISA for the determination of indole-3-acetic acid

*1.  Antiserum concentrations*

Appropriate amounts of antibodies (lyophilized IgG fraction, cf. Protocol C. 3) are dissolved in 50 mmol litre$^{-1}$ Na carbonate buffer (pH 9.6). If the optimal antibody (and enzyme tracer!) concentrations for the phytohormone assay are not known, they can be easily found by two-dimensional titre determinations. These assays are run in microtitre plates with decreasing antiserum concentrations (e.g. dilution 1:1000 in column 1*, 1:2000 in column 2, 1:4000 in column 3, etc.) and decreasing enzyme tracer concentrations (e.g. dilution 1:1000 in row A, 1:2000 in row B, 1:4000 in row C, etc.). For further optimization in the ELISA (cf. below) with respect to detectability and sensitivity antibody–tracer combinations are recommended which yield absorptions above 0.6 in the absence of the phytohormone.

---

*For labelling the columns and the row, cf. Fig. 8.

## 2.  Methylation of IAA

Methylation of IAA decreases the instability of IAA and at the same time increases the affinity of the antibodies towards the analyte, resulting in lower detection limits of the assay.

The following procedure is recommended (cf. Levitt, 1973). A 20 ml methylation vessel is filled with a spatula tip of $N$-nitroso-toluene-4-sulphomethylamide, 5 ml dry diethylether, and 2 ml diethylglycolmonoethylether. After mixing and dissolving, 1 ml 60% KOH is underlayered. $N_2$ is slowly bubbled through for 1 min (use 3 min for the standards). Diazomethane is produced which derivatizes IAA to its methyl ester. After evaporation over $N_2$, 1 ml TBS-G (Tris-buffered saline containing 50 mmol litre$^{-1}$ Tris, pH 7.8, 10 mmol litre$^{-1}$ NaCl, 2.1 mmol litre$^{-1}$ MgCl$_2$, and 0.5 g litre$^{-1}$ gelatine) is added and kept at 4° C overnight to dissolve the auxin.

## 3.  ELISA

300 $\mu$l of the appropriate antibody dilution are pipetted into each well of the microtitre plate for coating. After incubating for 4 h at 37° C, the coated plates are washed 3 times with double-distilled water. A washer is convenient although the procedure can be carried out manually. Then, each cavity is filled with 50 $\mu$l TBS-G (cf. above). Subsequently, 200 $\mu$l per well of the methylated IAA standards and samples are added. Figure 8 proposes a convenient pipetting scheme, using four parallels for each standard or sample, respectively, except for the wells at the rim. This step should be carried out under the control of a stop-watch to guarantee equal incubation periods for the subsequent manipulations. After an incubation period of exactly 1 h on a shaker, the enzyme tracer (50 $\mu$l per well, cf. Protocol B) is added in the same order as the IAA samples. After exactly 30 min, the plates are washed three times with double-distilled water. Then 200 $\mu$l substrate (0.5 mg dinitrophenyl phosphate per ml diethanolamine buffer = 1 mmol litre$^{-1}$, pH 9.8) are added. The plates are incubated for 2 h at 37° C. It is advisable to carry out the incubations at constant temperature (e.g. 20° C) on a tempered plate and to use pre-tempered solutions. Finally the absorptions are read at 405 nm and at 620 nm as a reference in an EIA reader.

EIA washers and readers, as well as microtitre plates, are commercially available in a great assortment. The use of pipetting aids is recommended. Commercial programs are also available for data processing.

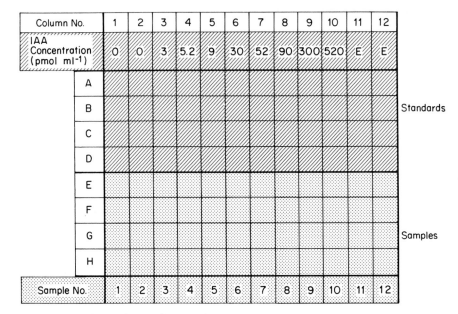

| Column No. | 1 | 2 | 3 | 4 | 5 | 6 | 7 | 8 | 9 | 10 | 11 | 12 | |
|---|---|---|---|---|---|---|---|---|---|---|---|---|---|
| IAA Concentration (pmol ml⁻¹) | 0 | 0 | 3 | 5.2 | 9 | 30 | 52 | 90 | 300 | 520 | E | E | |
| A | | | | | | | | | | | | | Standards |
| B | | | | | | | | | | | | | Standards |
| C | | | | | | | | | | | | | Standards |
| D | | | | | | | | | | | | | Standards |
| E | | | | | | | | | | | | | Samples |
| F | | | | | | | | | | | | | Samples |
| G | | | | | | | | | | | | | Samples |
| H | | | | | | | | | | | | | Samples |
| Sample No. | 1 | 2 | 3 | 4 | 5 | 6 | 7 | 8 | 9 | 10 | 11 | 12 | |

**Fig. 8** Pipetting scheme for an immunoassay. The standard concentrations (pmol ml⁻¹ of the methylated IAA) increase logarithmically. The zero and the excess samples (E) are duplicated.

## B.    Enzyme tracer synthesis

The procedure uses alkaline phosphatase for tracer synthesis. It follows the method of Weiler *et al.* (1981).

1.75 mg IAA (= 10 $\mu$mol) are dissolved in 0.2 ml 50% dimethylformamide and added to 3.8 mg EDC (= 1-ethyl-3[3-dimethylaminopropyl]carbodiimide-HCl, 20 $\mu$mol) in 0.1 ml $H_2O$. The pH is adjusted to 6.5 with 0.01 N NaOH (*c.* 12 drops). The reaction mixture is incubated for 1 h at 25° C with slow stirring. Then 100 $\mu$l alkaline phosphatase (Boehringer, 2500 U mg⁻¹) corresponding to 1 mg are diluted with 200 $\mu$l 50% dimethylformamide and added in 20 $\mu$l portions over a time period of at least 90 min to the IAA-EDC solution. Finally, $N_2$ is blown for 2–3 min into the reaction vessel, which is then tightly closed. The mixture is slowly stirred for 18 h at 25° C in the dark with a magnetic stirrer and then dialysed against 5% dimethylformamide for 8 h and subsequently for 3 days against TBS buffer, pH 7.8.

The tracer is mixed with glycerol in a ratio of 1:1.2 (v/v) and stored at −26° C.

## C.   Production of polyclonal antibodies

### 1.   Immunoconjugate synthesis

The synthesis of the immunoconjugate is carried out in a similar manner to the tracer synthesis.

10 mg IAA are dissolved in 0.1 methanol and made up with $H_2O$ to 8.0 ml. During the dilution, precipitates occur, which are redissolved during the volume increase. The pH is adjusted to $c$. 6.5 with 1 N NaOH in 10 $\mu$l portions. 20 mg EDC are dissolved in 0.5 ml $H_2O$ and the pH is again brought to 6.5 with 0.01 N NaOH. Both solutions are combined and reacted for 1 h at 25° C in the dark with slow stirring. 50 mg BSA (bovine serum albumin) are dissolved in 1 ml $H_2O$ and the pH is adjusted with 0.1 N NaOH to 6.5. Over a time period of 15 min, 20 $\mu$l aliquots of the BSA solution are pipetted into the IAA-EDC solution. $N_2$ is blown into the reaction mixture for 2–3 min, which is then slowly stirred at 25° C in the dark for 18 h. The conjugate is then dialysed for 3 days at 4° C against 20 mmol litre$^{-1}$ sodium phosphate buffer (pH 7.1) with 0.9% (w/v) NaCl (= PBS).

### 2.   Immunization

Six-month-old female rabbits (e.g. New Zealand White) are bled from the ear vein by making a $c$. 4 mm longitudinal cut with a sharp scalpel in order to obtain control serum ($c$. 15 ml), which is collected in a centrifuge tube or a small beaker. Then the animals are injected three times in weekly intervals subcutaneously into the neck area and intracutaneously into the caudal area with an emulsion prepared by sonication* of 0.5 ml of an immunoconjugate dilution ($c$. 0.5 mg) + 0.5 Freund's adjuvant, complete. A final booster injection of 0.5 ml immunoconjugate (without adjuvant) is injected into the ear vein.

Two weeks later, $c$. 30 ml blood is drawn from the ear vein for the preparation of antiserum. If necessary, additional samples can be obtained at monthly intervals after further booster injections as described above.

### 3.   Serum and immunoglobulin G preparation

The collected blood is kept for 1 h at room temperature. After separating the clot from the wall with a syringe needle, the tube is kept

---

*As an alternative procedure the mixture can be squeezed several times between two 1-ml syringes, which are tightly connected by a silicone tubing, until a white emulsion is obtained.

overnight at 4° C for further clotting. Then, the entire contents (serum- + clot) are transferred to a small glass funnel in order to collect the serum in another centrifuge tube. After a further 12 h at 4° C, the serum is centrifuged for 10 min at 6000 rpm (Sorvall centrifuge RC-5B, SS 34 rotor) and the supernatant is collected.

For the preparation of the immunoglobulin G (IgG) fraction, an ammonium sulphate (AS) precipitation is carried out. An aliquot of 6 ml saturated AS solution is slowly pipetted into 10 ml serum with gentle stirring. The pH is adjusted to 7.4 with 2N NaOH. After slow stirring for a further 30 min, the suspension is centrifuged for 30 min (20° C) at 10 000 rpm. The supernatant is discarded and the pellet is gently dissolved in 10 ml saline (0.9% NaCl, pH 7.4). The AS precipitation is repeated with 5 ml saturated AS solution as described. The pellet is again taken up in 10 ml saline and dialysed for 2–3 days against saline with two changes. It is terminated when a test for sulphate with 1% (w/v) $BaCl_2$ yields no precipitate.

The IgG fraction is divided into aliquots and then deep-frozen or lyophilized.

## D. Clean-up of mycorrhizal extracts

The clean-up procedure is based upon earlier work by Bandurski and Schulze (1977) and Weiler et al. (1981) and may be used to distinguish between free and bound auxin. Similar procedures have been described by Ek et al. (1983) and Frankenberger and Roth (1987) for fungal mycelia and culture media.

Samples of 0.1–0.5 g fresh weight are frozen in liquid nitrogen immediately after collecting the samples in aluminium foil and kept at −70° C for further analysis. The homogenization and extraction of auxin is performed with a pre-chilled mortar and pestle after the addition of washed quartz sand (Merck), 3 ml 70% acetone and 1 ml butylated hydroxytoluene. The crushed material is then transferred into glass vials and slowly shaken for 4 h in the dark. After centrifugation at 6000 rpm for 10 min, the supernatant is poured into a measuring cylinder and after adjusting the volume if required, partitioned into four equal parts into reagent vessels. Two of these are used for the determination of free IAA and two for the determination of total IAA (bound + free IAA). Figure 9 summarizes the further procedure, which is performed within 1 day.

IAA is methylated as described in Protocol A.2. The extracted and methylated auxin, which is dissolved in buffer, can be measured direct in the ELISA.

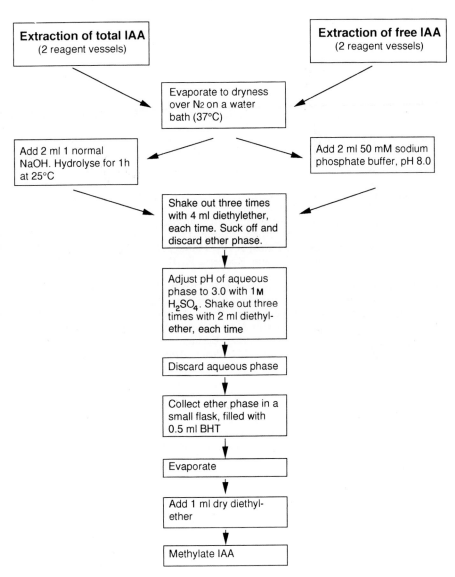

**Fig. 9** Clean-up of mycorrhizal extracts for the subsequent determination of auxins in the ELISA.

Care has to be taken to recognize contamination of mycorrhizal roots by other microorganisms and their contribution to phytohormone production. This applies especially to field grown material. In critical cases, it might be necessary to isolate contaminants and check for their ability to produce phytohormones.

## References

Bandurski, R. S. and Schulze, A. (1977). *Plant Physiol.* **60**, 211–213.

Bartal, A. H. and Hirshaut, Y. (eds) (1987). *Methods of Hybridoma Formation.* Humana Press, Clifton, NJ. 480 pp.

Beyrle, H., Penningsfeld, F. and Hock, B. (1991). *New Phytol.* **117**, 665–672.

Chan, W. D. (1987). In *Immunoassay. A Practical Guide* (W. D. Chan and Perlstein M. T., eds), pp. 1–13. Academic Press, Orlando.

Chard, T. (1982). In *Laboratory Techniques in Biochemistry and Molecular Biology* (T. S. Work, and Work, E., eds), 2nd edn, Vol. 6, Pt. II, p. 284. Elsevier, Amsterdam and New York.

Crozier, A., Sandberg, G., Monteiro, A. M. and Sundberg, M. (1986). In *Plant Growth Substances 1985* (M. Bopp, ed), pp. 13–21. Springer-Verlag, Berlin.

Dudley, R. A., Edwards, P., Ekins, R. P., Finney, D. J., McKenzie, I. G. M., Raab, G. M., Rodbard, D. and Rodgers, R. P. C. (1985). *Clin. Chem.* **3**, 1264–1271.

Ek, M., Ljungquist, P. O. and Stenström, E. (1983). *New Phytol.* **94**, 401–407.

Ekins, R. P. (1970). In *In vitro Procedures with Radioisotopes in Medicine.* International Atomic Energy Agency, Wien, Austria.

Ernst, D. (1986). In *Immunology in Plant Sciences* (H. F. Linskens, and Jackson J. F., eds) pp. 18–49. Springer-Verlag, Berlin.

Frankenberger Jr, W. T. and Roth, M. (1987). *Appl. Environm. Microbiol.* **53**, 2908–2918.

Kemeny, D. M. and Challacombe, S. J. (1988). *ELISA and Other Solid Phase Immunoassays. Theoretical and Practical Aspects*, J. Wiley & Sons, Chichester. 367 pp.

Levitt, M. J. (1973). *Anal. Chem.* **45**, 618–620.

Liebmann, S. and Hock, B. (1989). *Plant Physiol. (Life Sci. Adv.)* **8**, 99–104.

Maggio, E. T. (1980). *Enzyme-immunoassay.* CRC Press, Boca Raton, FL. 295 pp.

Mertens, R., Deus-Neumann, M. and Weiler, E. W. (1983). *FEBS Lett.* **160**, 269–272.

Münzenberger, B., Heilemann, J., Strack, D., Kottke, I. and Oberwinkler, F. (1990). *Planta* **182**, 142–148.

Pengelly, W. L. (1986). In *Plant Growth Substances 1985* (M. Bopp, ed.), pp. 35–43. Springer-Verlag, Berlin.

Peters, J. H., Baumgarten, H. and Schulze, M. (1985). *Monoklonale Antikörper. Herstellung und Charakterisierung.* Springer-Verlag, Berlin. 268 pp.

Rodgers, R. P. C. (1984). In *Clinical Immunoassay: The State of the Art* (W. R. Butt, ed.). Marcel Dekker, New York.

Schneider, P., Horn, K., Lauterbach, R. and Hock, B. (1990). *Environ. Pollut.* **64**, 347–351.

Sembdner, G., Schneider, G. and Schreiber, K. (1988). *Methoden der Pflanzenhormonanalyse.* VEB Gustav Fischer-Verlag, Jena. 296 pp.

Slankis, V. (1951). *Symbolae Bot. Upsal.* **11**, 1–63.

Slankis, V. (1958). In *The Physiology of Forest Trees* (K. V. Thimann, ed.), pp. 427–443. The Ronald Press, New York.

Tijssen, P. (1985). *Laboratory Techniques in Biochemistry and Molecular Biology* (R. H. Bordon and P. H. van Knippenberg, eds), Vol. 15. Elsevier, Amsterdam. 549 pp.

Wang, T. L., Griggs, P. and Cook, S. (1986). In *Plant Growth Substances 1985* (M. Bopp, ed.), pp. 26–34. Springer-Verlag, Berlin.

Weiler, E. W. (1981). *Planta* **153**, 319–325.

Weiler, E. W. (1990). In *Chemistry of Plant Protection* (G. Haug and Hoffmann, H., eds), Vol. 3, pp. 145–220. Springer-Verlag, Berlin and Heidelberg.

Weiler, E.W., Jourdan, P. S. and Conrad, W. (1981). *Planta* **153**, 561–571.

Wittmann, C. and Hock, B. (1990). *Food Agric. Immunol.* **1**, 211–224.

Zola, H. (1987). *Monoclonal Antibodies: A Manual of Techniques.* CRC Press, Boca Raton, FL. 214 pp.

# 14

# The Quantification of Arbuscules and Related Structures Using Morphometric Cytology

RONALD TOTH

*Department of Biological Sciences, Northern Illinois University, DeKalb, IL 60115-2861, USA*

## I. Introduction

The arbuscule is the most characteristic structure of the vesicular-arbuscular mycorrhizal complex. This "little tree" begs to be considered the site of host–fungus metabolite exchange. The large number of hyphal branches and concomitant cortical protoplast surface area make the sum total of the arbuscules as important as the entire external mycelium. They are the business end of the system.

METHODS IN MICROBIOLOGY
VOLUME 24   ISBN 0-12-521524-X

Copyright © 1992 by Academic Press Limited
All rights of reproduction in any form reserved

Many vesicular-arbuscular mycorrhiza have been characterized cytologically and, in general, arbuscular structure and development are very similar across host–fungus combinations (Bonfante-Fasolo, 1984). In all cases so far studied, the arbuscule appears a likely candidate for metabolite exchange. However, little direct evidence links arbuscules to such a function. The difficulties involved in studying phosphorus transport through individual arbuscules have prevented us from obtaining this much needed information. Physiological studies on vesicular-arbuscular mycorrhiza have been primarily concerned with host growth and phosphorus uptake in whole plants.

By their nature, data from physiological studies are numerical and yield information on quantities of nutrients and host growth rates. In contrast to this, most cytological and ultrastructural studies on arbuscules have been largely qualitative. Using the techniques of morphometric cytology, arbuscular parameters can be quantified. Cytological parameters can then be correlated with physiological processes. Correlations and possibly cause and effect relationships can be established if cytological and physiological data can be integrated. Morphometric analysis makes this possible.

Morphometry is the quantification of different structures or phases of a body such as the proportions of different minerals in a geological specimen or organelles in a cell (James, 1977). Not only relative proportions (percentages) but absolute volumes, surface areas, numbers of particles, etc., can be calculated. Morphometry is but one small part of the larger discipline known as stereology, which is the three-dimensional interpretation of two-dimensional (flat) images such as very thin sections or projections (Elias et al., 1971). It can be used in the study of structures independent of their size or composition, such as the distribution of stars in space or an estimation of the proportions of soil and roots beneath trees (James, 1977).

Many questions concerning the "cost" to the host in bound carbon used by the fungus and the amount and rates of phosphorus transported from the fungus remain unanswered. For example, does arbuscular and/or host protoplast surface area in large or fine roots correlate with phosphorus transport? The quantification of arbuscules and host cells can help to answer such physiologicaly related questions.

The same techniques of morphometric cytology used to quantify arbuscular parameters can also be used to quantify intercellular hyphae, vesicles, and even the external mycelium. This chapter will deal primarily with cytological parameters of arbuscules but will briefly consider total fungal biovolume and surface area.

## II.  The arbuscular cycle: description

It is not the intention of this chapter to review arbuscular anatomy, development or degeneration, since this has been done most adequately by Bonfante-Fasolo (1984, 1987, 1988) and others (Bonfante-Fasolo, 1983; Strullu, 1985; Smith and Douglas, 1987; Gianinazzi-Pearson and Gianinazzi, 1988; Bonfante-Fasolo and Perotto, 1990; Bonfante-Fasolo and Scannerini, 1990; Smith and Smith, 1990). We will only consider those aspects of the cycle which may relate to physiological parameters.

A portion of the intercellular hypha penetrates the host cortical cell wall and later, branches dichotomously resulting in the formation of a trunk (2–10 $\mu$m in diameter) and a series of branches (0.3–1 $\mu$m in diameter). Although the cell is entered, the protoplast is not. For our purposes, the cell is defined as the outermost limits of the cell wall. As the trunk and branches develop, the host cell membrane (plasmalemma) invaginates to accommodate the ingrowing fungus.

Cell size (wall) remains the same, but the protoplast decreases in size and increases in surface area. Concurrently there is an increase in both nuclear and cytoplasmic volume. Many other changes occur such as nuclear shape, plastid starch content, Golgi vesicle size, size and number of vacuoles (Bonfante-Fasolo, 1984). From a physiological point of view, only some of these changes are relevant and need to be quantified. For example, in a study of phosphorus transport and host growth, the phosphorus released by the arbuscule (assuming it is) must be absorbed through the host plasmalemma surrounding the branches. Likewise, the bound carbon given to the fungus probably passes through this same membrane. To correlate physiology with anatomy, the surface area of this membrane must be quantified. The surface area of the arbuscule could substitute just as well. It is not important to quantify the number of branches. The total biovolume of fungus may relate to host carbon lost, and branches are a part of this total. The volume of the arbuscule would, therefore, be a useful parameter to quantify.

It has been assumed that host cells with arbuscules are metabolically more active. Changes in nuclear size and shape, increased cytoplasmic volume, Golgi hypertrophy, etc. (Bonfante-Fasolo, 1984) have been noted and used to justify this assumption. However, none of these changes would be as important to quantify as mitochondrial volume and the surface area of the inner membrane. When doing morphometric cytology, as with any study, one must define the relevant parameters beforehand.

Arbuscular volume, host cytoplasm and plasmalemma have been

quantified in several vesicular-arbuscular mycorrhiza (Cox and Tinker, 1976; Toth and Miller, 1984; Alexander *et al.*, 1988, 1989). These are probably the most significant parameters although, as already mentioned, mitochondria should also be quantified and probably vacuole volume and surface area also. Since cortical cells do not develop wall ingrowths as do transfer cells (an indicator of large-scale transport) (Gunning and Steer, 1975), it is possible that phosphorus is stored in the vacuoles and transferred out of the cell more slowly at a later time. The vacuoles may absorb and store phosphorus.

## III.  Morphometric cytology: methods for quantifying cellular structures

### A.  Words of encouragement

Do not allow the fancy words to scare you away. Using nothing more complicated than a micrograph, a clear plastic sheet with horizontal and vertical lines drawn on it and a pocket calculator, volumes and surface areas can be calculated for any cellular structure. The mathematics required is simple arithmetic with a little algebra thrown in for spice. The number of structures in a given volume (e.g. number of mitochondria in the cell) is more complicated to determine but fortunately rarely needs to be quantified. Volume and surface area of an organelle, or in our case an arbuscule, relate to function. Knowing the number is more useful when studying population dynamics.

There are many excellent texts on morphometry; the two best are by Steer (1981) and Williams (1977). Both contain excellent examples in the actual analysis of micrographs. A shorter introductory article for beginners (Toth, 1982) may be a good starting point. The extensive yet readable text by Weibel (1979) is an excellent reference. Those more adventurous and mathematically inclined to consider the theory of quantitative stereology are encouraged to consult Hilliard (1968), James (1977), Underwood (1970) and Weibel (1980).

### B.  Background

When considering a model system it is often easier to imagine relationships on a macroscopic level using materials and shapes we are familiar with. The same principles will apply to microscopic structures of unknown shape. For our example of morphometric principles let us consider two square cement blocks of side length 1 m in each of which are embedded 100 steel balls. In block A the balls are 5 cm in diameter and in block B the balls are 25 cm in diameter. Obviously the proportion of the total solid (steel and cement) occupied by the steel balls in block A will be less than in block B. Likewise the ratio of the surface area of the steel balls to the volume of the cement block and the absolute surface area of the steel balls in block A will be less than in block B. However, the number of steel balls would be the same.

How would you discover these facts? A physiologist or biochemist would probably measure and weigh the blocks, digest away the cement using mineral acids and analyse the salts made for type and amount. In this way they could determine the composition and quantity of the matrix. They might be surprised to discover the corroded steel balls at the bottom of the acid container (one of the joys of science). They would weigh and measure and count the balls and probably analyse their composition also. Their analysis would yield quantitative data.

Cytologists might get a large diamond-impregnated cutting wheel and saw the blocks in half. The cut surfaces would produce flat planes of two dimensions. They would see a homogeneous grey matrix material of unknown composition (the cement) and scattered about shiny circular profiles (the steel) of various sizes. In block A the maximum sized circles are smaller than in block B. These data are qualitative. The size differences in the shiny circles would quickly (and correctly) be interpreted as artefacts of cutting spheres across random planes and they would be correctly interpreted as spheres in the original solid since no elongated profiles were seen.

In order for the data obtained by the biochemist and the cytologist to be of maximum usefulness, they must correlate. One way to do this is to quantify the cytological data. The techniques of quantitative stereology (morphometry) allow us to do this.

Morphometry can be carried out on any system that can be viewed in two dimensions whether it is a projection, cut surface or very thin section, so thin as to be viewed as a surface (thin section for electron microscopy). Morphometry yields quantitative information such as volumes, surface areas and number of particles in a whole structure. As seen in our example, when a cut surface is exposed from a three-dimensional structure, three-dimensional structures (spheres = steel balls) present in the object are observed as two-dimensional structures, in our case, as circles of various sizes. Two-dimensional structures (planes = the real original surfaces of the steel balls) present in the object are observed as one-dimensional structures (lines = edges of circles). Any one-dimensional structures (lines) would be reduced to points. There were none of these in our example. The numerical reconstruction of the original three-dimensional structure proceeds in the reverse manner.

Data collected in one dimension are used as an exact predictor of two-dimensional properties of the structures, and so on (James, 1977).

### C. Volume fraction, $V_V$

*1. Theory*

Cuts through the block will produce artificial surfaces—the cut faces (viewed as planes of two-dimensions). On average, smaller circles will be produced from cuts through the first block (A) than from the second (B) (Fig. 1). These circles represent artificial surfaces, but they will relate (be proportional) to the amount or volume of steel in the concrete. The more steel (larger balls), the larger the circles. In other words the area of the circles is proportional to the volume of the steel in the cement. The more area exposed in the cut surfaces, the more volume that part or phase occupied in the whole structure (Fig. 1).

This is a direct proportion and is often expressed by the relationship:

$$A_A = V_V \qquad\qquad (1)$$

where $A_A$ is the areal fraction and $V_V$ is the volume fraction. In the morphometric literature, symbols such as $V_V$, $A_A$, $B_V$, $S_V$, etc.

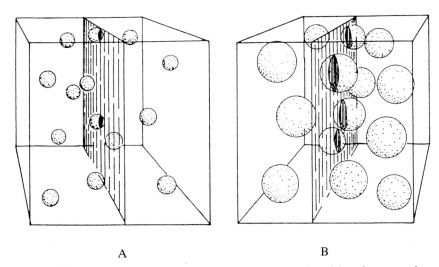

A                                          B

**Fig. 1.** Diagramatic representation of cut surfaces produced by planes passing through blocks containing spherical inclusions. Note that in block B with the larger spheres, more artificial "surfaces" are produced. From Weibel, E. R. (1963). *Laboratory Investigation* **12**, 131–155. Reproduced with permission of the US and Canadaian Academy of Pathology.

represent fractions. With a symbol like $A_A$ the upper letter represents the area of steel circles (for example in $cm^2$) and the lower letter represents the area of the whole cut block face (steel and cement). For example, if in block A, $150\ cm^2$ of steel is measured on a cut face and the whole block face has an area of $10\,000\ cm^2$ ($1\ m^2$), then $A_A = 150\ cm^2/10\,000\ cm^2 = 0.015$. In block B, with the 25 cm diameter balls, suppose that $4000\ cm^2$ of steel circles are produced on a cut surface, then $A_A = 4000\ cm^2/10\,000\ cm^2 = 0.4$. Since $A_A = V_V$, then in block A the steel balls occupied 1.5% of the volume of the block, and in block B the steel balls occupied 40% of the volume.

This principle was first developed and applied by Delesse (Weibel, 1973) using geological specimens. The mathematical justification for $V_V = A_A$ has been elucidated in some detail and can be found in many references (Underwood, 1970; Weibel, 1980).

At this point the reader may be thinking that this does not sound at all easy, because in practice it is very difficult to measure areas. As Weibel (1969) states, "it was therefore a great advance when the Russian geologist Glagolev demonstrated in 1933 that $V_V$ could be estimated by superimposing a regular point lattice on the section." In practice, a clear plastic sheet is laid over the micrograph (Fig. 2a). On the plastic is drawn a lattice of points. It is hard for the eyes to follow a series of points (see Steer, 1981, p. 85 for a good example of this phenomenon) and most often a lattice of lines is used. The intersections of the lines are used as points (Fig. 2b). If the distance between intersecting lines is $d$ and the intersections of the lines are considered as "points" for point counting, it is easily seen that there are as many squares of area $d^2$ centred around each point (Weibel, 1973, p. 261). In other words each "point" on the lattice represents an area around it. The accuracy of this technique increases as $d$ decreases. Each point then represents a smaller area around it. The limit would be an infinite series of points counted over the micrograph. In practice 100–500 does quite nicely.

By counting the number of points superimposed over the arbuscule and dividing that number by the number of points over the entire cell, the areal fraction can be calculated. Likewise the volume fraction. Therefore

$$P_P = A_A = V_V \tag{2}$$

In order to simplify notations and aid in identification of phases to which fractions like $V_V$ $S_V$, $B_V$, etc. pertain, a standard system using lower case subscript letters has been adopted (Weibel, 1979). These lower-case letters are attached to the parameter symbol, for example:

$$V_{steel} = V_{(s)}$$

for the volume of the steel balls, and

$$V_{block} = V_{(b)}$$

for the volume of the entire block. The volume fraction is then written as:

$$\frac{V_{steel}}{V_{block}} = \frac{V_{(s)}}{V_{(b)}} = V_{V(s,b)}$$

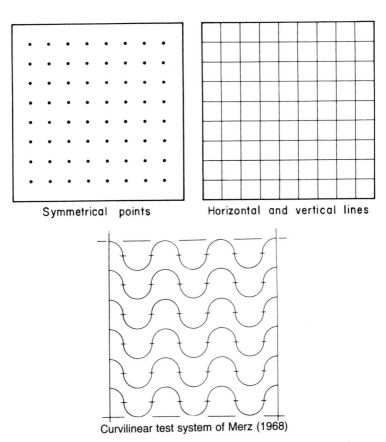

Symmetrical  points

Horizontal  and  vertical  lines

Curvilinear test system of Merz (1968)

**Fig. 2.** Common examples of probes. The classic series of points on which much current theory is based is in practice harder to use than a lattice of lines where the intersections are used as points for $P_P$ determinations and lines for $S_V$ analysis. For specimens with highly regular and ordered phases (muscle fibres, annulate lamella, etc.) a system of curved lines (Merz style) should be used.

where $V_V$ is the volume fraction of the steel (s) in the total volume of the steel and cement, the block (b).

Since we are dealing with fractions and the dimensions are of equal value, they cancel out. In our example of the concrete block B, the areal fraction, $A_A$ = 4000 cm$^2$/10 000 cm$^2$ = 0.4, is dimensionless and so is the volume fraction $V_V$ = 0.4. This is fortunate because it eliminates the need to measure accurately specimen dimensions (as will be necessary for surface area analysis).

### 2.  Quantifying arbuscular volume fraction, $V_V$

The accuracy of this and other methods relies on the assumption that data are gathered from a truly two-dimensional view. There is some error associated with viewing sections, especially when structure size approaches section thickness. Fortunately, the basic problems of section stereology have been solved (Elias *et al.*, 1971) and most of the structures we are interested in are much larger than the section thickness. In most cases technical errors will not occur.

This is the case for most large cell structures (1 $\mu$m and larger) when thin sections for electron microscopy are used. Due to the small diameter of arbuscular branches, studies on arbuscules require electron micrographs. Let us proceed now from our steel ball and cement model to a cortical cell containing an arbuscule.

The first thing to notice is that we are presented with profiles of very different proportions, the large ones from trunks and the small ones from branches. In theory each profile should only be sampled once (Weibel, 1973). If a point lattice is chosen so that there are enough points to have a high probability of "hitting" a large number of branch profiles, the trunk profiles would be "hit" many times. Clearly a different point lattice would need to be used for each structure.

In practice, it is easier to use a "coherent double lattice" test grid (Weibel, 1973, 1979). This is a point lattice with two different point densities superimposed. The easiest way to construct one is to use a lattice with every third or fourth line wider, but the wider lines often obscure fine detail. A more efficient system employs a 3/4 circular arc "around" every third or fourth point (Weibel, 1979) (Fig. 3). The precise test point is taken to be the intersection of the line edges in the open quadrant of the circle. The arcs mark the "coarse" points which are used for large structures (trunks) and "fine" points are used for small structures (branches). In the example (Fig. 3) there are sixteen fine points for every coarse point. There are 228 total points over the cell, 42 "fine" points over the arbuscular branches and three "coarse"

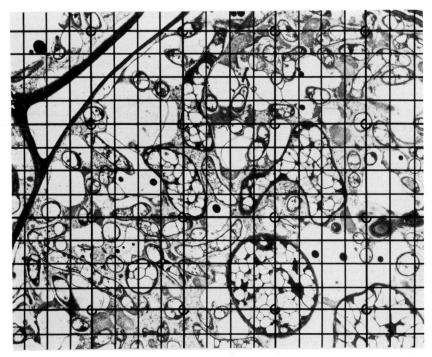

**Fig. 3.** Micrograph of *Zea mays* L. containing a mature arbuscule of *Glomus fasciculatum* with a superimposed grid pattern. Magnification, 5600 ×.

points over the trunk profiles. Therefore, Pp for branches = 42/228 = 0.184. Since the point fraction equals the volume fraction (Equation 2), $V_{V(b,c)}$ = 0.184. There are 3 × 16/228 points over the trunk, so that $P_{P(t,c)}$ = 0.27 and therefore $V_{V(t,c)}$ = 0.27. The use of the double lattice saves a great deal of time since it is not necessary to change plastic cover sheets or to count all the fine points over the large structures.

Much has been made of "digitizer boards", computerized morphometric programs, etc. Arbuscule branches are too small and too numerous for these to be helpful. The error introduced by trying to outline each branch would be enormous and the task extremely tedious. In an early study on onion mycorrhiza (Cox and Tinker, 1976) copies of micrographs showing host cytoplasm in silhouette were made and analysed by computer. Eight cells were quantified. In the same time 50 micrographs can be analysed with an overlaid point-counting grid, without the expense of a computer and with the same or greater accuracy. The great advantage of the single grid overlay is that counting can be done at any time and in any place—using a clip board, the overlay is held over the micrograph (a little tape or a rubber band at the bottom helps), and counting can begin. Very little data from very few types of research can be gathered in such a flexible way.

Another advantage is that the micrographs do not have to be finished prints. Chatter, knife marks and stain contaminations are no problem. Time need not be wasted obtaining better negatives, nor does uneven density of prints present a problem. All that is needed is a micrograph with a density and contrast that can be seen clearly enough to allow counting to be done. Compression during sectioning (Weibel, 1973) as well as shrinkage of the specimen (Steer, 1981) during dehydration and polymerization of the plastic can be ignored for now (these will be considered later). Estimations of volume fractions are probably the easiest morphometric data to collect and an important parameter when comparing anatomy with physiology.

## D.  Surface area to volume fraction, $S_V$

### 1.  Theory

Many biochemical and physiological processes (hormone reception, electron transport, ATP production, etc.) occur on, or are regulated by, a membrane. These membranes surround cells and provide the boundaries of organelles. In electron micrographs they appear as lines (at low magnification) and can be quantified using morphometric techniques. When making correlations between anatomy and physiology, probably the most useful parameter to quantify is the surface area of cells and organelles. This parameter is expressed as $S_V$ where S is the surface area of the phase in question and V is the volume of the containing structure. For our purposes a good example would be the surface area of the host protoplast in a cell with an arbuscule.

Using our analogy of steel balls in cement again, we can see that when a cut is made in block B with the 25 cm spheres, large circles are produced. These have longer circumferences than the smaller circles in block A with the 5 cm spheres. If a series of test lines, either straight or curved (Fig. 2), is laid over the cut face of the block, some of these will intersect these boundaries. The more lines, and the more boundaries, the more intersections. The boundaries (circumferences of the circles) represent the surface areas of the original spheres. By counting the number of intersections we can gain information about the original surface areas.

There is a direct proportionality between the surface area of a phase in a unit volume of sample and the number of intersections that a unit length of test line makes with the boundary exposed in sections. This was first shown by Saltykov in 1945 (Rhines, 1968). The basic equations for obtaining the area of surfaces in a volume (Underwood, 1970; Weibel, 1973) are:

$$S_V = 2I_L \ \mu m^2/\mu m^{3*} \tag{3}$$

$$S_V = 4N_L \ \mu m^2/\mu m^3 \tag{4}$$

where $S_V$ is the surface area per unit volume, $I_L$ is the number of times the test lines *intersect* the boundary being studied per unit length of test line, and $N_L$ is the number of times the test lines *intercept* the phase being studied, per unit length of test line (Fig. 4).

## 2. Quantifying $S_V$ for host protoplasts and arbuscules

A cortical cell with a mature arbuscule is a perfect candidate for demonstrating the use of both equations. Once the wall is breached, the protoplast invaginates to accept and surround the ingrowing trunk and branches. The relationship $S_V = 2I_L$ can be used for determining the surface area to volume ratio ($S_V$) of the protoplast (p) to the cell (c)

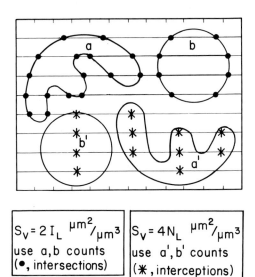

**Fig. 4.** Diagram representing phases in a test area with an overlying series of lines for $S_V$ determinations (from Toth, 1982).

*Any units can be used; $\mu m$ are convenient for most cytological studies.

volume, while the relationship $S_V = 4N_L$ can be used to determine the $S_V$ of the arbuscule (a) to the cell volume. The equation $S_V = 2I_L$ is most easily applied to membrane systems (plasmalemma, endoplasmic reticulum, etc.) whereas $S_V = 4N_L$ is more easily used for structures (mitochondria, chloroplasts, arbuscules). In our example (Fig. 5), using only the horizontal lines, there are four test lines which measure 11.3 cm each, for a total of 45.2 cm. There are 42 cm of test line over the cell. These intersect the host protoplast plasmalemma 63 times. Substituting 63 for I and 42 for L in the equation $S_{V(p,c)} = 2I_L$ yields 3.0 cm$^2$/cm$^3$ or 3.0 cm$^{-1}$. The magnification of the micrograph is 5600 ×. This means that 1 cm on the micrograph corresponds to 1.8 $\mu$m on the sample, so $S_{V(p,c)} = 3.0/1.8\ \mu$m$^{-1} = 1.66\ \mu$m$^2$/$\mu$m$^3$.

Using the same example, the ratio of the surface area of the arbuscule to the volume of the cell can also be calculated. In this case it is easier to use the number of times a test line intercepts (Fig. 5) a branch or trunk profile. There are 30 intercepts and the same 42 cm of test line.

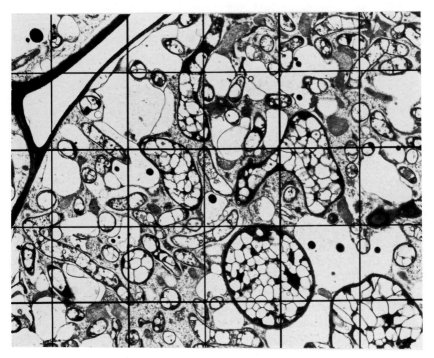

**Fig. 5.** Micrograph of *Zea mays* L. containing a mature arbuscule of *Glomus fasciculatum* with a superimposed grid pattern. Example in text uses only the horizontal lines for $S_V$ determinations. Magnification, 5600 ×.

Using the equation $S_V = 4N_L$ yields $4 \times 30/42$ cm. Therefore $S_{V(a,c)} = 2.86$ cm$^{-1}$. Converting cm to $\mu$m (1 cm = 1.8 $\mu$m at 5600 $\times$) yields 1.59 $\mu$m$^2$/$\mu$m$^3$.

Equations (3) and (4) differ by a factor of 2 and their relationship to one another is obvious (Fig. 4). Choosing between them is a matter of convenience. You should note that these equations do *not* determine the surface area of the phase in relation to its own volume but rather to the test volume. In our case the test volume was the cell. Test line lengths were measured over the cell. The host cell's plasmalemma surface area was calculated in relation to the cell volume and not to the volume of the protoplast. In order to obtain the surface area of the plasmalemma (protoplast) in relation to the volume of the protoplast, the surface area to volume ratio ($S_V$) provided by those equations must be divided by the fraction of the test volume (whole cell) occupied by the protoplast. For example, if using the equation $P_P = V_V$ and a point-counting grid, we determine that the volume fraction of the protoplast in the cell $V_{V(p,c)}$ is 0.5 $\mu$m$^3$/$\mu$m$^3$ and using the equation $S_V = 2I_L$ we determine that the ratio of the surface area of the plasmalemma to a unit test volume (the cell) is $S_{V(p,c)} = 1.66$ $\mu$m$^2$/$\mu$m$^3$, then there are (1.66/0.50) 3.32 $\mu$m$^2$ of plasmalemma surface area per $\mu$m$^3$ of protoplast ($S_{V(p,p)} = 3.32$ $\mu$m$^2$/$\mu$m$^3$).

## 3. Precautions

Surface area to volume fractions are not as easy to obtain as volume fractions. As previously stated, $V_V$ is a dimensionless quantity. Assuming that the arbuscule and host cell shrink uniformly as the epoxy plastic shrinks (which is a logical assumption), no correction is necessary. It is not even necessary to know the magnification of the micrograph! This is not the case for $S_V$ fractions which do have a dimension ($\mu$m$^2$/$\mu$m$^3$ = $\mu$m$^{-1}$).

Cell dimensions must be measured accurately. This is often difficult, due to the artefacts of tissue preparation. During dehydration most specimens shrink. In onion roots the length decreases only 5% but the diameter 20% (R. Toth, unpubl. obs.). This results in a considerably smaller root volume since the radius is squared (volume = $\pi r^2 l$). When calculating volume even greater losses can result in other tissues where all dimensions shrink uniformly.

During polymerization, shrinkage of the epoxy resin introduces more error. Tissues must be accurately measured before processing and after embedding in all three dimensions to calculate a shrinkage factor since (as in roots) not all sides shrink the same amount. All of the factors taken together can easily result in a 25% decrease in all three dimensions (75% of original length, width, and depth remaining). The resulting volume would only be 0.42 ($0.75^3$) of the original! During sectioning, compression in the direction of cutting introduces more error. Sections must also be assessed for this compression (Steer, 1981; Toth, 1982). Both of these factors will introduce a considerable error into any morphometric calculation but can easily be corrected, once the amount of shrinkage and compression is known (Toth, 1982). Fortunately, these values only need to be determined once, assuming the same tissues and type of plastic remain constant from experiment to experiment.

These problems arise because we are looking at a micrograph of a cell which is actually converted to a smaller volume and therefore presents a smaller area of cut surface upon sectioning (our micrograph). However, the probes (test lines) will intersect the same number of profile edges (membranes, arbuscular branches, etc.). Therefore, the number of intersections will remain the same but the length of the test line measured over the micrograph will be less and $S_V$ will be greater than it should be. Remembering that $S_V = 2I_L = 4N_L$, it is easily seen that if I and N remain constant and are divided by smaller values of L, the result ($S_V$) will be larger than it should be. This point is stressed here because it is often overlooked and can introduce a large error since shrinkage effects are cubed.

Another consideration is that all of the magnifications are known accurately. Three things must be considered. Electron microscopes do not often produce images at the stated magnification: they need to be tested with a calibration grid, easily obtainable from any standard supply firm. If the magnification at which negatives are taken is reached differently by either increasing or decreasing magnification to reach it, the actual magnification will be different ($\pm$ 10%) due to lens hysteresis (Meek, 1976). It is advisable always to approach the taking magnification from a higher magnification, and not to go past (lower than) the taking magnification. The negatives must be carefully catalogued and all printed at the same enlargement. It is a dangerous practice to print micrographs at different magnifications. Printing procedures should be standardized.

Several other factors must be considered, such as the optimum magnification of the final micrographs (in order to see clearly the structures being quantified), the line spacing of the probes, sampling procedures, number of micrographs to take and numbers of points or line intersections to count for each micrograph and for the total study.

Space limits our discussion of these considerations but standard references on morphometric techniques may be consulted for more details (Williams, 1977; Weibel, 1979; Steer, 1981; Toth, 1982).

In most cases it will be found that a magnification of 5000 × is adequate for quantifying most structures. Even though the branches are small, their darkly stained walls and the clear separation from the host protoplast by the "interfacial zone" renders them easy to see and therefore to quantify. For endoplasmic reticulum and mitochondrial membranes a magnification of 25 000 × would be recommended. For most 8″ × 10″ (20 cm × 25 cm) micrographs, a line spacing of 2/3 to 1 cm works well.

## IV.  Approximation of the arbuscular cycle using morphometric techniques

Arbuscules must be quantified if anatomical data are to be related to physiological data in a precise way, but what parameters should be quantified? Clearly $V_V$ and $S_V$ for arbuscules are useful parameters, but should they be calculated for "mature" arbuscules or averaged for the whole cycle?

In addition to amount ($V_V$) and rates of change for the arbuscule, the $S_V$ of the arbuscule and host protoplast should relate to phosphorus transfer. The proportion of the cycle during which the mature branches exist would be an important parameter to quantify. For example, if total $V_V$ arbuscule is equal in two different mycorrhiza but different levels of phosphorus are transferred, the amount of time in the cycle during which branches occur may be very important. If branches develop quickly, the "mature" arbuscule may exist for a longer part of the development phase in one mycorrhiza than in another.

What controls the rates at which parameters develop and what causes the development phase to end? If these factors are under the control of the fungus, one would expect to see similar patterns of development in different host cells with arbuscules of the same fungus. If the control processes are modified by each host, then variability across host species would be expected. Obviously, this is an oversimplification of a complex group of host–fungus interactions but does provide a starting point.

Values for $V_V$ (arbuscule), $V_V$ (host cytoplasm), $S_V$ (host) and the rates of change of these parameters have been studied in six host species mycorrhizal with an Illinois isolate of *Glomus fasciculatum* (Toth and Miller, 1984; Alexander *et al.*, 1988, 1989). The rate of branch development and the duration of the development phase were constant across hosts, and therefore assumed to be primarily under the control of the fungus. The number and size of "clumps" of collapsed branches, and their rates of disappearance (digestion?) varied. Host cytoplasm values and rates of change also varied (as one might expect) across host species and all of these parameters were assumed to be primarily under the control of the host cell.

There are, however, problems with defining and quantifying a mature arbuscule (Cox and Tinker, 1976; Alexander et al., 1989). How accurately can any parameter be estimated in such a dynamic system as arbuscule development? How accurately can the rate of change for any parameter be determined? As it turns out it is very difficult to estimate accurately some of these parameters. Although the basic techniques of morphometry are well understood and can yield accurate estimates of volume and surface area fractions, studies of developing arbuscules present special difficulties.

The system of asynchronous arbuscule development in adjacent cortical cells and the rapid changes in arbuscular parameters within a cell make it impossible to quantify accurately the arbuscular cycle. The non-random location of arbuscular branches in the cell further complicates the situation. Only an approximation can be hoped for. To explain this we must first consider several assumptions about sampling procedures used in morphometry.

Morphometric cytology is based on several assumptions. The most important is that the specimen is sampled randomly. This implies that no bias is introduced in selecting the view to study, the probes randomly intersect phases, and no one specimen is sampled more than once (one micrograph per cell).

When a specimen is sampled it is cut by random planes and artificial surfaces are created and exposed. Some cuts will expose a greater proportion of a phase than is really in the whole specimen. It will, therefore, be over-represented. Likewise, chances are equal that another cut (on another cell) will produce a view where that phase is under-represented. Given a large number of random cuts these will average out. This is an underlying principle of morphometry. In geological specimens or mature cells where shape is not changing and organelle turnover has reached equilibrium, this would be the case. An accurate estimation of $V_V$, $S_V$, etc. can easily be determined.

How does one sample and deal with a population of cells which are developing (changing)? Let us use as our example the mitochondrial (m) volume fraction in root cortical cells (c), $V_{V(m,c)}$. In a developing cortical tissue, cells of different ages will have different $V_{V(m,c)}$ values. For example, cells sampled near the apex will be small, their $V_V$ values for cytoplasm will be high and their $V_{V(m,c)}$ values will also be high. Mature cortical cells, 1–2 cm from the root apex, are extremely large (40 $\mu$m × 240 $\mu$m in onion), have large central vacuoles, little cytoplasm and low $V_{V(m,c)}$ values. Because of our knowledge of root growth and development we "know" that the cells at the apex are young (outside of the quiescent centre) and those further from the apex are older. In practice, if we want to study the dynamics of how $V_{V(m,c)}$ changes during a cell's life, it is much easier to sample cell populations from the root apex rearward, along an age gradient, than to sample cell populations at different times. It is difficult to keep track of where a cell population is, in a growing root, and to time its age precisely. There are many difficulties. Ideally we would want to sample the same cell throughout its life. A cell in tissue culture can be monitored with a light microscope for a large part of its life but it is not possible to do electron microscopic morphometry on a cell and have it survive!

So, in practice, we would sample populations of cortical cells, from areas near the apex and progressively further away. In a section of meristematic tissue, profiles of 20 cells might be viewed. These are quantified and $V_{V(m,c)}$ determined. Each cell is in a slightly different stage of development but they are so close that we accept (define) them as being at the same stage. Any over- or under-estimations cancel out. We do this for local cell populations along our age gradient. Knowing the growth rate of the root we can then construct a curve plotting the rate of change for $V_{V(m,c)}$ in cortical cells over time. This sounds relatively easy, but there are problems associated with arbuscules.

As already mentioned, any random cut may over- or under-represent a phase. Cells nearby are in the same stage of development and if many cells in the same stage of development are sampled, errors will cancel out. The problem with developing arbuscules is that each cortical cell with an arbuscule may be in a very different stage of development. There is no nearby cell in the same stage of development, which when sampled will cancel out error.

But if many are sampled, why do they not average out? They do average out for an overall value for arbuscular (a) volume in the cell (c) volume, $V_{V(a,c)}$ for the entire cycle, but not for any one stage (corresponding to a time-course). We need to study many cells (arbuscules) at the same stage of development, at each point in time, to cancel out errors inherent in the sampling system. This is simply not possible—each cell can only be sampled once. Morphometry does not work in this way.

Another problem relates to data grouping to construct the rate curves. Using the analogy of a cell going through mitosis we can analyse the difficulties. Using a root tip squash and a light microscope we can see whole nuclei going through mitosis. We can easily assign each to a "phase" (prophase, etc.). The proportion of nuclei in each phase to the total number of nuclei is equal to the proportion of the total cell cycle time taken by the nucleus to pass through that stage. For example, if 2% of the nuclei are observed to be in anaphase, then anaphase only takes 2% of the cycle time.

Using this analogy, the rates at which arbuscular parameters developed or degenerated were determined (Toth and Miller, 1984; Alexander, 1988, 1989). All cells were first quantified for all parameters and the values for a "mature" arbuscule were calculated. Five of the most mature arbuscules, from cells with the highest total fungus, cytoplasm and largest $S_V$, were averaged to obtain an estimation of the characteristics of a mature arbuscule. The values obtained from each parameter of a mature arbuscule were then divided into 10 size classes (five for development and five for degeneration). Histograms were constructed for each of the parameters, showing the number of cells in each size class (analogous to number of cells in each mitotic phase) (Alexander *et al.*, 1988, Fig. 1). From each histogram a curve was constructed to show the velocity at which each parameter changed during the arbuscular cycle. By plotting cycle time (number of cells) on the $x$-axis and percentage of the way through the cycle (size classes) on the $y$-axis, a velocity curve for each parameter was constructed (Alexander *et al.*, 1988, Figs. 2–9).

The problem is that from a single micrograph one cannot be sure to which size class (analogous to a mitotic phase) to assign any one arbuscule. A random cut through a highly developed arbuscule at the edge of the branches will yield a section showing a cell with a few branch profiles (Fig. 6, line C). It is often impossible to know if this image represents a mature arbuscule or an early stage of a developing arbuscule (the whole cell is not visible). A section may pass through the middle of a cell where a large concentration of branches is found (Fig. 6, line B). The arbuscule is often centrally located, or if not, the branches do not often "fill" the cell uniformly, and at the edges of the cell few branches will be encountered. A view of the cell centre will over-represent the actual $V_{V(a,c)}$. In theory, this will be averaged out by views near the edges of the cell with few branches.

However, if these views are interpreted as being young developing arbuscules they would not be averaged with the mature arbuscule data. The data for mature arbuscules would be over-represented ($V_V$ too large). More micrographs (arbuscules) would be assigned to the early developing stages, and put in these size classes in the histograms. This would over-represent the length of this phase in the total cycle time.

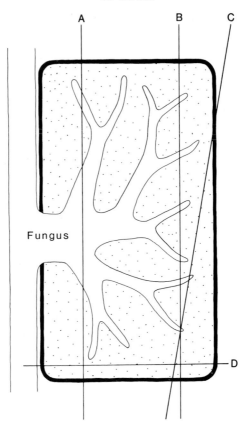

**Fig. 6.** Diagramatic representation of numbers of arbuscular branches intersected by planes of section.

Using our mitosis example, it would be as if an unknown number of anaphase views were wrongly interpreted as prophase views. The length of the cycle taken to complete anaphase would be incorrectly decreased and the time the cell spends in prophase increased. We would think that the arbuscule develops more slowly at the early stages and is only mature for a very short time. If phosphorus exchange takes place through the arbuscular branches, it is very important to know the length of time the mature arbuscule exists. This is the stage which may most closely correlate with phosphorus transport to the host.

Clearly there are problems. The curves thus far generated and the estimations for maximum levels of all the parameters are probably in error by ± 10%. Attempts to section all roots longitudinally and to use low magnification views have kept the error this low. In this way large parts of cells were seen and the approximate stage of arbuscule development could be determined. Cell ends where few branches exist were not sectioned (as would happen if cross-sections were taken) (Fig. 6, line D). In the case of bean cells, which are very short and where branches more evenly "fill" available spaces, the error is less. In onion, the arbuscule occupies the smallest space and therefore the error is probably greatest (Alexander *et al.*, 1989).

Even if every cell were sectioned perfectly, through its midpoint, along its long axis, the problem of the ever-changing values for $V_V$, $S_V$, etc. would still persist. There is an asynchronous population of cells with arbuscules. Truly accurate data can only be obtained by averaging all micrographs and finding a value for $V_V$ over the entire cycle. Development and degeneration can be clearly distinguished so an average for each is also possible. Unfortunately, earlier work was more concerned with the cycle dynamics so that this was not done (Toth and Miller, 1984; Alexander *et al.*, 1988, 1989). More recently a number of arbuscular parameters averaged for the whole cycle and for development and degeneration have been calculated (Table I).

## V.   Estimation of total membrane surface area

Knowing the $S_{V(p,c)}$ and cell size, absolute values for the surface area of a cortical cell protoplast with a mature arbuscule can be calculated. Values for development, degeneration or the entire "cycle" are also obtainable (Table I). In order to correlate phosphorus uptake to surface area, the entire surfaces in the root system involved in arbuscular phosphorus transport must be calculated.

In order to do this, some estimation of the percentage of cells in the root system containing arbuscules in needed. This can be done by careful observations of whole roots, root squashes or sections. Recently the number of cells containing arbuscules has been correlated with per cent colonization as determined by the gridline-intersect method (Toth *et al.*, 1990). This provides an easy way to estimate arbuscular number and, therefore, the total amount of surface area of host membrane in contact with arbuscules.

## VI.   Estimation of total fungal biovolume

As pointed out in the previous section, the relative values (volume fractions, etc.) produced by morphometry can be converted into

**TABLE I**

Parameters of arbuscules and related host cell structures for six species mycorrhizal with *Glomus fasciculatum* averaged over one cycle

| Parameter | Host species | | | | | |
|---|---|---|---|---|---|---|
| | *Maize* | *Wheat* | *Oats* | *Onion* | *Bean* | *Tomato* |
| Cell dimensions Length × diameter ($\mu m$) | $42.5 \times 125$ | $44 \times 140$ | $50 \times 113$ | $42.5 \times 240$ | $54 \times 64$ | $45 \times 93$ |
| Cell volume ($\mu m^3$) | 177 328 | 212 874 | 221 875 | 340 468 | 146 574 | 147 910 |
| Host cytoplasm (%) Average for arbuscule development | 15.5 | 16.8 | 15.0 | 5.0 | 9.0 | 9.3 |
| Host cytoplasm (%) Average for whole cycle | 10.2 | 9.2 | 9.5 | 3.7 | 5.6 | 5.7 |
| Average $S_V(p, c)$ $\mu m^2/\mu m^3$ for arbuscule development | 0.77 | 0.66 | 0.56 | 0.26 | 0.52 | 0.34 |
| Average $S_V(p, c)$ $\mu m^2/\mu m^3$ for whole arbuscular cycle | 0.45 | 0.39 | 0.33 | 0.19 | 0.34 | 0.29 |
| Average arbuscule (%) for development | 22.9 | 21.5 | 19.0 | 5.2 | 12.5 | 11.0 |
| Average arbuscule (%) for whole cycle | 14.2 | 12.07 | 11.6 | 5.1 | 9.7 | 10.3 |

Relative values (percentages) can be converted to absolute values using cell volume data. Values for uninfected cortical cells and those containing mature arbuscules can be found in Alexander *et al.* (1988, 1989).

absolute amounts if cell dimensions are known. The same techniques described earlier can be used to quantify intra- and intercellular hyphae and vesicles. Using 0.5–1 $\mu$m plastic sections and a light microscope at a magnification of 1000 × would be adequate for this. In short, $V_V$ for all fungal structures in the root can be quantified. Knowing the volume of the root system an estimation of total fungal biovolume can be calculated. Root length can be calculated using a line-intersect method (Ambler and Young, 1977) and average diameter can easily be obtained from measurements using a light microscope.

This approach has been used to determine total fungal biovolume in onion roots (Toth et al., 1991). Obviously such a tedious analysis would prohibit its use in a study primarily concerned with host growth or phosphorus uptake. Fungal biovolume has also been correlated with per cent whole root colonization using the gridline-intersect method, enabling quick estimations to be made (Toth et al., 1991). The correlation between fungal biovolume and per cent colonization determined using morphometric analysis in onion was similar to that obtained using a chitin assay in western wheat grass (Toth et al., 1991). Biochemistry and morphometric cytology do correlate!

## VII. Conclusion

Few vesicular-arbuscular mycorrhiza have so far been quantified. Arbuscules of five host species mycorrhizal with one fungus and one host species mycorrhizal with two fungi have been quantified. This is a very small sample size. The techniques are available and it is hoped that, with the precautions that have been listed concerning an accurate interpretation of cycle data, many more mycorrhiza will be studied. Making cycle curves may prove not to be cost effective but average values for $V_V$ (fungus—especially branches), $V_V$ (host cytoplasm), $S_V$ (protoplast) and $V_V$ (mitochondria) etc. should be determined.

More mycorrhiza need to be studied, and data from individual arbuscules need to be used to calculate values for the entire root system in order to make meaningful correlations. For example, onion had the lowest levels of $S_{V(p,c)}$ increases and final $V_{V(a,c)}$ levels of six mycorrhiza (Cox and Tinker, 1976; Alexander et al., 1989), yet is is considered extremely mycorrhiza-dependent. Maize, on the other hand, generally considered a more mycorrhiza-independent plant, had the highest levels of $V_{V(a,c)}$ and the greatest $S_{V(p,c)}$ increases (Alexander et al., 1989). These data seem to be just the opposite of what one would expect. Clearly the total number of arbuscules in the root system must be known and also some ratio of the extent of the external mycelium to the root system. Many more factors control phosphorus transport to the host other than arbuscular parameters, but they are an important part of the overall picture.

Arbuscules need to be quantified in other fungus-host combinations, and over a wide range of phosphorus levels. It is possible that the characteristics of the cycle $V_{V(a,c)}$ averaged for development, and mature arbuscules are very different at low and high phosphorus levels. Ericoid and orchid mycorrhiza, and also ectomycorrhiza, all need to be quantified. The basic techniques of morphometry can and should be used on these systems to provide us with quantitative data.

## Acknowledgements

I wish to thank Drs P. Bonfante-Fasolo and R. M. Miller for advice and F. Altepeter for technical assistance during the preparation of this chapter.

## References

Alexander, T., Meier, R., Toth, R. and Weber, H.-C. (1988). *New Phytol.* **110**, 363–370.
Alexander, T., Toth, R., Meier, R. and Weber, H.-C. (1989). *Can. J. Bot.* **67**, 2505–2513.
Bonfante-Fasolo, P. (1984). In *VA Mycorrhiza*, (C. L. Powell and D. J. Bagyaraj, eds), pp. 5–34. CRC Press, Boca Raton, Fl.
Bonfante-Fasolo, P. (1987). *Symbiosis* **3**, 249–268.
Bonfante-Fasolo, P. (1988). In *Cell to Cell Signals in Plant, Animal, and Microbial Symbiosis* (S. Scannerini, D. C. Smith, P. Bonfante-Fasolo, and V. Gianinazzi-Pearson, eds), pp. 219–235. Springer-Verlag, Berlin.
Bonfante-Fasolo, P. and Perotto, S. (1990). In *Electron Microscopy Applied in Plant Pathology* (K. Mendgen and D. E. Lesemann, eds), pp. 265–275. Springer-Verlag, Berlin.
Bonfante-Fasolo, P. and Scannerini, S. (1990). In *Functioning in Mycorrhizae* (M. Allen, ed.). Academic Press, New York (in press).
Cox, G. and Tinker, P. B. (1976). *New Phytol.* **77**, 371–378.
Elias, H., Henning, A. and Schwartz, D. E. (1971). *Physiol. Rev.* **51**, 158–200.
Gianinazzi-Pearson, V. and Gianinazzi, S. (1988). In *Cell to Cell Signals in Plant, Animal, and Microbial Symbiosis* (S. Scannerini, D. C. Smith, P. Bonfante-Fasolo and V. Gianinazzi-Pearson, eds), pp. 73–84. Springer-Verlag, Berlin.
Gunning, B. E. S. and Steer, M. W. (1975). *Plant Cell Biology: An Ultrastructural Approach*. Edward Arnold, London.

Hilliard, J. E. (1968). In *Quantitative Microscopy* (R. T. DeHoff and F. N. Rhines, eds). McGraw-Hill, New York.

James, N. T. (1977). In *Analytical and Quantitative Methods in Microscopy* (G. A. Meek and H. Y. Elder, eds), Society for Experimental Biology, Seminar Series 3, pp. 9–28. Cambridge University Press, Cambridge.

Meek, G. A. (1976). *Practical Electron Microscopy for Biologists*, 2nd edn. John Wiley & Sons, Chichester.

Rhines, F. N. (1968). In *Quantitative Microscopy* (R. T. DeHoff and F. N. Rhines, eds), pp. 1–10. McGraw-Hill, New York.

Scannerini, S. and Bonfante-Fasolo, P. (1983). *Can. J. Bot.* **61**, 917–943.

Smith, D. C. and Douglas, A. E. (1987). *The Biology of Symbiosis*. Edward Arnold, London.

Smith, S. E. and Smith, F. A. (1990). *New Phytol.* **114**, 1–38.

Steer, M. W. (1981). *Understanding Cell Structure*. Cambridge University Press, Cambridge.

Strullu, D. G. (1985). *Les Mycorrhizes. Encyclopedia of Plant Anatomy*, pp. 1–198. Gebruder Borntraeger, Berlin, Stuttgart.

Toth, R., Doane, C., Bennet, E. and Alexander, T. (1990). *Mycologia* **82**, 519–522.

Toth, R. (1982) *Am. J. Bot.* **69**, 1694–1706.

Toth, R. and Miller, R. M. (1984). *Am. J. Bot.* **71**, 449–460.

Toth, R., Miller, R. M., Jarstfer, A., Alexander, T. and Bennet, E. (1991). *Mycologia* (in press).

Underwood, E. E. (1970). *Quantitative Stereology*. Addison-Wesley, Reading, MA.

Weibel, E. R. (1969). *Int. Rev. Cytology* **26**, 235–302.

Weibel, E. R. (1973). In *Principles and Techniques for Electron Microscopy: Biological Applications* (M. A. Hayat, ed.), Vol. 3. Van Nostrand Rheinhold, New York.

Weibel, E. R. (1979). *Stereological Methods*, Vol. 1, Practical Methods for Biological Morphometry. Academic Press, New York.

Weibel, E. R. (1980). *Stereological Methods*, Vol. 2, Theoretical Foundations. Academic Press, New York.

Williams, M. A. (1977). *Practical Methods in Electron Microscopy* (A. M. Glauret, ed.) Vol. 6, pt. 2. North Holland, Amsterdam.

# 15

# Methods for Studying Vesicular-arbuscular Mycorrhizal Root Colonization and Related Root Physical Properties

SRIYANI RAJAPAKSE*

*Department of Soil Science, Massey University, Palmerston North, New Zealand*


J. CREIGHTON MILLER, Jr

*Department of Horticultural Sciences, Texas A&M University, College Station, TX 77843, USA*

*Present address: Department of Horticulture, Clemson University, Clemson, SC 29634, USA.

METHODS IN MICROBIOLOGY
VOLUME 24   ISBN 0-12-521524-X

Copyright © 1992 by Academic Press Limited
All rights of reproduction in any form reserved

# I. Introduction

## A. The need

Mycorrhiza are a common but special type of root, resulting from a mutualistic symbiotic association between a plant and a fungus. This association is so integrated that complete separation of the plant roots and fungus is virtually impossible; therefore, mycorrhizal fungi should be studied as a part of the plant root system.

The primary purpose of studying root colonization in many vesicular arbuscular mycorrhizal experiments is to observe and confirm mycorrhiza formation and to quantify development of mycorrhiza within the root system. Other methods which have been developed to assess formation of vesicular-arbuscular mycorrhiza include the determination of: fungal biomass measured by colorimetric assay of glucosamine produced from fungal chitin (Hepper, 1977); production of yellow pigment in mycorrhizal onion roots by colorimetry (Becker and Gerdemann, 1977); and length of extramatrical hyphae (Abbott et al., 1984; Bethlenfalvay et al., 1982; Schubert et al., 1987). These methods have not been used extensively to quantify formation of mycorrhiza because each has its own limitations.

The study of root colonization is also necessary to observe and describe the morphology of specific mycorrhizal structures formed within the root, which is not possible with other quantification methods. Root colonization is also measured to evaluate the responses to various treatments such as the effectiveness of different inocula or inoculation methods and the extent of host specificity in different host–fungus combinations. Furthermore, because the anatomical features of some of the fungal structures inside the root are diagnostic for certain species, it is sometimes possible to differentiate the vesicular-arbuscular mycorrhizal fungi responsible for root colonization. Using this approach Lopez-Aguillon and Mosse (1987) were able to recognize *Glomus fasciculatum* E3, *Glomus tenue* and *Gigaspora margarita* inside a single root.

Study of physical properties of the root system is often needed to quantify formation of vesicular-arbuscular mycorrhiza in plants, for example measurement of total root length is needed if the total amount of colonized root length is to be determined. Study of roots is also necessary to observe how root growth and other root properties are affected directly by formation of mycorrhiza and indirectly by other treatments.

## B. Root physical properties of interest

The physical properties of the root that need to be determined in individual experiments are dependent on the specific objectives of the study, resources available and the plant species involved. At present, the most commonly used root physical properties are: total root length, specific root length, specific root mass and root dry weight. Other properties, such as root architecture, root abundance and characteristics of root hairs have been used occasionally. Estimation of most of these properties will be discussed in this chapter.

## II. Collection of roots

Regardless of the type of root system, fine terminal roots should be collected to observe root colonization, since these are the roots that harbour mycorrhizal fungi. These roots which are widely referred to as "feeder roots", are usually <0.5–2.0 mm in diameter and provide much of the extensive surface area of a root system. However, they comprise only a small portion of total root biomass (Zobel, 1986). Mycorrhiza are not formed in roots which have secondary thickenings.

When collecting roots, soil cores, spades, hand forks, etc., are used depending on the size of the plant, whether the plants are pot- or field-grown, and on the additional root parameters that need to be measured. When collecting roots from large trees grown in the field, the rooting pattern and depth of rooting should be determined. Usually vesicular-arbuscular mycorrhizal roots are observed in the upper 0–30 cm of the soil profile. Special care is required in collecting smaller roots because they easily break away from the main roots. For pot experiments, sandy soil is the best potting medium for collecting fine roots. If vermiculite is incorporated into the mix, roots can grow through or adhere to the expanded mica particles. For ease of handling, initially only lateral roots to which smaller roots are attached should be removed in both field- and pot-grown plants. Larger roots should be collected separately. It is often necessary to separate soil particles containing smaller roots and remove the particles by washing in a 0.05 mm sieve. Failure to include sufficient numbers of feeder roots in the sample can lead to underestimation of root colonization if larger roots constitute most of the sample.

## A. Sampling

Sampling is usually necessitated because of the large size or extensive growth of most root systems and of time and labour constraints. Root studies are often time-consuming and labour intensive compared to shoot studies. Therefore the number, size and method of sampling to be employed should be considered carefully. Very often, these sampling decisions vary from one situation to another.

In small seedlings, entire root systems can be collected and used in the measurement of root colonization. In pot studies, most parts of the root system can be recovered and sampling can be accomplished by taking at least four samples from different locations on the root, and pooling them to obtain one or more subsamples. For large field-grown trees, samples of roots or soil cores of adequate size from 5–10 different sites around the root should be collected, depending on the spread of the root system. The soil from all cores should be mixed thoroughly and the subsamples can be obtained either by using a subsampler or by repeated quartering and pooling opposite quarters (Ambler and Young, 1977).

Whatever the method employed, representative root samples must be obtained to estimate accurately root colonization. Since sample to sample variation is greater for roots than for shoots, usually more samples are required for root studies (Taylor, 1986). Sampling strategies in measuring vesicular-arbuscular mycorrhizal root colonization have been presented by Reich and Bernard (1984). They determined different variance components for assessing percentage of root colonization in four root samples taken from each of three apple trees. Each sample was divided into three subsamples. The variance component for samples was greater than that for subsamples or entire trees. While the conditions of other experiments vary, these results suggest that precision of the root colonization assay is improved by increasing the number of samples per plant, rather than by increasing the number of subsamples.

## B. Sample preparation and preservation

To prepare samples for processing, large particles of organic matter, debris and leaves should be removed manually and roots should be washed gently in water to remove soil particles. Soil dispersion agents such as sodium pyrophosphate ($Na_4P_2O_7$) have been used to recover fine roots from hard soil (Ambler and Young, 1977). If the entire root system or larger roots to which smaller roots are attached have been collected, they should be washed in a sieve before sampling for easy and rapid handling. Roots should then be cut into 1–2 cm pieces and stored in individual vials containing a formalin, acetic acid and alcohol (FAA) solution. FAA solution is prepared by 1:1:18 (v/v/v) combination of formalin, glacial acetic acid and ethanol. Root samples should be stored at room temperature until processing can be completed.

## III.    Observation and assessment of root colonization

Unlike ectomycorrhiza, which ensheath the roots, vesicular-arbuscular mycorrhiza do not induce obvious changes on the root surface. Therefore, vesicular-arbuscular mycorrhizal root colonization must be observed and assessed through internal microscopic examination.

### A.    Clearing and staining of roots

The standard method of clearing and staining roots for observing vesicular-arbuscular mycorrhizal root colonization (Nicolson, 1959; Hayman, 1970) was first improved by Phillips and Hayman (1970), and later modified by other researchers. A major modification was that of Kormanik and McGraw (1982).

In this method, cytoplasm and nuclei of plant roots are cleared with a KOH solution and the remaining fungal structures are visualized using a stain. Choices of stains include trypan blue in lactophenol, cotton blue, acid fuchsin, aniline blue or chlorozol black E. Chlorozol black E was found to be superior to most other stains for studying the structural details of arbuscules and internal hyphae (Brundrett et al., 1983), while some researchers prefer other stains.

The procedure of clearing and staining roots using the method of Phillips and Hayman (1970) with some modifications is briefly outlined below.

1. Removing FAA: root samples stored in FAA should be washed with tap water several times and placed in small glass vials.
2. Clearing: 10% (w/v) KOH solution should be added to cover roots and heated at 90 °C for about 1 h. Alternative methods are to autoclave samples for 15 min at 120 °C or to leave at room temperature overnight. Clearing time and temperature depend on the type of root being evaluated. For example, delicate roots will be cleared in shorter periods or at lower temperatures. Therefore, clearing time and method should be adjusted to suit the specific root material.

3. Washing: KOH solution should be poured off and roots should be rinsed with tap water at least 3 times.
4. Bleaching: bleaching is necessary only for heavily pigmented root material which is not cleared adequately in KOH alone. Roots should be immersed from 10 min to 1 h in alkaline $H_2O_2$ solution at room temperature until roots are bleached. Alkaline $H_2O_2$ is prepared by adding 3 ml of $NH_4OH$, 30 ml of 10% $H_2O_2$ to 567 ml of tap water (Kormanik and McGraw, 1982). Roots should be rinsed thoroughly in water to remove all $H_2O_2$.
5. Acidifying: roots should be covered with 1% HCl, which is poured off after 3 min. Roots should not be washed after this step as an acidic medium is required by the stain.
6. Staining: an equal volume of 0.05% trypan blue in lactophenol stain should be added to roots and autoclaved for 10 min at 120 °C or heated at 90 °C for 10–15 min or left unheated for several hours. Lactophenol is prepared by combining 250 ml lactic acid, 300 g phenol, 250 ml glycerin and 300 ml water.
7. Destaining: roots should be immersed in lactophenol for destaining.
8. Roots should be stored in lactophenol or poured into a Petri dish for immediate microscopic observation.

Slight variation in the above steps might be necessary for different root material. Kormanik and McGraw (1982) discussed a method of clearing and staining with acid fuchsin in detail. They have used acid fuchsin without phenol and have obtained staining and clarity sufficient for quantifying root colonization, but insufficient for observing morphology of fungal structures. Phenols are toxic and can cause irritations or other side effects if proper precautions are not taken. It is not known whether the same clarity could be obtained by trypan blue without phenol. Clearing and staining many samples are laborious and time-consuming processes and require careful attention. Kormanik et al. (1980) have developed equipment and procedures for handling large numbers of root samples at one time.

**B. Microscopic observation**

Examination of stained roots in lactophenol in a Petri dish under a good dissecting microscope at 10–40 × magnification is adequate for detecting the presence or absence or for rapid assessment of root colonization. For some methods of quantification and observation of structures inside the root, root samples must be mounted on microscope slides, either temporarily in glycerol or lactophenol, or permanently in PVA (polyvinyl alcohol resin)-lactophenol. They are then observed under a compound microscope at 100–400 × magnification. Brundrett *et al.* (1983) have obtained clear and detailed views with chlorozol black E stain combined with Nomarski interference microscopy. Occasionally, when observation of the ultrastructure of the fungus root is required, root tissues should be sectioned in epoxy resin followed by electron microscopy. Refer to Cox and Sanders (1974) or Gianinazzi-Pearson *et al.* (1981) for details of this procedure.

For quantification, root segments to be mounted on slides should be randomly selected from the Petri dishes. Giovannetti and Mosse (1980) suggested that it would be better to mount more segments and make one observation per segment rather than to make close observations on only a few mounted roots.

## C. Methods of assessing root colonization

There is no standard method for quantification of root colonization in cleared and stained root samples. Researchers have used various assessment methods to meet their requirements. Giovannetti and Mosse (1980) summarize these methods and have compared several, which will be discussed later. Most of the common assessment methods are described below.

### 1. Detecting the presence or absence of colonization

This should be done by scanning the whole stained root samples in a Petri dish for the presence of any mycorrhizal structures, i.e. hyphae, arbuscules, vesicles or internal spores and rated positive (+) or negative (−) on a per sample or per plant basis. This is the most basic and rapid assessment method of those discussed here. This method is not quantitative, but adequate for some types of work such as checking host–fungus specificity and observing non-inoculated "control" plants for root colonization.

### 2. Calculating the percentage of root segments colonized

After examining roots for the presence or absence of vesicular-arbuscular mycorrhizal structures, the number of root segments colonized should be counted and expressed as a percentage of total root segments in the sample. Root segments can be observed in Petri dishes or mounted on slides. If Petri dishes are used, roots should be spread uniformly in the dish and all root segments should be visible without having to move other segments. A good light source and illumination is required for this work. Replicated assessments can be made by rearranging the roots and counting again (Giovannetti and Mosse, 1980). If mounted on slides, 30–100 random root segments per sample should be mounted in groups of 10 (Giovannetti and Mosse, 1980). Some drawbacks of this method will be discussed in Section III.D below.

### 3.  Categorizing root colonization

Root colonization can be broadly categorized into several classes on the basis of the percentage of root length colonized, after scanning either whole root samples or mounted subsamples without counting. Kormanik and McGraw (1982) reported the following classes as used at the Institute of Mycorrhizal Research and Development in Athens, Georgia, USA. Class 1, 0–5%; Class 2, 6–25%; Class 3, 26–50%; Class 4, 51–75%; and Class 5, 76–100%. Although this is a subjective assessment, Giovannetti and Mosse (1980) stated that these subjective assessments can give reliable results with only a few hours of training.

### 4.  Estimating the percentage of root length colonized

In this method, the length of colonized roots is expressed as a percentage of the total length of roots in the sample. This is measured by either the gridline-intersect method carried out with root segments in Petri dishes or by the direct estimation of percentage of root length colonized in mounted root segments.

(a) Gridline-intersect method. This method, described by Giovannetti and Mosse (1980), has been the most extensively used procedure for assessing root colonization in vesicular-arbuscular mycorrhizal research. Both the proportion of root length colonized and total root length can be estimated using this method. Estimation of total root length using this method is described in Section IV.

Root samples should be evenly distributed in the Petri dish and the dish placed on a plastic sheet ruled with gridlines. The dimension of the grid squares is important for measuring the total root length but not for estimating the percentage of root length colonized. For the latter purpose, gridlines only serve as a device for systematic selection of observation points. The vertical and horizontal gridlines should be scanned under a dissecting microscope and the presence or absence of root colonization should be recorded at each point where a root intersects a gridline. The percentage of root length colonized is expressed as the number of intersections with root colonization out of 100 total intersections counted.

Giovannetti and Mosse (1980) compared the standard error of percentage of root length colonized by counting 100, 200, 300 and the total number of intersections, which was sometimes as many as 600 with up to 15 replications in each instance. They found that counting 100 intersections on three sample rearrangements (replicates) gave a smaller standard error than counting 300 intersections on one arrangement. It was also shown that little, in terms of reducing standard error, was gained by counting 300 rather than 200 intersections. In order to obtain an accuracy of ±4% it was only necessary to count 100 intersections once, whereas an accuracy of ±1% required a count of all intersections with 3–5 arrangements.

*(b)  Direct estimation of percentage of root length colonized.* Although direct estimation of percentage of root length colonized is normally achieved using randomly selected and mounted root segments, some researchers have used entire root samples (Giovannetti and Mosse, 1980) or random subsamples in Petri dishes (Biermann and Linderman, 1981) for observation. The proportion of the length of each segment which contains vesicles, arbuscules or hyphae is estimated to the nearest 10% and the average is calculated for the sample observed, with the help of a frequency table.

## 5.   Total mycorrhizal root length per plant

Total mycorrhizal root length per plant can be calculated to assess the extent of mycorrhiza formation in the entire root system. The extent of root colonization, estimated by any of the above methods, should be multiplied by the total root length per plant.

## 6.   Total dry weight of mycorrhizal roots

As with the length of mycorrhizal roots, total dry weight of mycorrhizal roots should be calculated by multiplying the estimated percentage of root colonization by the dry weight of roots.

## D.   Comparison of methods of assessing root colonization

Both the estimation of percentage of root segments colonized and the direct estimation of percentage of root length colonized are time-consuming and difficult when large numbers of samples are involved. The latter method provides a more accurate assessment compared to estimation of percentage of root segments colonized by rating root segments positive or negative (Biermann and Linderman, 1981). Quantification of root colonization by estimating the percentage of root segments colonized can provide overestimates of the extent of root colonization (Giovannetti and Mosse, 1980; Biermann and Linderman, 1981; Thomson and Wildermuth, 1988). As Biermann and Linderman (1981) pointed out, the major reason for the overestimation is that a segment which is counted as positive may not be mycorrhizal for its entire length, whereas those which are counted as negative are completely non-mycorrhizal. The percentage of root segments colonized is also dependent on the size of root segments. Shorter root segments with colonization will result in overestimation of the number of root segment colonized, while longer root segments may provide an underestimation of root colonization. Estimation of percentage of root length colonized does not result in these inaccuracies, and is no more time-consuming than determining the percentage of root segments with colonization (Biermann and Linderman, 1981). Since root segments are observed closely, fungal structures such as arbuscules, vesicles and entry points can also be observed with this method.

In an attempt to progress towards standardization of assessment methods, Biermann and Linderman (1981) recommended that percentage of root length colonized be used as a standard measurement of vesicular-arbuscular mycorrhizal root colonization, so that comparisons could be made between results of different researchers. However, the standard error involved in assessing percentage of root length colonized was found to be only slightly lower than that obtained when the percentage of root segments colonized was estimated from mounted slides (Giovannetti and Mosse, 1980). One drawback in observing mounted roots on slides is that three-dimensional roots are viewed as two-dimensional objects. As a result, when a colonized area of root is overlying an uncolonized area in the same position, the estimation will not be any less than when colonized areas overlap in the same position of the root segment.

Giovannetti and Mosse (1980) compared four ways of estimating root colonization in cleared and stained roots. These are: (1) visual estimation of percentage of root cortex colonized observed in Petri dishes; (2) the gridline-intersect method; (3) estimation of percentage of root length colonized in mounted root segments; and (4) estimation of presence or absence of colonization in root segments mounted on slides and expressed as a percentage of root segments colonized. Giovannetti and

Mosse (1980) calculated the standard error of root colonization measured by these four methods on the same root sample. For mounted methods, root segments were randomly selected from the samples. They found that the gridline-intersect method produced the smallest standard error, followed by the visual estimation of percentage of root length colonized observed in Petri dishes.

Stained mycorrhizal root segments are visible as darker root segments when picking root segments for mounting on slides. Giovannetti and Mosse (1980) stated that some bias may be introduced into the estimation of the extent of root colonization at this point. Some selection method, for example, selecting every fifth root along an arbitrary axis in the Petri dish or picking from pre-determined positions on the grid sheet in which roots are randomly spread, could help prevent this.

Thompson and Wildermuth (1988) compared vesicular-arbuscular mycorrhizal root colonization measured by four methods including the gridline-intersect method, percentage of root length colonized measured in root segments mounted on slides, percentage of root cortex colonized measured in root segments mounted on slides and percentage of root segments colonized measured in root segments mounted on slides. They found that the results of all four methods correlated well, indicating a general agreement between methods in ranking plant species in the order of the extent of root colonization. However, it is not known which method best estimates or reflects true root colonization.

Researchers working on vesicular-arbuscular mycorrhiza do not use one method exclusively, as no method is unequivocally superior. When the assessment has to be made rapidly, or if total root length also has to be estimated as part of the experiment, it would be wise to select the gridline-intersect method. If the fungal structures are to be observed and studied, the method of direct estimation of the percentage of root length colonized in mounted root segments would be advantageous.

Root colonization, measured quantitatively by using any of the methods, does not necessarily reflect the effectiveness of vesicular-arbuscular mycorrhizal fungi in nutrient transfer capacity. It is not known that such functions are even related to the proportion of arbuscules present. One should be cautious in interpreting the functional basis of differences in the level of root colonization between any two sets.

## IV.  Measurement of root physical properties

The extent of the root system is often estimated to obtain a comprehensive picture of mycorrhiza formation as well as of the possible influence of root colonization on root growth. Traditionally, the most widely used variable in measuring the extent of root systems has been root length. Other variables which have been used are specific root length and specific root mass. Root weight, which has been used occasionally, is more related to the thickening of roots than their ability to form mycorrhiza or their absorptive capacity.

Entire root systems collected from smaller plants or root samples collected from larger plants should be used in measuring these variables. Estimation of these variables should precede observation and estimation of root colonization.

## A.  Root length

### 1.  Direct measurement of root length

In the direct measurement, roots should be spread over graph paper which has been covered with a glass plate or plastic sheet. Roots are straightened and their length estimated to the nearest mm (Reicosky *et al.*, 1970). Except in young plants, this method is time-consuming and tedious.

In a modification of this method, roots are cut into 1 cm segments and evenly spread over the graph paper. The grid size and area of paper most commonly used are 1 cm and 100 cm$^2$, respectively. The number of roots in three columns and three rows of graph paper is counted, averaged per grid and multiplied by the total number of grids in order to estimate the total root length. This method involves less time than the previous method.

### 2.  Gridline-intersect method

Root length is usually estimated by the gridline-intersect method manually or by using a scanner. Newman (1966) developed a formula, $R = \pi NA/2H$, for estimating the total length of roots ($R$) spread out in a given area ($A$) when randomly placed straight lines having a total length of $H$ in the area intersect with root $N$ number of times. The longer the root, the more intersections it makes with the superimposed randomly arranged lines. Marsh (1971) simplified the method of Newman by arranging the random lines in the form of a grid and, because there is a simple relationship between $A$ and $H$, he calculated that when the grid size is 1.25 cm, $\pi A = 2H$ and $R$ (in cm) $= N$. When grid dimensions are different from the above, root length can be calculated by the modified formula of Tennant (1975), $R =$ grid size in cm $\times N/1.25$. The 1.25 of this equation can be combined with the grid size to give a length conversion factor.

Roots should be spread over a flat surface and a grid, usually made on a plastic sheet or plexiglass, is placed on the area. The number of intersections between roots and random lines should be counted manually or automatically. Manual counting is time-consuming and can impose considerable strain on the eye. Rowse and Phillips (1974) developed a photoelectric instrument to count the number of intersections between the root sample and a set of parallel lines. Intersections are counted mechanically by moving the root sample beneath a modified binocular microscope fitted with a photoelectric counting device (Rowse and Phillips, 1974). A similar machine used by Richards *et al.* (1979) is capable of measuring root samples as long as 200 cm. Voorhees *et al.* (1980) used computer-controlled scanning and digitizing of photographic images of roots. The digitized light brightness was used to calculate root length based on Newman's line intersect approach.

### 3. Root length estimation devices

Instruments have been developed and are commercially available to estimate root length (Taylor, 1986). In addition to these instruments, area meters with video display monitors are also used to measure root length since the widths of roots are very small. A good contrast between roots and background is needed for smaller roots to form an image in the monitor.

## B. Specific root length and specific root mass

Usually, root length is measured from roots collected from soil samples rather than from the entire root system. Specific root length and specific root mass (also known as root length density and root mass density, respectively) are widely measured when root samples are collected.

Specific root length is calculated as length of root per unit volume of soil. Specific root mass is calculated as weight of root per unit volume of soil. Root length could be measured by any one of the methods described earlier. Root weight should be measured after oven-drying. For these methods, roots should be collected in a known volume from a soil core. Specific root length is preferred over specific root mass, because the formation of mycorrhiza is not directly related to weight of roots, as mentioned earlier.

## C.  Shoot : root ratio

The proportion of shoots to roots is compared using dry weights or by using height of shoots and total length of the root system or length of the tap root.

## D.  Root abundance

Root abundance is expressed as the number of roots present in a plane 5, 10 or 15 cm below the shoot–root interface, depending on the size of the root system. The intact root system is laid over a covered grid sheet and the shoot–root interface is placed on the baseline of the grid. The number of roots that cross the selected plane, e.g. the 5 cm line, is counted.

## E.  Characteristics of root hairs

The most important characteristics of root hairs are average length, number per cm of secondary root and average diameter. Measurements of these variables is time-consuming and is limited to studies that have specific objectives, such as investigating the influence of root hairs on formation of vesicular-arbuscular mycorrhiza and growth effects.

The number of root hairs present should be counted by observing under a binocular microscope 5 cm long pieces of secondary roots carefully removed from soil- or sand-grown plants. Average length is calculated by recording the length of root hairs in the same sample using a rotatable gridded eye-piece. The diameters of the root hairs could be measured by using a digital micrometer or by direct observation under the microscope.

## V.  Conclusions

A number of methods for estimating vesicular-arbuscular mycorrhizal colonization and related root physical properties are currently used. Among the more popular are, percentage of root segments colonized, percentage of root length colonized, total root length, and specific root length. These, as well as other methods described, have their advantages and disadvantages. Each researcher should be aware of the various methods and should select the most appropriate method after considering the objectives of the study, the plant species under study, and available resources. Researchers should also not limit their methods to those described here. New methods as well as improvements to those already used should be developed.


## Acknowledgement

We gratefully acknowledge the review of this manuscript by Professor Ruth A. Taber.


## References

Abbott, L. K., Robson, A. D. and De Boer G. (1984). *New Phytol.* **97**, 437–446.
Ambler, J. R. and Young, J. L. (1977). *Soil Sci. Soc. Am. J.* **41**, 551–556.
Becker, W. N. and Gerdemann, J. W. (1977). *New Phytol.* **78**, 289–295.
Bethlenfalvay, G. J., Brown, M. S. and Pacovsky, R. S. (1982). *New Phytol.* **90**, 537–543.
Biermann, B. and Linderman, R. G. (1981). *New Phytol.* **87**, 63–67.
Brundrett, M. C., Piché, Y. and Peterson, R. L. (1983). *Can. J. Bot.* **62**, 2128–2134.
Cox, G. and Sanders, F, (1974). *New Phytol.* **73**, 901–912.
Gianinazzi-Pearson, V., Morandi, D., Dexheimer, J. and Gianinazzi, S. (1981). *New Phytol.* **88**, 633–639.
Giovannetti, M. and Mosse, B. (1980). *New Phytol.* **84**, 489–500.
Hayman, D. S. (1970). *Trans. Br. Mycol. Soc.* **54**, 53–63.
Hepper, C. M. (1977). *Soil Biol. Biochem.* **9**, 15–18.
Kormanik, P. P. and McGraw, A. C. (1982). In *Methods and Principles of Mycorrhizal Research* (N. C. Schenck, ed.), pp. 37–45. The American Phytopathological Society, St. Paul, MN.
Kormanik, P. P., Bryan, W. C. and Schultz, R. C. (1980). *Can J. Microbiol.* **26**, 536–538.
Lopez-Aguillon, R. and Mosse, B. (1987). *Plant Soil* **97**, 155–170.
Marsh, B. a'B. (1971). *J. Appl. Ecol.* **8**, 265–267.
Newman, E. I. (1966). *J. Appl. Ecol.* **3**, 139–145.
Nicolson, T. H. (1959). *Trans. Br. Mycol. Soc.* **42**, 421–438.
Phillips, J. M. and Hayman, D. S. (1970). *Trans Br. Mycol. Soc.* **55**, 158–161.
Reich, L. and Bernard, J. (1984). *New Phytol.* **98**, 475–479.
Reicosky, D. C., Millington, R. J. and Peters, D. B. (1970). *Agron. J.* **62**, 451–453.
Richards, D., Goubran, F. H., Garwoli, W. N. and Daly, M. W. (1979). *Plant Soil* **52**, 69–76.
Rowse, H. R. and Phillips, D. A. (1974). *J. Appl. Ecol.* **11**, 309–314.
Schubert, A., Marzachi, C. and Mazzitelli, M., Cravero, M. C. and Bonfante-Fasolo, P. (1987). *New Phytol.* **107**, 183–193.
Taylor, H. M. (1986). *Hort. Sci.* **21**, 952–956.

Tennant, D. (1975). *J. Ecol.* **63**, 995–1001.
Thompson, J. P. and Wildermuth, G. B. (1988). *Can. J. Bot.* **69**, 687–693.
Voorhees, W. B., Carlson, V. A. and Hallauer, E. A. (1980). *Agron J.* **72**, 847–851.
Zobel, R. W. (1986). *Hort. Sci.* **21**, 956–959.

# 16
# Wet-sieving and Decanting Techniques for the Extraction of Spores of Vesicular-arbuscular Fungi

GIOVANNI PACIONI

*Dipartimento di Scienze Ambientali, Università, I-67100 L'Aquila, Italy*

## I. Introduction

The extraction of small living structures from soil and other substrata originally presented a problem for nematologists. The methods they developed have been adapted to enable the separation of spores of Endogonaceous fungi, the structures of which were initially confused with cysts of *Heterodera,* a common pest of roots. Both spores and cysts can be separated by flotation methods (Triffitt, 1935). One such method was successfully developed by Gerdeman and Nicolson (1963) and this is now the most widely used procedure for the study of spores of Endogonaceous fungi in soil. The method is described here and a few modifications are proposed to improve its effectiveness.

METHODS IN MICROBIOLOGY
VOLUME 24   ISBN 0-12-521524-X
Copyright © 1992 by Academic Press Limited
All rights of reproduction in any form reserved

## II.  Description of the technique

A suspension of soil in water is passed through a series of metallic sieves arranged in decreasing order of mesh width. The method is illustrated in Fig. 1. The steps are as follows:

1. Suspend 250 ml of soil in 1 litre of water, or more if necessary.
2. Wait for heavier particles to settle.
3. Pass the suspension through a sieve with a 1 mm wide mesh, saving the filtrate.
4. Debris retained by this sieve must be re-suspended into the saved suspension, stirred again, or washed under a stream of water, saving the suspension.

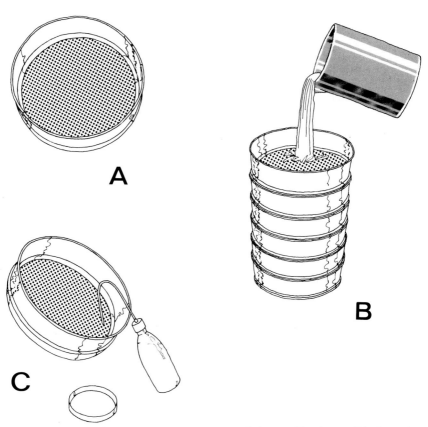

**Fig. 1.**   Steps in the wet-sieving technique: (A) metallic sieve; (B) decanting through a sieve series; (C) removal of debris to a Petri dish.

5. Decant all suspensions through a sieve series ranging from 1 mm to 40 $\mu$m mesh.
6. With the aid of a jet of water directed at both sides of the sieve, the contents of each sieve is transferred to a Petri dish.
7. The shallow suspension can be observed with a stereo-microscope and spores and microsporocarps picked up with a flattened needle or a Pasteur pipette.
8. Small concave watch-glasses or slides can be used for further observations.

### III.   Modifications and improvements

Experience has led to the introduction of some modifications to the original method.

### A.   Preparation of the suspension

The following modifications to the method for preparing the suspension give better results:

(1) An optimal ratio of soil/water was found to be 1/10, that is 100 ml of soil in 1 litre of water.
(2) The suspension should be stirred with a magnetic stirrer or by hand using a rod. Magnetic stirring gives better results, using a water-driven stirrer that does not heat. Stirring time varies according to the nature of soil—usually 10 min. is sufficient.
(3) To avoid foam that could retain debris, add an anti-foam agent, for example Tween 80, 0.1–0.5%.
(4) If the soil is a clay that yields a suspension which blocks the sieve, precipitate the particles in 0.1 M sodium pyrophosphate (Fogel and Hunt, 1979).

The last two procedures are sometimes indispensable when using the nylon filters recommended by Pacioni and Rosa (1985). Particularly when working with either humic or clay soils, the efficiency of the filter can be altered by the presence of foam or suspended particles.

### B.   Filtering apparatus

The removal of spores from metallic sieves can be difficult and, particularly when working with soil where a predominance of single spores occurs, significant losses of spores can occur. An effective modification of the filtering apparatus was introduced and satisfactorily tested by Pacioni and Rosa (1985).

Nylon filters (supplied by RI.MA., Via Gramsci 2/C, Bologna, Italy, fax. 39-51-553761) with standard meshes of the kind used in palynology or with cellular cultures for the separation of protoplasts were used. The method is illustrated in Fig. 2.

(1) A series of filters, decreasing in pore size (1 mm to 40 $\mu$m), are attached to each other by plastic tubing of 20 cm diameter, cut in lengths of 20 cm.

(2) Once filled, the filters are placed onto a transparent plexiglas grid-lined sheet and placed under a stereoscope.

The spores are supported by the filter meshes where they can be counted accurately and manipulated easily. Using a transmitted light stereoscope, the transparency of the supports (nylon filter and plexiglas sheet) allows easy observation of spores.

## C. Technique for a large amount of soil

The techniques described above are suitable for small-scale studies, involving up to 100 g of soil at a time. If kilogram quantities of soil are to be examined, a drum, for example a petrol barrel without cover but with a lateral over-flow pipe, can be utilized. The filtering apparatus is made up of a metallic sieve with meshes of 1 mm for retaining larger debris and a nylon filter bag held by a fixing band. The size of the mesh is selected to be appropriate for the dimensions of the spores required. The method is illustrated in Fig. 3.

(1) Place a plastic pipe or insert a fixed water source at the bottom of the drum.

(2) Place the filtering apparatus to catch the water suspension.

(3) Fill the drum, suspend the soil, stir with a rod, treat as necessary and then turn on the water, leaving the suspension to flow through the filtering apparatus.

## IV. Conclusion

Of the various methods proposed for the recovery of Endogonaceous spores from soil, the wet-sieving and decanting technique is the simplest and most effective. However, using the procedures described spores with a diameter less than 40 $\mu$m will be lost. Unfortunately these smaller spores often occur in compact soils, such as clays, that quickly block a

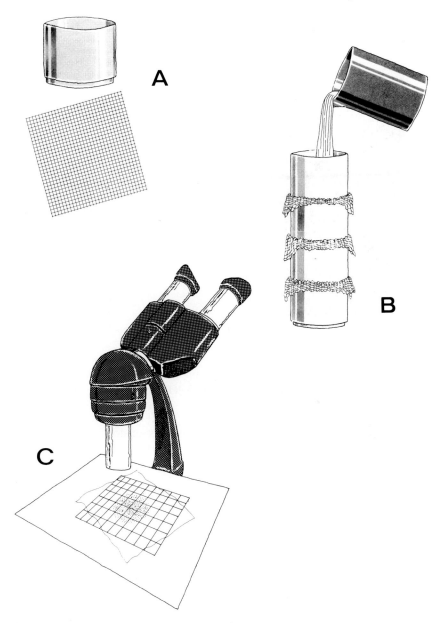

**Fig. 2.** Steps in the modified wet-sieving technique: (A) section of wide plastic tubing and nylon filter; (B) soil suspension being poured onto a column containing nylon filters; (C) filter on plexiglas grid-lined sheet under stereoscope.

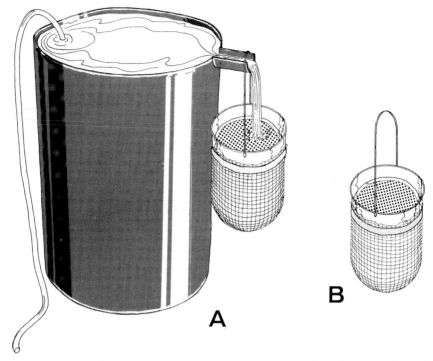

**Fig. 3.** Wet-sieving technique for a large amount of soil: (A) drum with soil
suspension; (B) filtering apparatus.

cellulosic filter (Pacioni and Puppi, 1989). If the presence of these
smaller spores is suspected, save the passed suspension, precipitate the
clay particles still in suspension and pass again through a cellulosic filter
under vacuum. Observe the filter directly under a stereoscope with
indirect light.

### References

Fogel, R. and Hunt, G. (1979). *Can. J. For. Res.* **9,** 245–256.
Gerdeman, J. W. and Nicolson, T. H. (1963). *Trans. Br. Mycol. Soc.* **46,**
    235–244.
Pacioni, G. and Puppi, G. (1989). *Micol. Veg. Medit.* **3,** 133–142.
Pacioni, G. and Rosa, S. (1985). *Bull. Br. Mycol. Soc.* **19,** 66–68.
Triffitt, M. J. (1935). *J. Helminth.* **13,** 59–66.

# 17
# Usefulness of the Pinnule Technique in Mycorrhizal Research

MITIKU HABTE

*Department of Agronomy and Soil Science, University of Hawaii, Honolulu HI 96822, USA*

## I. Introduction

Numerous investigations have demonstrated that most species of plants in nature carry out vital processes under the influence of vesicular-arbuscular mycorrhizal fungi (Mosse, 1981; Kormanik *et al.*, 1982; Giannazzi-Pearson and Diem, 1982). Global interest in the fungi has now reached a point whereby any discussion of agricultural biotechnology that does not include the role of vesicular-arbuscular mycorrhizal fungi is considered incomplete.

The fungi improve plant growth largely by enhancing the uptake of diffusion-limited nutrients, particularly phosphorus. The presence of the fungi in plant roots has also been associated with improved uptake of other essential nutrients, tolerance to drought, protection from plant

METHODS IN MICROBIOLOGY
VOLUME 24   ISBN 0-12-521524-X

Copyright © 1992 by Academic Press Limited
All rights of reproduction in any form reserved

pathogens, resistance to salinity, and increased plant chlorophyll content. Because of the positive ways by which vesicular-arbuscular mycorrhizal fungi influence plants, researchers and practitioners have been interested in harnessing the vesicular-arbuscular mycorrhizal symbiosis for increased plant productivity in agriculture and in natural ecosystems. This goal, however, is far from being achieved, partly because the biotic and abiotic factors that regulate the symbiosis have yet to be clearly understood. A significant aspect of this problem is the lack of simple, rapid and reliable techniques for monitoring the development of vesicular-arbuscular mycorrhizal symbiotic effectiveness across host, endophyte and environmental variables. The usual procedures of evaluating vesicular-arbuscular mycorrhizal activity such as spore counts (Gerdemann and Nicolson, 1963), counts of infective propagules (Porter, 1979), assessment of root infection levels (Giovannetti and Mosse, 1980) and detection and quantification of extramatrical hyphae (Pacovsky and Bethlenfalvey, 1982) are tedious, lack precision, and often are not well correlated with the symbiotic effectiveness that may develop (Mosse, 1972; Kormanik and McGraw, 1982; Graham et al., 1982; Menge, 1983).

Because growth response to vesicular-arbuscular mycorrhizal infection of host species largely results from increased uptake of P, one of the best ways of determining vesicular-arbuscular mycorrhizal symbiotic effectiveness would be to monitor the P status of host plants as the symbiosis develops. This chapter is concerned with the determination of pinnule P status as a rapid and precise technique for monitoring the development of symbiotic effectiveness in the vesicular-arbuscular mycorrhizal association.

## II.  Methodology

### A.  Criteria for selecting indicator plants

The key criteria to be considered in selecting indicator plants for the pinnule technique are: (1) they must have compound leaves; (2) they must grow reasonably rapidly; (3) they must be moderately to very highly dependent on vesicular-arbuscular mycorrhizal fungi; and (4) they must have pinnules or subleaflets that detach readily from leaflets. Species that are marginally dependent on vesicular-arbuscular mycorrhizal fungi could serve as indicator plants, but the range of soil solution P levels at which they would be useful is limited (M. Habte and A. Manjunath, in press). The species used in the initial development of the pinnule technique was *Leucaena leucocephala* var. K8 (Habte et al.,

1987). Subsequently, other species such as *Acacia koa*, *L. diversifolia*, *L. trichoides* and *Sesbania grandiflora* were found to be useful as indicator plants (M. Habte and A. Manjunath, in press; M. Habte, unpubl. res.)

## B.  Growth conditions

A good rule of thumb is that the health of the vesicular-arbuscular mycorrhizal symbiosis is intimately linked to the health of the host plant. Growth conditions must, therefore, be adjusted such that indicator plants will develop and grow normally in the presence of effective vesicular-arbuscular mycorrhizal endophytes. Phosphorus level is particularly critical, since P appears to be the most important regulator of the symbiosis. High levels of P are inhibitory to the development of vesicular-arbuscular mycorrhiza, and the effectiveness of vesicular-arbuscular mycorrhiza may not be adequately expressed if there is too little P in the soil solution (Ross, 1971; Habte and Manjunath, 1987; Aziz and Habte, 1987; Habte *et al.*, 1988). The optimum level of P for vesicular-arbuscular mycorrhizal activity in host species highly to very highly dependent on vesicular-arbuscular mycorrhiza lies between 0.01 and $0.02 \, mg \, P \, litre^{-1}$ (Habte and Manjunath, 1987; Manjunath and Habte, 1990). If available, rock phosphate could replace superphosphate, and perhaps is a better source of P since it seldom inhibits vesicular-arbuscular mycorrhizal activity (Powell, 1980). A rock phosphate containing 16% P could be applied at the rate of $5.44 \, g \, kg^{-1}$ of soil or higher (Manjunath *et al.*, 1989a; M. Habte, unpubl. data). Other essential nutrients must be supplied in sufficient amounts by blanket application to prevent these nutrients from limiting growth. Several blanket nutrient formulae have been successfully used for mycorrhizal effectiveness studies in my laboratory (Habte and Manjunath, 1987; Habte *et al.*, 1989).

For acid-sensitive indicator plants, soil pH must be adjusted to around 6.0. Vesicular-arbuscular mycorrhizal development also requires adequate light and moisture supply (Graham *et al.*, 1982; Menge, 1984). In Hawaii, we routinely grow indicator plants in a glasshouse under natural light with soil moisture content maintained at approximately field capacity level.

## C.  Sampling and sample preparation

Figure 1 represents a photocopy of a pressed *Leucaena leucocephala* leaf showing leaflets and pinnules (subleaflets). Pinnule sampling can begin

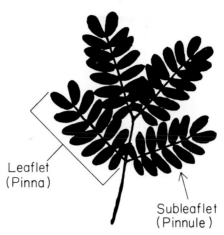

**Fig. 1.** A photocopy of pressed *L. leucocephala* leaf showing pinna and pinnules (Habte *et al.*, 1987) (courtesy of Marcel Dekker Inc.).

as early as 10 days from planting. Subsequent sampling can be carried out as frequently as once every 5 days. The youngest fully open leaf is selected, and one or two pinnules may be removed per leaf at each sampling day from a fixed position on a leaf. Because P is mobile within the plant, the youngest pinnule on the youngest fully open leaf is the most sensitive indicator of vesicular-arbuscular mycorrhizal effectiveness. However, the youngest pinnule on a leaflet is often difficult to remove intact. Any other convenient pinnule position on the youngest fully open leaf will do, provided that this position is maintained throughout the duration of an investigation. In our studies, we have used the second, third or fourth pinnule from the tip or base of a leaflet without any noticeable sacrifice in information (Habte and Manjunath, 1987; Habte *et al.*, 1987, 1988; M. Habte, unpubl. res.). I personally prefer the fourth pinnule from the tip of a leaflet, because it is relatively easy to remove.

Pinnules from a given sample are deposited in labelled plastic vials and brought to the laboratory for drying (70 °C for 4 h). Pinnules are then weighed (if P concentration calculations are going to be made), and transferred into 18 × 50 mm Pyrex test tubes for ashing.

## D.  Analysis of samples

Pinnules are ashed in a muffle furnace at 500 °C for 3 h. The ash is then dissolved and colour developed according to the molybdenum blue technique (Murphy and Riley, 1962). To achieve this, 2.5 ml of reagent

B is added into the test tube containing the ashed sample; 10 ml of distilled or de-ionized water is added to the mixture. The contents are mixed thoroughly using a vortex mixer. After standing 20 min, the intensity of the colour that develops is measured in a spectrophotometer at a wavelength of 840 nm. Reagent B is prepared by dissolving 0.428 g of ascorbic acid in 100 ml of reagent A. Reagent A is prepared by dissolving 0.35 g of antimony potassium tartrate in 2.7 litres of distilled/de-ionized water, adding to it 168 ml of concentrated sulphuric acid and bringing the total volume to 3 litres with distilled/de-ionized water. This solution is stored in a dark bottle and used as needed. Reagent B should be made daily.

The concentration of P in a sample is determined by referring to a standard curve prepared by plotting the absorbance of standard P solutions against P concentration. Results may be expressed either as total P content per pinnule or as P%. Both expressions give comparable results for pinnule samples taken from *L. leucocephala* (Figs 2 and 3). It is, however, advantageous to express results as total P content as this will not require pinnule weight determination. Moreover, recent observations made using *Acacia koa* suggest that P% values may be highly variable in some plant species (S. Miyasaka and M. Habte, unpubl. res.).

### E.  Calibration

Vesicular-arbuscular mycorrhizal effectiveness measured in terms of pinnule P content and that measured by monitoring shoot P content of host plants in sequential harvest experiments have been found to be highly correlated (Habte *et al.*, 1987). Other studies have also clearly demonstrated the strong relationship existing between vesicular-arbuscular mycorrhizal effectiveness values obtained by employing the pinnule technique and those determined using response variables such as root dry matter yield, shoot dry matter yield or shoot P concentration (Aziz *et al.*, 1991a; M. Habte, unpubl. data).

### III.  Application

### A.  Monitoring the progress of symbiotic effectiveness

Because the pinnule technique allows many more samplings than is possible in multiple-harvest experiments, it provides a better resolution of the phases of development that are involved in the vesicular-arbuscu-

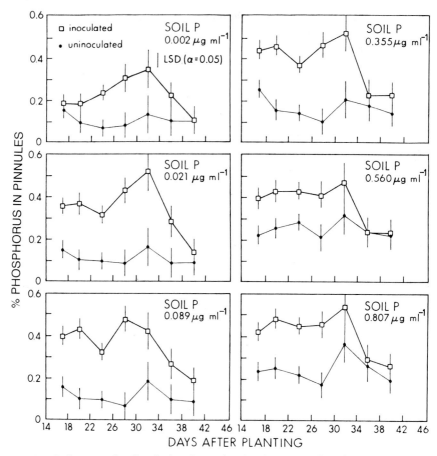

**Fig. 2.** Influence of soil solution P on the development of vesicular-arbuscular mycorrhizal effectiveness measured as P concentration of *L. leucocephala* pinnules (Habte and Manjunath, 1987).

lar mycorrhizal symbiosis (see Fig. 2). During the first few days of sampling, pinnule P content in both vesicular-arbuscular mycorrhizal and non-vesicular-arbuscular mycorrhizal plants declines as P in the seed reserve is depleted. This is the phase during which vesicular-arbuscular mycorrhizal endophytes are becoming established on host roots, but have not yet developed adequate fungal structures to translocate sufficient P from soil to host plants. This phase is followed by a phase of rapid P accumulation in pinnules of mycorrhizal but not of non-mycorrhizal plants. This activity subsequently reaches a peak value after which time pinnule P content declines to a level which corresponds to the

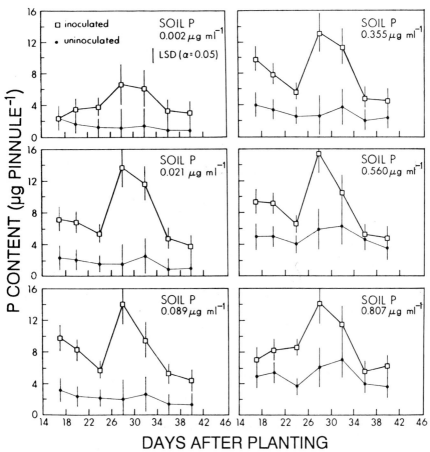

**Fig. 3.** Influence of soil solution P on the development of vesicular-arbuscular mycorrhizal effectiveness measured as total P content of *L. leucocephala* pinnules (Habte and Manjunath, 1987).

initial pinnule P content. This decline probably reflects the operation of a host control mechanism which limits the supply of P coming to the host (Pacovsky and Fuller, 1986) and the translocation of P from pinnules to other plant parts (Habte and Manjunath, 1987). In experiments of longer duration, this phase may be followed by a rise in vesicular-arbuscular mycorrhizal activity, yielding another peak which is smaller than the first one (Fig. 4) (M. Habte, unpubl. data). It is conceivable that vesicular-arbuscular mycorrhizal effectiveness is expressed in a series of cycles corresponding to cycles of arbuscule formation and digestion.

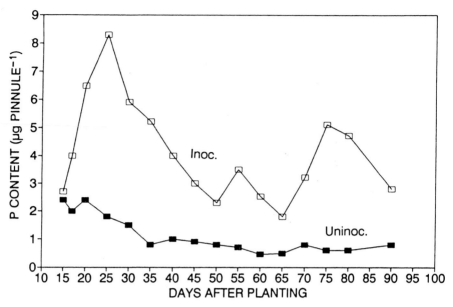

**Fig. 4.**   Cycles of vesicular-arbuscular mycorrhizal effectiveness demonstrated in
terms of pinnule P content of *L. leucocephala* pinnules.

## B.   Evaluation of introduced and indigenous endophytes

Ability to use the vesicular-arbuscular mycorrhizal technology in a
predictable manner will depend, among other things, on the extent to
which efforts are made towards the selection of inoculum species with
high levels of symbiotic effectiveness. Figure 5 illustrates the usefulness
of the pinnule technique for achieving this purpose. To obtain the data
summarized in Fig. 5, a subsurface (50–55 cm) Wahiawa soil (Tropeptic
Eutrustox) devoid of vesicular-arbuscular mycorrhizal endophytes was
optimized for mycorrhizal activity (Habte and Manjunath, 1987). The
soil was either uninoculated or inoculated with a 6-month-old crude
inoculum of *Glomus aggregatum*, *G. deserticola* or *G. mosseae* at
comparable densities of infective propagules. *L. leucocephala* was
grown on the soil, and the development of vesicular-arbuscular mycorrhizal effectiveness was monitored as a function of time. Figure 5 clearly
shows that *G. aggregatum* was the endophyte with the highest symbiotic
effectiveness, followed by *G. deserticola* and *G. mosseae*. These findings confirm the earlier observations of Huang and colleagues (1983),
who found that inoculation of fumigated field plots with *Glomus
aggregatum* (at that time misidentified as *G. fasciculatum*) gave the

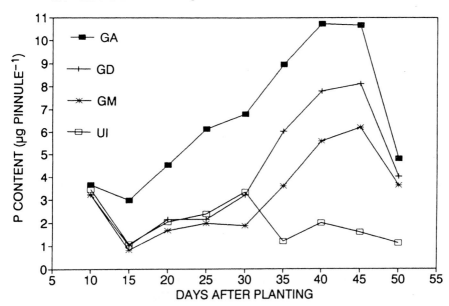

**Fig. 5.**    Effectiveness of selected vesicular-arbuscular mycorrhizal endophytes indicated by pinnule P status of *L. leucocephala* leaves.

highest *L. leucocephala* dry matter yield compared with inoculation of the plots with either *G. etunicatum* or *G. mosseae*.

Another important variable which determines the extent to which a given host plant responds to vesicular-arbuscular mycorrhizal inoculation is the symbiotic status of native vesicular-arbuscular mycorrhizal endophytes. Habte *et al.* (1987) developed a simple procedure which utilized the pinnule technique for determining the symbiotic effectiveness of endophytes contained in different soils relative to the effectiveness of a known vesicular-arbuscular mycorrhizal inoculum. In this approach, a fumigated sand–soil medium was optimized for vesicular-arbuscular mycorrhizal activity (Habte and Manjunath, 1987). The medium was either uninoculated or inoculated with *G. aggregatum* or with 50 g portions of a test soil of interest. *L. leucocephala* was grown on the medium, and vesicular-arbuscular mycorrhizal effectiveness was monitored using the pinnule technique. Figure 6 represents a summary of the data collected using the procedure. The results indicate that the support medium inoculated with the Kaneohe soil (Humoxic Tropohumult) showed no evidence of vesicular-arbuscular mycorrhizal activity. There was vesicular-arbuscular mycorrhizal activity in the Halii soil (Typic Gibbsihumox), but symbiotic effectiveness was expressed after a pro-

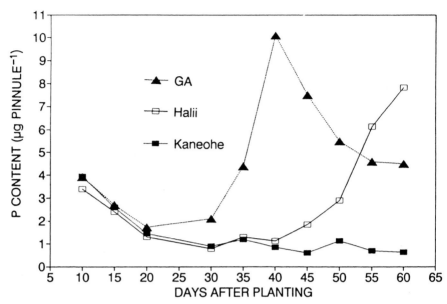

**Fig. 6.** Effectiveness of indigenous vesicular-arbuscular mycorrhizal fungi determined in terms of P content of *L. leucocephala* pinnules.

longed lag phase. In contrast, in the support medium inoculated with *G. aggregatum*, symbiotic effectiveness was expressed relatively rapidly and at a high level. Most probable number counts of infective propagules in the test soils revealed that there were no viable propagules in the Kaneohe soil. This soil was collected from a severely eroded surface. The Halii soil had a high density of infective propagules ($650 g^{-1}$), but expressed their symbiotic effectiveness very slowly. Since in the soils evaluated vesicular-arbuscular mycorrhizal effectiveness was either nonexistent or was expressed after a prolonged lag, the likelihood of getting a good response to inoculation of these soils with a highly effective endophyte such as *G. aggregatum* is great. These findings demonstrate the usefulness of the pinnule technique for predicting host response to vesicular-arbuscular mycorrhizal inoculation. In fact, a slight modification of this approach was recently used to predict with reasonable accuracy the magnitude of response of *L leucocephala* to inoculation of five tropical soils with *G. aggregatum* (Habte and Fox, 1989).

## C.  Evaluation of soil conditions and environmental stresses

The pinnule technique can be used to study the influence of any variable on the development of the vesicular-arbuscular mycorrhizal symbiosis. When soil variables are not the subject of an investigation, the confounding effect of these variables must be minimized by optimizing soil conditions for vesicular-arbuscular mycorrhizal activity. The pinnule technique has been particularly useful in this optimization process (Habte and Manjunath, 1987; Habte et al., 1987; Manjunath and Habte, 1990).

The technique has also been tested for usefulness in the evaluation of the influence of natural and imposed erosion on several occasions. As noted above, a severely eroded ultisol (Kaneohe series), which was found not to have vesicular-arbuscular mycorrhizal activity when tested by the pinnule technique, was also found not to contain vesicular-arbuscular mycorrhizal propagules (Habte and Manjunath, 1987). Figure 7 represents a summary of a segment of pinnule P data collected from an experiment designed to determine the influence of simulated erosion on vesicular-arbuscular mycorrhizal effectiveness. Studies like this one have revealed that losses of symbiotic effectiveness accompanying erosional soil losses, contrary to existing opinion, cannot be restored through mere re-infestation of eroded soils with vesicular-arbuscular mycorrhizal fungi (Aziz and Habte, 1987; Habte et al., 1988; Aziz and Habte, 1990). The pinnule technique has also been useful in the identification of critical soil components other than vesicular-arbuscular mycorrhizal propagules which are lost during erosional soil losses, and in the elucidation of the restorative inputs required for establishing mycorrhizal host species on eroded soils (Habte et al., 1988; Aziz and Habte, 1990; Aziz et al., 1991a).

Many pesticidal compounds control pest populations by hampering biochemical processes that are common to pest and non-pest species. The effect that pesticides, particularly fungicides, might have on the vesicular-arbuscular mycorrhizal symbiosis can be conveniently and accurately evaluated using the pinnule technique. Recently, M. Habte and A. Manjunath (unpubl. res) used the technique to evaluate the influence of the fungicide Thiram (Tetramethylthiuram disulphide) on the development of the vesicular-arbuscular mycorrhizal symbiosis 71 days after the pesticide was incorporated into a subsurface (30–35 cm) Wahiawa soil. The mycorrhizal inoculum used was G. aggregatum and the indicator host was L. leucocephala. The results are presented in Figs 8 and 9. Figure 8 represents pinnule P data observed when soil P level was optimum for mycorrhizal activity. At this level of P, the first increment of Thiram had no effect on vesicular-arbuscular mycorrhizal effectiveness, while at the higher levels tested the chemical either prolonged the time required for the expression of vesicular-arbuscular

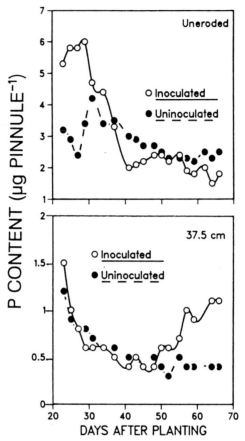

**Fig. 7.** Vesicular-arbuscular mycorrhizal effectiveness measured in terms of pinnule P content of *L. leucocephala* grown in uneroded Wahiawa soil compared to that of *L. leucocephala* grown in the soil after 37.5 cm of the top soil had been removed (from Habte *et al.*, 1988).

mycorrhizal effectiveness or suppressed it completely. At soil P level sufficient for non-mycorrhizal host growth, pinnule P contents of mycorrhizal and non-mycorrhizal *L. leucocephala* were similar; Thiram treatment thus had no influence on P content. The adverse effect of the fungicide, therefore, was confined to the vesicular-arbuscular mycorrhizal endophytes. Other pesticides evaluated using the pinnule technique include phenamiphos (Habte *et al.*, 1988), Fosetyl-Al (Aziz *et al.*, 1991b), and Chlorothalonil (Aziz *et al.*, 1991a).

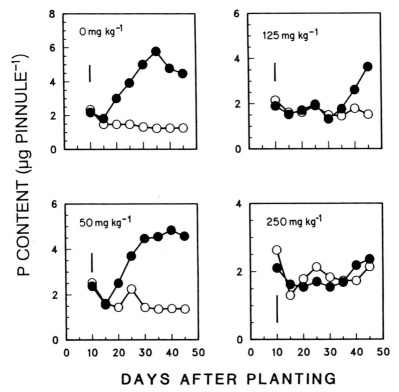

**Fig. 8.**    Influence of Thiram on pinnule P status of *L. leucocephala* grown on a Wahiawa soil uninoculated (○) or inoculated (●) with *G. aggregatum* at soil solution P level optimum for vesicular-arbuscular mycorrhizal activity (M. Habte and A. Manjunath, unpubl. data).

## D.   Evaluation of vesicular-arbuscular mycorrhizal dependency

The vesicular-arbuscular mycorrhizal dependency of a given species could be dependent, among other things, on its stage of development, on the stage of development of vesicular-arbuscular mycorrhizal endophytes and on the soil solution P status. If vesicular-arbuscular mycorrhizal dependency data are going to form part of the knowledge base that is required for predictable management of the symbiosis in practical agriculture, it is essential to define vesicular-arbuscular mycorrhizal dependency of host plants as a function of critical soil solution P levels and at a particular stage of the development of the vesicular-arbuscular mycorrhizal symbiosis. Pinnule P data are particularly useful in determining the critical stage of vesicular-arbuscular mycorrhizal development at which vesicular-arbuscular mycorrhizal dependency of host species is

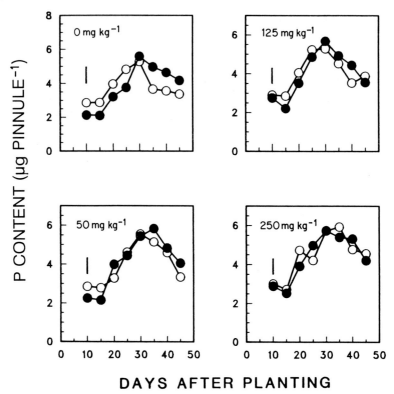

**Fig. 9.**   Influence of Thiram on pinnule P status of *L. leucocephala* grown on a Wahiawa soil uninoculated (○) or inoculated (●) with *G. aggregatum* at soil solution P level sufficient for non-mycorrhizal host growth (M. Habte and A. Manjunath, unpubl. data).

established. Such approaches were recently used by Manjunath and Habte (1991) to define categories of vesicular-arbuscular mycorrhizal dependency of host species and to determine the root morphological characteristics associated with the vesicular-arbuscular mycorrhizal dependency categories proposed.

## IV.   Conclusion

Data derived from single-harvest experiments at best provide limited understanding of vesicular-arbuscular mycorrhizal symbiosis, particularly with regard to its developmental aspects. This is so because the response of host plants to vesicular-arbuscular mycorrhizal infection varies not only with requirement for immobile nutrients, but also with the stage of development of host and endophyte species. The pinnule technique gives researchers the opportunity to monitor vesicular-arbuscular mycorrhizal effectiveness as a function of time without the tedium associated with multiple-harvest experiments and without the requirements of large amounts of time and space found in these kinds of experiments. From the data reviewed in this chapter it is clear that the pinnule technique enables one to assess rapidly and conveniently the influence of almost any biotic or abiotic variable on the development of the vesicular-arbuscular mycorrhizal symbiosis. It is simple, reliable and sensitive. These properties also make it attractive for use in instructional and training settings.

## References

Aziz, T. and Habte, M. (1987). *Can. J. Microbiol.* **33**, 1097-1101.

Aziz, T. and Habte, M. (1990). *Commun. Soil Sci. Plant Anal.* **21**, 493–505.

Aziz, T., Habte, M. and Yuen, J. E. (1991a). *Plant Soil* **131**, 47–52.

Aziz, T., Yuen, J. E. and Habte, M. (1991b) *Commun. Soil. Sci. Plant. Anal.* **21**, 2309–2317.

Bethlenfalvay, G., Pacovsky, R. S. and Brown, M. S. (1981). *Soil Sci. Soc. Am. J.* **45**, 871–875.

Gerdemann, J. W. and Nicolson, T. H. (1963). *Trans. Br. Mycol. Soc.* **46**, 234–244.

Gianinazzi-Pearson, V. and Diem, H. G. (1982). In *Microbiology of Tropical Soils and Plant Productivity* (Y. Dommergues and H. G. Diem, eds), pp. 209–259. Martinus Nijhoff/Dr. W. Junk Publishers, The Hague.

Giovannetti, M. and Mosse, B. (1980). *New Phytol.* **84**, 489–500.

Graham, J. H., Linderman, R. G. and Menge, J. A. (1982). *New Phytol.* **91**, 183–189.

Habte, M. and Aziz, T. (1991). *J. Plant Nutr.* **14**, 429–442.

Habte, M. and Fox, R. L. (1989). *Biol. Fertil. Soils* **8**, 111–115.

Habte, M. and Manjunath A. (1987). *Appl. Environ. Microbiol.* **53**, 791–801.

Habte, M. and Manjunath, A. (1988). *Biol. Fertil. Soils* **5**, 313–316.

Habte, M., Fox, R. L. and Huang, R. S. (1987). *Commun. Soil Sci. Plant. Anal.* **18**, 1403–1420.

Habte, M., Fox, R. L., Aziz, T. and El-Swaify, S. A. (1988). *Appl. Environ. Microbiol.* **54**, 945–950.

Huang, R. S., Yost, R. S., Fox, R. L., Habte, M. and Murdoch, C. L. (1983). *Leucaena Res. Rpts.* **4**, 83–85.

Kormanik, P. P. and McGraw, A. C. (1982). In *Methods and Principles of Mycorrhizal Research* ( N. C. Schenck, ed.), pp. 37–45. American Phytopathology Society, St. Paul, MN.

Kormanik, P. P., Shultz, R.C. and Bryan, W. C. (1982). *Forest Sci.* **28**, 531–539.

Manjunath, A. and Habte, M. (1990). *Commun. Soil Sci. Plant Anal.* **21**, 557–566.

Manjunath, A. and Habte, M. (1991). *Can. J. Bot.* **69**, 671–676.

Manjunath, A., Hue, N. V. and Habte, M. (1989a). *Plant Soil* **114**, 127–133.

Manjunath, A. *et al.* (1989b). *J. Plant Nutr.* **12**, 755–768.

Menge, J. A. (1983). *Can. J. Bot.* **61**, 1015–1024.

Menge, J. A. (1984). In *VA Mycorrhiza* (C. Ll. Powell and D. J. Bagyaraj, eds), pp. 187–203. CRC Press, Boca Raton, FL.

Mosse, B. (1972). *Rev. Ecol. Biol. Soc.* **9**, 529–537.

Mosse, B. (1981). Hawaii Institute of Tropical Agriculture and Human Resource, Research Bulletin **194**, pp. 5–82. Honolulu, HI.

Murphy, J. and Riley, J. P. (1962). *Anal. Chim. Acta.* **27**, 31–35.

Pacovsky, R. S. and Fuller, G. (1986). *Plant Soil* **95**, 361–377.

Pacovsky, R. S. and Bethlenfalvey, G. J. (1982). *Plant Soil* **68**, 143–147.

Porter, W. M. (1979). *Austral. J. Soil Res.* **17**, 515–519.

Powell, C. Ll. (1980). *Soil Biol. Biochem.* **12**, 247–250.

Ross, J. P. (1971). *Phytopathology* **61**, 1400–1403.

# 18

# Inoculum Production of Vesicular-arbuscular Mycorrhizal Fungi for Use in Tropical Nurseries

FALKO FELDMANN and ELKE IDCZAK

*Institut für Angewandte Botanik, Universität Hamburg, Marseiller Strasse 7, D-2000 Hamburg 36, Germany*

## I.  Introduction

The use of vesicular-arbuscular mycorrhizal fungi in the production of plants for tropical use has already been tested and described by some authors (e.g. Sieverding and Saif, 1984; Hayman, 1987). Under tropical conditions (poor soils, extreme temperatures and high humidity or drought, presence of many pathogens) the use of vesicular-arbuscular mycorrhizal fungi can be especially beneficial to perennial crops which are produced in nurseries and then planted in the fields. Vesicular-arbuscular mycorrhizal fungi not only give improved growth of different tropical crops, but also show the potential to increase resistance of the

METHODS IN MICROBIOLOGY
VOLUME 24   ISBN 0-12-521524-X

Copyright © 1992 by Academic Press Limited
All rights of reproduction in any form reserved

host to root pathogens (Zambolim, 1986) and to reduce the severity of foliar disease (Feldmann, 1990).

Most tropical perennial crops are propagated from seeds or cuttings in nurseries before they are transferred to the plantations. These crops are normally hosts for vesicular-arbuscular mycorrhizal fungi. A list of tropical plants that act as hosts for vesicular-arbuscular mycorrhizal fungi can be found in Janos (1987). In the tropical countries Colombia and Brazil vesicular-arbuscular mycorrhizal fungi are already in some cases an important factor in commercial cultivation systems, diminishing the cost of plant production (e.g. M. T. Lin, pers. commun.) not only because the plants can be produced using smaller amounts of fertilizers, but also because there are fewer losses during production. In general mycorrhizal plants are of better quality than non-mycorrhizal ones. They are more resistant to many stress factors when they are transferred to the plantations (see e.g. Schönbeck, 1987).

Why is the use of vesicular-arbuscular mycorrhizal fungi in plant production systems not yet a common procedure? Certainly the difficulties involved in the selection of the fungi which can be used are an important factor. But probably the most important reason for their current lack of use is that potential users of vesicular-arbuscular mycorrhizal fungi are not able to handle the inoculum production themselves and there are no producers of inoculum who can guarantee a standard quality and low cost of production and transport to the users. These problems can be solved by the use of an inorganic material as a carrier for the infection units of the vesicular-arbuscular mycorrhizal fungi—so-called "expanded clay" (see Fig. 1). The method of producing inoculum with expanded clay was first described by Dehne and Backhaus (1986). In this chapter we describe our experience with this method, and how we adapted it to the conditions of the humid tropical region of Amazonas, Brazil.

With expanded clay as the carrier for propagules of vesicular-arbuscular mycorrhizal fungi it is easily possible to produce an inoculum of high quality on a large scale. High quality means first that the inoculum causes a rapid colonization of the plant root system by the vesicular-arbuscular mycorrhizal fungi. This is guaranteed by an inoculum of high infectivity. In the case of expanded clay as a carrier for propagules of vesicular-arbuscular mycorrhizal inoculum, infectivity does not necessarily mean large numbers of spores. Mycelia are very often as important as spores for root colonization (see Grunewaldt-Stöcker and Dehne, 1989). Because of this it is necessary to include a method for the estimation of all infection units (spores and mycelia). Only by such a method does the control of the infectivity of the produced inoculum

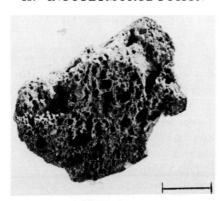

**Fig. 1.**  Expanded clay particle. For inoculum production of vesicular-arbuscular mycorrhizal fungi broken expanded clay can be used as an organic carrier material for the fungal propagules. The fungi are sporulating inside the cavities of the clay particles. The infectivity of the inoculum depends not only on the fungal spores, but also on fungal mycelia on the surface of the particles. Bar, 0.2 cm.

become convincing. For this purpose the use of the most probable number method (MPN test), described below, is recommended.

The second very important characteristic of an inoculum of high quality is that it should be uncontaminated or contain only small quantities of plant pathogenic micro-organisms. Depending on the purpose for which the inoculum is produced, a low level of contamination can be ignored. If, for example, it is to be used for the inoculation of young cuttings for seedlings in nurseries, it has to be pure because these plants are in most cases extremely susceptible to pathogenic fungi during the first weeks of growth. The use of expanded clay as a carrier for infection units of vesicular-arbuscular mycorrhizal fungi makes it possible to avoid contamination of the substrate by pathogens during inoculum production and to decontaminate it selectively if contamination has taken place during production.

The method described here can also be used in temperate climates. Some of the special adaptations to tropical conditions are normally not necessary in this case (e.g. reduction of evaporation from the pots). Inoculum produced by the procedure described below proved to be very suitable for successful colonization of the following tropical perennial crops (annual crops are not listed): *Hevea brasiliensis* (rubber, seringueira) *Carica papaya* (papaya, mamao), *Manihot esculenta* (cassava, manioca), *Citrus* spp. (e.g. orange, laranja), *Theobroma cacao* (cacao), *Theobroma grandiflorum* (cupuacu), *Elaia guianensis* (oil palm, dende), *Guilielma gassipaes* (pupunha).

## II.  Inoculum production with inorganic carrier material

### A.  Materials

Broken expanded clay (Leca) was used as the carrier material for the
infection units of the vesicular-arbuscular mycorrhizal fungi, as proposed
by Dehne and Backhaus (1986). The particle size was between 4 and
8 mm in diameter. Other expanded clay material seems not to work so
well, perhaps because of a different surface structure.

The host plant used was normally corn (*Zea mays*, variety "Badischer
Landmais" or the Brazilian hybrid BR 5102).

The vesicular-arbuscular mycorrhizal fungi available in mass production with expanded clay are listed in Table I.

The infectivity of the inoculum depends on the chosen isolate of the
vesicular-arbuscular mycorrhizal fungus; isolates with between 0 and 5
infection units per $cm^3$ in expanaded clay showed higher numbers of
propagules when cultivated in sand.

### B.  Initial phase of inoculum production

The first step in inoculum production of vesicular-arbuscular mycorrhizal
fungi is the colonization of host roots by the selected fungi. The plants
are grown in a sandy substrate and planted into the clay substrate later
in the second phase of mass production. Small quantities of defined
fungal material or small samples of undefined soil are sufficient for the

**TABLE I**

Inoculum infectivity of different vesicular-arbuscular mycorrhizal fungal species
in expanded clay as a carrier for the infection units

| Vesicular-arbuscular mycorrhizal fungus | Isolate | Source | Inoculum infectivity[a] |
|---|---|---|---|
| *Glomus etunicatum* | D13 | Dehne | *** |
| | T 6 | Dehne | *** |
| | 476 | Schenck | * |
| *Glomus intraradices* | 208 | Schenck | ** |
| | 267 | Schenck | *** |
| *Glomus manihotis* | W 9 | Weritz | ** |
| *Gigaspora margarita* | 185 | Schenck | * |

[a]The infectivity of the inoculum was assessed by the MPN-method (described below).
The meaning of the symbols is: ***, more than $30\,IU\,cm^{-3}$; **, $5-30\,IU\,cm^{-3}$; *,
$0-5\,IU\,cm^{-3}$. IU are infection units (spores, vesicles and mycelia (see below)).

colonization of the host plants. Pure defined isolates of vesicular-arbuscular mycorrhizal fungi can be produced by single-spore cultivation, using material received from other laboratories or bought from specialist companies.

The procedure is as follows:

- Seeds of the host plants are prepared for use in inoculum production by washing for 12 h under running water to remove all pesticides. After this period the seeds are thoroughly soaked and can be sown.
- Small pots of size 7 cm × 10 cm × 10 cm are cleaned intensively, especially if they have been used previously.
- A 3 cm layer of sand is added to the bottom of the pot. A 1 cm depth of substrate containing the start inoculum is then layered on the sand.
- The prepared seeds are sown and covered with sand. The optimal number of plants is 10 per pot.

After a few days the plant roots grow through the inoculum layer and later fill the whole volume while the fungi intensively colonize them. After three weeks the initial phase of inoculum production normally ends and the host plants are ready to be prepared for the second phase of mass production using expanded clay as substrate (see Fig. 2).

*Remarks*

If fungal material of low infectivity is used in this initial phase of inoculum production, the roots of the host plants may not be well colonized after 21 days of growth. In this case it is necessary to leave the host plants in the small pots for five weeks before mass production starts. If mixed populations of vesicular-arbuscular mycorrhizal fungi are used as start inoculum in the initial phase the fungi which will be multiplied in the following mass production may be selected by the speed with which they colonize the roots.

The cultivation parameters (e.g. irrigation, fertilization, etc.) in the initial phase can be deciding factors for the future quality of the inoculum. In our experience a substrate temperature of between 22 and 24 °C is optimal. In growth chambers the air temperature may be 24 °C while that of the substrate is much lower because of the cooling stream of air! Mycorrhizal fungi normally are less infective under such conditions.

The humidity of the substrate should be a little below field capacity.

The young plants require fertilizer, especially if they are grown for

F. FELDMANN and E. IDCZAK

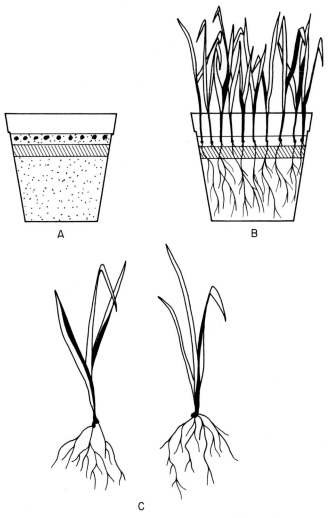

**Fig. 2.** The initial phase of inoculum production. (A) Small pots are partially filled with sand, the starter inoculum layered on the top. Corn seeds are then sown and covered with sand. (B) The corn seedlings are grown for 3–4 weeks in the small pots until their roots are colonized by the vesicular-arbuscular mycorrhizal fungi. (C) When the roots are colonized they are separated from each other by washing them thoroughly under running water. The roots can then be surface sterilized.

more than two weeks in small pots. We use the nutrient solution described below for mass production but diluted 1:5 and applied twice a week at 20 ml per pot (or more, if the plants show symptoms of deficiency).

An important factor in the initial phase of inoculum production is the period and intensity of irradiance. We obtain the best colonization under irradiance of an intensity of 12 000–15 000 lx from mercury vapour lamps for 14 h day$^{-1}$ (200–260 $\mu$E m$^{-2}$ s$^{-1}$). Dehne and Backhaus (1986) recommended an intensity of 5000 lx for 16 h day$^{-1}$. We find this to be insufficient for optimal root colonization by mycorrhizal fungi.

## C.   Mass production

### 1.   Separation of the host plants

In the first step of mass production the plants from the initial phase have to be separated from each other. This is easily achieved if they are washed carefully under running water.

### 2.   Sterilization of the host plants

The root systems of host plants from the initial phase of inoculum production can be surface sterilized after separation to avoid the transfer of pathogenic micro-organisms into the expanded clay substrate. Dehne and Backhaus (1986) propose ethanol (70%) for 2 min against *Pythium ultimum*; we use—if necessary—a solution of sodium hypochlorite (2%) for 2 min. If the initial phase of inoculum production is quite short (up to three weeks) and the conditions controlled (as in growth chambers) it is normally not necessary to surface sterilize the root systems of the host plants before planting them into the expanded clay.

### 3.   Preparation of expanded clay for mass production

Before the host plants are planted into the expanded clay the substrate should be washed. Pasteurization can be an advantage, but is not always necessary. Only if a test with agar plates shows contamination with pathogenic or saprophytic fungi is fumigation essential. We sterilize the expanded clay twice for 4 h with steam. Between the two fumigations the clay is cooled down to 20 °C.

After washing and sterilization the expanded clay is placed into 5-litre pots. Larger containers are also practical. We prefer the 5-litre size because the root systems of two corn plants grow intensively through the whole volume of the pots during the period of inoculum production.

### 4. Planting of the host plants

Two host plants whose root systems have been colonized by the vesicular-arbuscular mycorrhizal fungi in the initial phase are planted into each pot which has been filled with 5 litres of expanded clay. This has to be done very carefully to avoid damage to the roots by the very rough clay particles. Afterwards the substrate is watered thoroughly to fix the plants.

### 5. Cultivation parameters

Five basic requirements must be fulfilled to allow the production of inoculum of high quality:

(a) The illumination of the plants must be optimal.
(b) The optimal temperature range of the fungi in the substrate should be exceeded only for short periods of time.
(c) Long-term drought and waterlogging of the substrate must be avoided.
(d) The plant nutrition must be optimal for the symbiosis.
(e) The plants must be treated against pathogens.

*(a) Optimal illumination.* In temperate climates mass production independent of environmental conditions is possible in greenhouses with additional illumination. Dehne and Backhaus (1986) used sodium vapour high-pressure lamps with an intensity of 5000 lx for 16 h per day as an additional light source. We found that 14 h of light per day with an intensity of 10 000 lx or more ($200$–$260 \ \mu E \ m^{-2} \ s^{-1}$) generated by mercury vapour lamps improved the quality of inoculum produced in the greenhouse. In temperate climates during the summer we successfully produced inoculum of high quality and in large quantities in small plastic greenhouses. The covering material of the greenhouse was a film (Wepelen GF3, Werra-Plastik, Philippsthal, FRG) which allows sufficient light transmission above 350 nm (Idczak *et al.*, 1988). Therefore an additional light source was not necessary.

In the humid tropics, however, several complications with regard to interactions between environmental parameters and symbiont growth occur which require special adaptations. For example, intensive irradiation results in high temperatures in the greenhouses, which are not normally air-conditioned. This has a detrimental effect on plant growth. Shading the greenhouses results in decrease of irradiation. Therefore mass production outdoors is advisable. With varieties of corn adapted to these conditions the whole solar spectrum can be utilized. When the pots with expanded clay are placed outside (if necessary an acclimatisation period of the plants to the outdoor conditions must be taken into account) they must not have any contact with the ground so that mud splashes and soil fauna cannot contaminate them and the purity of the produced inoculum can be assured (see Fig. 3).

*(b) Temperature of the substrate.* Under direct irradiation the temperature of the substrate might attain a level which has a negative influence on development of the symbiosis. To avoid this shading of the pots is recommended, e.g. by placing them into a wide, well-aerated trough (see Fig. 3).

*(c) Irrigation.* A problem in the use of expanded clay as a substrate is the low potential for water transport from the bottom of the container to the upper layers of the substrate (larger particle size in the upper layers). On the other hand, much water can be held in the lower layers of the substrate, which consists mainly of smaller particles. Because of this the lower layers may be temporarily too wet. Regular drying out can be achieved by the use of clay pots instead of plastic containers. To reduce evaporation from the surface of the substrate the pots should be

**Fig. 3.** Trough made of wood. A trough like the one shown guarantees that contamination of the inoculum by soil fauna and splashing mud is largely avoided during inoculum production. It can be constructed to provide shade for the pots while retaining adequate air flow around them.

covered with a white plastic film with holes which limits gas exchange between substrate and air but does not prevent it completely. In this way the plants need watering only once a day or less. The amount of water necessary for the plants depends on the prevailing transpiration and evaporation. Substrate humidity should always be kept a little below the field capacity of the expanded clay.

*(d) Plant nutrition.* Adequate plant nutrition is guaranteed by use of a fertilizer (modified after Hoagland and Arnon, 1938) added to the water three times a week. In the nutrient solution the proportion between the main macronutrients should be as follows:

$$N:P:K \quad as \quad 1:0.85:3.35$$

with a concentration of

$$83 \text{ mg N litre}^{-1}, \quad 31 \text{ mg P litre}^{-1} \quad and \quad 232 \text{ mg K litre}^{-1}$$

Two to three times a month a stock solution is prepared which contains

| | |
|---|---|
| $KNO_3$ | 50.0 g litre$^{-1}$ |
| $Ca(NO_3) \cdot 4 H_2O$ | 11.6 g litre$^{-1}$ |
| $KH_2PO_4$ | 12.4 g litre$^{-1}$ |
| $MgSO_4 \cdot 7 H_2O$ | 49.3 g litre$^{-1}$ |
| Fe-citrate/Fe EDTA | 8.0 g litre$^{-1}$ |
| $MnSO_4 \cdot 1 H_2O$ | 0.15 g litre$^{-1}$ |
| $ZnSO_4 \cdot 7 H_2O$ | 0.02 g litre$^{-1}$ |
| $CuSO_4 \cdot 5 H_2O$ | 0.08 g litre$^{-1}$ |
| $Na_2B_4O_7 \cdot 10 H_2O$ | 0.2 g litre$^{-1}$ |
| $(NH_4)_6Mo_7O_{24} \cdot 4 H_2O$ | 0.2 g litre$^{-1}$ |
| $CoNO_3 \cdot 6 H_2O$ | 0.25 g litre$^{-1}$ |

The pH of the nutrient solution is adjusted to 5.5 with HCl.

A commercially available fertilizer (Flory 9) with a similar composition of nutrients as given above can be applied to the plants for inoculum production with expanded clay. It may be anticipated that other commercial fertilizers, too, are suitable to circumvent the time-consuming preparation of the nutrient solution. It is critical to make sure that the concentrations of P and N do not exceed maximum levels of 70 ppm for P and 50 ppm for N in the substrate as this might be detrimental to the colonization of the roots.

*(e) Treatment against pathogens.* Once a month the plants are watered with an added fungicide (Previcur, 0.15%) selective for oomycetes such as *Pythium*, an important plant pathogen which can occur in the

substrate during inoculum production. No effect of Previcur on the mycorrhizal fungi has been detected. Other fungicides can be used, only with caution! Negative effects have been shown for Captan, Captafol, Thiabendazol, Formaldehyde, Maneb, Chlorothalonil (Nemec, 1980) and Benomyl (Fitter and Nichols, 1988). Our preliminary investigations show negative effects of the fungicides Bayleton, Cercobin and (in higher concentrations) of the herbicides Gramoxon and Round up (for further data, see Nemec, 1987).

Sometimes contamination of the substrate by saprophytic or pathogenic fungi can be observed during inoculum production. The presence of saprophytes and pathogens can be revealed by a growth test on agar. If contamination has taken place it is possible to treat the expanded clay with ethanol (70%) as Dehne and Backhaus (1986) propose. We have observed good results by using sodium hypochlorite (2%) for 2–5 min and washing afterwards. However, this procedure also reduces the infectivity of the inoculum. It is much less complicated and cheaper to avoid contamination during inoculum production by maintenance of clean conditions and appropriate irrigation.

## 6.   Growth period and harvest

Mass production of inoculum in containers with expanded clay as substrate takes 3–4 months, depending on the growth of the host plants. A macroscopical check on root colonization by vesicular-arbuscular mycorrhizal fungi is possible as corn roots often show an increasing yellowish colouring as the symbiosis develops. But to make sure that the roots really are colonized they should be cleared and stained for detection of the fungi (see below). If the roots are well colonized no further irrigation is necessary. The drought stress will stimulate spore production. After the first week of the drought period the plants are cut. Three weeks later the substrate should be spread out in a thin layer to be air-dried rapidly (see Fig. 4).

An important step is the separation of the roots from the expanded clay. We have repeatedly noticed increasing occurrence of pathogenic and saprophytic fungi if the roots remain in the substrate. If the inoculum produced shows high purity, i.e. no or little contamination by pathogens, the remaining roots increase the infectivity of the inoculum as they act as an additional source of infection units. Therefore we recommend testing the contamination of the inoculum by pathogens first and then deciding whether the roots should be separated from the expanded clay or not. Rapid drying of the substrate reduces growth of saprophytic fungi.

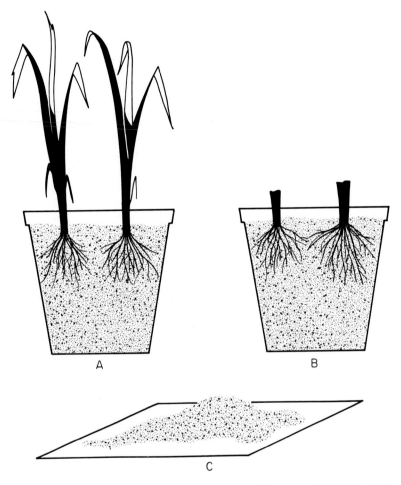

**Fig. 4.** The mass production of vesicular-arbuscular mycorrhizal fungal inoculum. (A) Two plants are grown in each pot (5-litre volume) for 3–4 months. (B) After the growth period the plants are not watered further. One week later they are cut. The substrate is dried for three weeks in the pots. (C) After the drought period the expanded clay is spread out in a thin layer and air-dried rapidly. The dry inoculum can be stored for a long time in a cool, dry place.

## 7. Storage of the inoculum

Storage of the inoculum under dry and cool conditions is possible for several years without significant loss of infectivity. As an example we found that the infectivity of an inoculum (*Glomus etunicatum*) which was stored at 20–23 °C and 30–50% relative humidity for three years was reduced by only 10–15% .

## III.   Control of inoculum infectivity

To determine the infectivity of a produced inoculum is of great importance for application in the field and in nurseries, as well as for scientific purposes. In practice one always has to deal with economic aspects. In the case of the application of vesicular-arbuscular mycorrhizal inoculum in the field and in nurseries costs can be minimized by using the minimum amount of inoculum needed to guarantee the colonization of the host plant. Therefore the infectivity of the inoculum has to be tested. If the inoculum is used for scientific research its infectivity must be known to standardize experimental conditions.

A method often used for the estimation of the infectivity is to determine the spore numbers in the substrate. This can be carried out quickly by a wet sieving and decanting procedure and a few centrifugation steps. When expanded clay is used as a carrier for the vesicular-arbuscular mycorrhizal inoculum even the determination of the spore number in the substrate is impossible as spores are mainly formed inside the numerous cavities of the expanded clay, from which they cannot readily be removed (see Grunewaldt-Stöcker and Dehne, 1989). This circumstance and the fact that many other propagules (mycelia, vesicles) are attached to the substrate require another method for determining the infectivity of the inoculum. The most probable number method (MPN test) has been adapted to the special conditions that exist when expanded clay is used as a carrier for fungal propagules.

### A.   The MPN test

This test determines the infectivity of different dilutions of an inoculum and finally calculates the number of propagules in the original sample by mathematical means.

To determine the most probable number of propagules per $cm^3$ of expanded clay the following are required: 20 pots (volume $70 cm^3$); 40 pre-soaked corn seeds; sterile expanded clay; and the expanded clay inoculum to be tested. From the original inoculum dilutions of 0, 1/10, 1/100 and 1/1000 are prepared. The 1/10 dilution is made by mixing one part of the inoculum and nine parts of sterilized expanded clay. One part of this mixture is mixed together with nine parts of sterilized expanded clay to give the next higher dilution, and so on. It is very important to mix the inoculum and the sterile expanded clay thoroughly to guarantee its homogeneous distribution.

The dilution samples are filled into pots (five pots per dilution). Then two pre-soaked corn seeds are planted in each pot. This bioassay is incubated for 30 days at 23 °C with 14 h of light per day ($>200 \ \mu E \ m^{-2} \ s^{-1}$). The plants are watered moderately. After the first 10 days 20 ml fertilizer of the nutrient solution described above (1/5 diluted) are added at weekly intervals. The plants are incubated for exactly four weeks and all root systems are then harvested separately. To allow microscopical assessment of root colonization by vesicular-arbuscular mycorrhizal fungi all the roots are cleared for 10 min at 90 °C in 10% KOH, rinsed with tap water and stained in 0.05% trypan blue in glycerol–lactic acid–distilled water (2:2:1) for 15 min at 90 °C.

## B.   Calculation of the MPN of propagules

Microscopical examination shows if a root system is colonized by vesicular-arbuscular mycorrhizal fungi. For the five replicates in each of the four dilutions one might obtain a combination of numbers such as:

$$5 \quad 5 \quad 3 \quad 2$$

This means that all root systems are colonized in dilutions 0 and 1/10, three root systems are colonized in the dilution 1/100, and two in the dilution 1/1000.

For the calculation of the MPN of propagules only three numbers of the given combination are required. The first number (N1) is that corresponding to the least concentrated dilution in which all (or the greatest number of) root systems are colonized. The two other numbers (N2 and N3) are those corresponding to the next two higher dilutions. In our example it would be the combination:

$$N1 \quad N2 \quad N3$$

$$5 \quad 3 \quad 2$$

Making use of Table II with these values of N1, N2 and N3, we find the value given for the combination 5 3 2 is 1.4. This is the most probable number of propagules in the second dilution of the combination used, which in our example is 1/100. To obtain the MPN for the original sample, the value is multiplied by the appropriate dilution factor. In our case the original sample would have an MPN of

$$1.4 \times 100 = 140 \text{ propagules cm}^{-3}$$

**TABLE II**

Most probable numbers for use with 10-fold dilutions and five pots per dilution
(Cochran, 1950)

| N1 | N2 | Most probable number for indicated values of N3 | | | | | |
|---|---|---|---|---|---|---|---|
| | | 0 | 1 | 2 | 3 | 4 | 5 |
| 0 | 0 | – | 0.018 | 0.036 | 0.054 | 0.072 | 0.090 |
| 0 | 1 | 0.018 | 0.036 | 0.055 | 0.073 | 0.091 | 0.11 |
| 0 | 2 | 0.037 | 0.055 | 0.074 | 0.092 | 0.11 | 0.13 |
| 0 | 3 | 0.056 | 0.074 | 0.093 | 0.11 | 0.13 | 0.15 |
| 0 | 4 | 0.075 | 0.094 | 0.11 | 0.13 | 0.15 | 0.17 |
| 0 | 5 | 0.094 | 0.11 | 0.13 | 0.15 | 0.17 | 0.19 |
| 1 | 0 | 0.02 | 0.04 | 0.06 | 0.08 | 0.10 | 0.12 |
| 1 | 1 | 0.04 | 0.061 | 0.081 | 0.10 | 0.12 | 0.14 |
| 1 | 2 | 0.061 | 0.082 | 0.10 | 0.12 | 0.15 | 0.17 |
| 1 | 3 | 0.063 | 0.10 | 0.13 | 0.15 | 0.17 | 0.19 |
| 1 | 4 | 0.11 | 0.13 | 0.15 | 0.17 | 0.19 | 0.22 |
| 1 | 5 | 0.13 | 0.15 | 0.17 | 0.19 | 0.22 | 0.24 |
| 2 | 0 | 0.045 | 0.068 | 0.091 | 0.12 | 0.14 | 0.16 |
| 2 | 1 | 0.068 | 0.092 | 0.12 | 0.14 | 0.17 | 0.19 |
| 2 | 2 | 0.093 | 0.12 | 0.14 | 0.17 | 0.19 | 0.22 |
| 2 | 3 | 0.12 | 0.14 | 0.17 | 0.20 | 0.22 | 0.25 |
| 2 | 4 | 0.15 | 0.17 | 0.20 | 0.23 | 0.25 | 0.28 |
| 2 | 5 | 0.17 | 0.20 | 0.23 | 0.26 | 0.29 | 0.32 |
| 3 | 0 | 0.078 | 0.11 | 0.13 | 0.16 | 0.20 | 0.23 |
| 3 | 1 | 0.11 | 0.14 | 0.17 | 0.20 | 0.23 | 0.27 |
| 3 | 2 | 0.14 | 0.17 | 0.20 | 0.24 | 0.27 | 0.31 |
| 3 | 3 | 0.17 | 0.21 | 0.24 | 0.28 | 0.31 | 0.35 |
| 3 | 4 | 0.21 | 0.24 | 0.28 | 0.32 | 0.36 | 0.40 |
| 3 | 5 | 0.25 | 0.29 | 0.32 | 0.37 | 0.41 | 0.45 |
| 4 | 0 | 0.13 | 0.17 | 0.21 | 0.25 | 0.30 | 0.36 |
| 4 | 1 | 0.17 | 0.21 | 0.26 | 0.31 | 0.36 | 0.42 |
| 4 | 2 | 0.22 | 0.26 | 0.32 | 0.38 | 0.44 | 0.50 |
| 4 | 3 | 0.27 | 0.33 | 0.39 | 0.45 | 0.52 | 0.59 |
| 4 | 4 | 0.34 | 0.40 | 0.47 | 0.54 | 0.62 | 0.69 |
| 4 | 5 | 0.41 | 0.48 | 0.56 | 0.64 | 0.72 | 0.81 |
| 5 | 0 | 0.23 | 0.31 | 0.43 | 0.58 | 0.76 | 0.95 |
| 5 | 1 | 0.33 | 0.46 | 0.64 | 0.64 | 1.1 | 1.3 |
| 5 | 2 | 0.49 | 0.70 | 0.95 | 1.2 | 1.5 | 1.8 |
| 5 | 3 | 0.79 | 1.1 | 1.4 | 1.8 | 2.1 | 2.5 |
| 5 | 4 | 1.3 | 1.7 | 2.2 | 2.8 | 3.5 | 4.3 |
| 5 | 5 | 2.4 | 3.5 | 5.4 | 9.2 | 16.0 | – |

## C.  Critical factors in the MPN test

### 1.  The host plant

The choice of the host plant used for the MPN test can have a dramatic influence on the results. A strict specificity between a vesicular-arbuscular mycorrhizal fungus and certain plant species does not exist but some preferences for plant species and for varieties of a particular species by defined vesicular-arbuscular mycorrhizal isolates have been found (Azcon and Ocampo, 1981). The best choice of a host plant for the MPN test would be a plant species for which the vesicular-arbuscular mycorrhizal inoculum is produced. In cases of trees and plants cultivated from cuttings this may not be practical because of the time taken for their development. Therefore fast-growing plants of low specificity such as corn or onion are frequently used for the bioassy. These are known to be colonized heavily by a wide range of vesicular-arbuscular mycorrhizal fungi. If comparing the infectivity of different vesicular-arbuscular mycorrhizal inocula, the same host plant must be used in any case.

### 2.  The incubation period

The propagules of the inoculum tested need some time to colonize the roots of their host plants—time for germination of the spores, time for growth of mycelium and time for penetration of the roots. The time period necessary for these developmental stages may vary between different vesicular-arbuscular mycorrhizal isolates. If the chosen incubation period for the MPN test is too short the number of propagules will be underestimated—if the incubation period is too long infections occur in the root systems of all plants in the MPN test, even in the higher dilutions, and the number of propagules is overestimated. A compromise is inevitable. We found the incubation period of 30 days recommended by Porter (1979) suitable for all purposes.

### 3.  The distribution of the inoculum

Because of the relatively large particle size of expanded clay (4–8 mm diameter) the homogeneous distribution of propagules in dilutions is difficult. As a consequence, the number of propagules may be underestimated if the root system does not grow through the whole pot and misses the few expanded clay particles in a high dilution which carry vesicular-arbuscular mycorrhizal propagules.

## IV. Discussion

Expanded clay as the carrier for the vesicular-arbuscular mycorrhizal propagules allows easy production of qualitatively excellent inoculum. Several advantages speak in favour of the use of expanded clay for inoculum production compared to conventional substrates such as soil or peat–vermiculite mixtures:

- The purity of the inoculum can be maintained without difficulty, contamination by pathogens can be largely avoided and decontamination afterwards is possible.
- The low weight of expanded clay allows easy long-distance transport and distribution of large quantities of inoculum.
- Inoculum can be stored for a long period without loss of infectivity.
- Application is very simple. The large particle size facilitates control of a homogeneous distribution of the inoculum in the substrate of inoculated plants.
- The high numbers of propagules (30–50 propagules per 8–10 particles of expanded clay, i.e. per $cm^3$ expanded clay) permit the use of low doses of inoculum. Successful inoculation of a host plant requires 2–5 vol% of expanded clay with an MPN of approximately 30 propagules per $cm^3$.

All these considerations highlight the possibility of introducing the use of mycorrhizal fungi easily where their effect is likely to be greatest—in tropical climates, where several physical stresses lead to poor conditions for the cultivation of plants. Users can be taught the method of application very quickly because the handling of the inoculum is very simple. The method of inoculum production described above can be used immediately by specialized staff of centralized agricultural institutions, which can also distribute the expanded clay inoculum to farmers without great effort.

The use of expanded clay inoculum is of most value in nurseries, where vesicular-arbuscular mycorrhizal fungi can be introduced to different plants successfully. Application in the field requires the use of selected and adapted fungi. Mycorrhizal plants which are produced in nurseries are less susceptible to transplantation stress than non-mycorrhizal plants (see e.g. Hayman, 1987). This is also the case if the fungi used are not adapted to the conditions prevailing at the final planting site, although these fungi are often suppressed by competitive indigenous fungi after some time (Feldmann, 1990).

Expanded clay inoculum is suitable for field application in both temperate climates (Baltruschat, 1987) and the tropics. Even smallest quantities of inoculum (less than $1 \, cm^3$ of expanded clay per two corn seeds) caused significant additional yield of mycorrhizal plants on Brazilian latosols (J. Weritz, pers. commun.) Expanded clay used in this way as starter inoculum offers the possibility for field production of large quantities of soil inoculum. In a yellow latosol the number of vesicular-arbuscular mycorrhizal propagules per $cm^3$ soil was increased from 4 to 75 with corn as the host plant. The fungi involved, *Glomus etunicatum* and *G. manihotis*, were bound to the expanded clay particles and could be harvested for use as vesicular-arbuscular mycorrhizal inoculum (Feldmann, 1990).

## Acknowledgements

We thank the Brazilian Centro de Pesquisa Agroflorestal da Amazonia, Manaus, Brazil, for the opportunity to carry out the experiments described. We thank Dr H.-W. Dehne, Bayer AG, Leverkusen, Germany, Dr N. C. Schenck, INVAM, Gainesville, FL, USA and Dipl.-Agr. J. Weritz, Symbionta, Gifhorn, Germany, who provided us with the starter inocula of the vesicular-arbuscular mycorrhizal fungi used in the experiments.

## References

Azcon, R. and Ocampo, J. A. (1981). *New Phytol.* **87**, 677–685.
Baltruschat, H. (1987). In *Proceedings of the 7th North American Conference of Mycorrhizae*, 3–8 May 1987, Gainesville, FL (D. M. Sylvia, L. L. Hung and J. H. Graham, eds), p. 16.
Cochran, W. G. (1950). *Biometrics* **6**, 105–116.
Dehne, H.-W. and Backhaus, G. S. (1986) *Z. Pflanzenkr. Pflanzenschutz* **93**, 415–424.
Feldmann, F. (1990). Die Mykorrhiza des Kautschukbaumes (*Hevea* spp.): Vorkommen am Naturstandort, Auswirkung auf das Resistenzverhalten, Verwendung im Plantagenbau. PhD thesis, Universistät Braunschweig, FRG.
Fitter and Nichols (1988). *New Phytol.* **110** (2), 201–206.
Grunewald-Stöcker, D. and Dehne, H.-W. (1989). *Z. Pflanzenkr. Pflanzenschutz* **96**, 615–626.
Hayman, D. S. (1987). *Ecophysiology of VA-mycorrhizal Plants* (G. R. Safir, ed.), pp. 171–192. CRC Press, Boca Raton, FL.
Hoagland, D. R. and Arnon, D. J. (1938). *The Water Culture Method for Growing Plants without Soil.* University of California College of Agriculture Circular No. 347.
Idcazak, E., Mattusch, P., Lieberei, R. and Crüger, G. (1988). *Nachrichtenbl. Deut. Pflanzenschutzd.* **40**, 183–186.

Janos, D. P. (1987). In *Ecophysiology of VA-mycorrhizal Plants* (G.R. Safir, ed.), pp. 107–134. CRC Press, Boca Raton, FL.

Nemec, S. (1980). *Can. J. Bot.* **58**, 522–526.

Nemec, S. (1987). In *Ecophysiology of VA-mycorrhizal Plants* (G. R. Safir, ed.), pp. 193–212. CRC Press, Boca Raton, FL.

Porter, W. M. (1979). *Austral. J. Soil Res.* **17**, 515–519.

Safir, G. R. (ed.) (1987). *Ecophysiology of VA-mycorrhizal Plants.* CRC Press, Boca Raton, FL.

Schönbeck, F. (1987). *Angew. Bot.* **61**, 9–13.

Sieverding, E. and Saif, S. R. (1984). CIAT Annual Review 02/1984.

Zambolim, L. (1986). In *ANAIS da I Reuniao Brasileira Sobre Microrrizas*, 11 a 14 de Novembro de 1985, Lavras, MG, 1986, pp. 76–99.

# 19

# Vesicular-arbuscular Mycorrhiza: Application in Agriculture

D. J. BAGYARAJ

*Department of Agricultural Microbiology, University of Agricultural Sciences,
Bangalore 560065, India*

## I. Introduction

Interest in vesicular-arbuscular mycorrhizal fungi has reached a peak in recent years. The ability of these fungi to produce dramatic responses in plant growth is well documented. However, the application of this technology to commercial production of food, fibre or fuel has been minimal. One of the main reasons for this is the difficulty of inoculum production, the fungi being obligate symbionts (Jeffries, 1987). There have been suggestions that we should not move too rapidly with

Copyright © 1992 by Academic Press Limited
All rights of reproduction in any form reserved

vesicular-arbuscular mycorrhizal inoculation of agricultural crops until more is known about the ecology and physiology of vesicular-arbuscular mycorrhizal endophytes (Carling and Brown, 1980; Hetrick, 1984; Bagyaraj, 1990) and indeed most field trials have been very empirical in the choice of the fungal inoculants, inoculation procedures, etc. This chapter deals with some of the methods used for exploiting the vesicular-arbuscular mycorrhizal symbiosis in agriculture.

## II.  Starter cultures, multiplication and storage

### A.  Starter cultures

The first step towards application of vesicular-arbuscular mycorrhizal technology is to obtain a good starter culture. Such cultures can be obtained from research workers in universities, research or industrial organizations. At present they can also be obtained from the International Collection of Vesicular-arbuscular Mycorrhizal Fungi (INVAM), University of West Virginia, 401 Brooks Hall, PO Box 6057, Morgantown, WV 26506-6057, USA.

Another approach is to isolate spores of vesicular-arbuscular mycorrhizal fungi from soil by wet sieving and decanting technique (Gerdemann and Nicolson, 1963). Such spores must be checked for the occurrence of mycoparasites. They are then surface sterilized and introduced into pot cultures by the use of the funnel technique (Nicolson, 1967). These pot cultures can be maintained in a greenhouse on a suitable host.

### B.  Multiplication

Techniques are available for the production of vesicular-arbuscular mycorrhizal inoculum in an almost sterile environment through nutrient film techniques, circulation hydroponic culture systems, aeroponic culture systems, root organ culture and tissue culture (see Menge, 1984; Nopamornbodi et al., 1988). For large-scale field trials, however, the only convenient method of producing large quantities of inoculum is by the traditional "pot culture" technique developed about 30 years ago by Mosse and Gerdemann (Wood, 1985).

Pot cultures are complex systems comprising host plants, mycorrhizal fungi, soil microflora and microfauna and supporting soil. Soil cultures can sometimes harbour root pathogens and thus can act as a source of disease. Soil inoculum, because of its bulk, can also create transport problems, and many countries restrict soil importation despite assurances of sterility. The problem is therefore to manipulate the host, substrate and the environment in such a way as to produce pot cultures yielding vesicular-arbuscular mycorrhizal inoculum with high inoculum potential, few contaminants, longer shelf life and easy transportation.

Several host plants including Sudan grass (*Sorghum bicolor* var. *sudanense*), Bahia grass (*Paspalum notatum*), Guinea grass (*Panicum maximum*), cenchrus grass (*Cenchrus ciliaris*), clover (*Trifolium subterraneum*), strawberry (*Fragaria* sp.), sorghum (*Sorghum vulgare*), maize (*Zea mays*), onion (*Allium cepa*), coprosma (*Coprosma robusta*) and coleus (*Coleus* sp.) have been studied for their suitability in producing vesicular-arbuscular mycorrhizal inoculum. Recently it was reported that Rhodes grass is the best host for mass production of *Glomus fasciculatum* (Sreenivasa and Bagyaraj, 1988a). In view of the disadvantages of using soil as the substrate for producing vesicular-arbuscular mycorrhizal inocula, it is desirable, where possible, to produce vesicular-arbuscular mycorrhizal inoculum in a partially, if not completely, artificial substrate. Recently we have found 1:1 perlite:soilrite (soilrite is a commercial preparation with perlite, vermiculite and peat moss of 1:1:1 proportion by volume) to be the best substrate for mass production of *G. fasciculatum* (Sreenivasa and Bagyaraj, 1988b). Dehne and Backhaus (1986) suggested the use of expanded clay for multiplying vesicular-arbuscular mycorrhizal fungi (see Feldmann and Idozak, Chapter 18, this volume). Moisture content and temperature of the substrate, as well as irradiance, mineral content, pot size, presence of mycoparasities and associated organisms can all influence mycorrhizal inoculum production (Bagyaraj, 1984; Menge, 1984) and hence should be evaluated for each fungal isolate.

## C.  Greenhouse sanitation

General greenhouse sanitation such as overall cleanliness, control of weeds, insects, fungal and bacterial pests and rodents is very important for pure inoculum production. Benches on the floor should be periodically washed with disinfectant solution. To prevent roots growing out of drainage holes and causing cross-contamination of cultures, it is advantageous to use double pots or pots with saucers. Care must be taken to prevent the hose pipe touching the potting mix when watering.

## D.  Storage

Pot cultures of vesicular-arbuscular mycorrhizal inoculum are usually harvested after about 2 months. The inoculum, consisting of the substrate and the roots chopped into small bits, is air-dried and packeted in polythene bags and stored. Ferguson and Woodhead (1982) observed that *Glomus fasciculatum* inoculum produced in this way was viable after four years storage. Attempts to freeze-dry vesicular-arbuscular mycorrhizal inoculum as a method of storage have not been successful (Crush and Pattison, 1975) but it can be preserved for up to six years using L-drying or cryopreservation techniques (Tommerup and Kidby, 1979).

## III. Mycorrhizal dependency of plants

Plants differ greatly in their need for and response to mycorrhizal infection. Gerdemann (1975) defined relative mycorrhizal dependency (RMD) as the degree to which a plant is dependent on the mycorrhizal condition to produce its maximum growth or yield at a given level of soil fertility. This was based on determining the extent of growth increase attributed to the mycorrhizal condition. Menge *et al.* (1978) proposed the following formula to calculate mycorrhizal dependency of crop plants using an inoculant vesicular-arbuscular mycorrhizal fungus in sterilized soil.

$$\text{RMD} = \frac{\text{Dry weight of mycorrhizal plant}}{\text{Dry weight of non-mycorrhizal plant}} \times 100$$

The drawback of this formula is that RMD values go beyond 100%, making if difficult to categorize crop plants.

Plenchette *et al.* (1983) proposed another formula to calculate relative field mycorrhizal dependency (RFMD) of crop plants under field conditions. They compared plants in fumigated and non-fumigated soils. This measures the extent of growth increase due to native endophytes. The calculated values lie between 0 and 100%.

$$\text{RFMD} = \frac{\text{Dry weight of mycorrhizal plant} - \text{Dry weight of non-mycorrhizal plant}}{\text{Dry weight of non-mycorrhizal plant}} \times 100$$

This formula is of great practical importance for ascertaining the response of crop plants to native mycorrhiza in fields before applying field inoculation to crops.

There have been many reports that the introduction of mycorrhizal fungi can improve plant growth under unsterile soil conditions in spite of the presence of native endophytes. Hence, Bagyaraj *et al.* (1988) proposed another formula which enables calculation of the mycorrhizal inoculation effect (MIE) to assess the growth improvement brought about by inoculation with a mycorrhizal fungus in unsterile soil with indigenous vesicular-arbuscular mycorrhizal fungi.

$$\text{MIE} = \frac{\text{Dry weight of inoculated plant} - \text{Dry weight of uninoculated plant}}{\text{Dry weight of inoculated plant}} \times 100$$

MIE is very useful for the assessment of the extent to which introduced fungi compete with native endophytes to bring about plant growth response. This information is of great value in practical agriculture, especially in developing countries, where farmers do not fumigate or sterilize either nursery or field sites.

## IV. Selection of efficient vesicular-arbuscular mycorrhizal fungi

Mycorrhiza workers rarely provide the rationale for selection of particular vesicular-arbuscular mycorrhizal endophytes in their experiments. There are a few "favourite" spore types for field inoculation including *Glomus fasciculatum*, *G. mosseae*, *G. etunicatum*, *G. tenue* and *Gigaspora margarita*. These fungi were probably chosen because of their ready availability. Ideally, vesicular-arbuscular mycorrhizal fungi selected for inoculation into agricultural soils must be able both to enhance nutrient uptake by plants and to persist in soils. Many of the characteristics required by inoculant fungi correspond with those essential for the success of inoculant strains of *Rhizobium* for legumes (Chatel *et al.*, 1968). These are infectiveness, effectiveness, extent of colonization and survival in the rhizosphere and in soil. An excellent discussion on the selection of vesicular-arbuscular mycorrhizal fungi for possible use in agriculture has been published by Abbott and Robson (1982) (see also Abbott *et al.*, Chapter 1, this volume).

Vesicular-arbuscular mycorrhizal fungi are known to improve plant growth by increasing nutrient uptake, primarily by increasing the volume of soil explored. Therefore, characteristics that could be associated with differences between them in their effectiveness and increasing nutrient uptake are: the ability to form extensive well-distributed mycelium in soil, the ability to form extensive colonization, the efficiency of absorbing phosphorus from soil solution, and the time that hyphae remain effective in transporting nutrients to the plant. It was suggested by Abbot and Robson (1982) that the initial selection of inoculant fungi may be conducted under controlled conditions; after that, selected vesicular-arbuscular mycorrhizal fungi can be evaluated in the field.

There have been a few attempts to select for efficient vesicular-arbuscular mycorrhizal fungi on a rational basis, such as ability to stimulate plant growth. Govinda Rao *et al.* (1983) suggested that several fungi can be screened for symbiotic response using a test host through pot culture, followed by microplot and then field trials. Seedlings were first raised in unsterile nursery soils containing indigenous mycorrhizal fungi to which different test vesicular-arbuscular mycorrhizal fungi had been added. It was important to add the same number of infective propagules of different fungi based on most probable number counts (Porter, 1979). Once the infection had developed, the seedlings were transplanted to experimental field plots and plant growth was monitored. This technique followed precisely the procedure used by farmers in India and many Third World countries. It is extremely good for selecting fungi for the pre-inoculation of transplanted crops raised in unsterile soils. The fungi thus selected will compete well with indigenous vesicular-arbuscular mycorrhizal fungi. In this selection process it is essential to include vesicular-arbuscular mycorrhizal fungi isolated from the root zone of the test host. This procedure for selecting efficient fungi has led to the selection of inoculant vesicular-arbuscular mycorrhizal fungi for many economically important forest tree species such as *Leucaena* spp. (Bagyaraj, *et al.*, 1989), *Tamarindus indica* (Reena and Bagyaraj, 1990a), *Acacia nilotica*, *Calliandra calothyrsus* (Reena and Bagyaraj, 1990b); root stocks of citrus (Vinayak and Bagyaraj, 1990) and mango (Balakrishna Reddy and D. J. Bagyaraj, unpubl. res.). Inoculation of citrus root stocks with efficient vesicular-arbuscular mycorrhizal fungi made the plants ready for budding 4–6 months early, a feature of great importance in citriculture.

Hall (1984), Gianinazzi-Pearson *et al.* (1985) and Schubert and Hayman (1986), for example, have also outlined procedures for selecting vesicular-arbuscular mycorrhizal fungi for crops sown directly in the field and pastures. These experiments can be criticised because fungi were screened in greenhouse pot experiments, but they are still useful. Before laying down a field trial with onions, it was determined in a pot experiment that a mixture of *Glomus* spp. was the best inoculant for the onion cultivar Pukekohe Long Keeper. This *Glomus* inoculum was then shown to be the best inoculant in a field trial (Powell and Bagyaraj, 1982), amply vindicating the effort of prior fungal selection even for a directly field-sown crop.

## V.  Concept of host preference

It is well known that vesicular-arbuscular mycorrhizal fungi are not host-specific. Any vesicular-arbuscular mycorrhizal plant species can be infected by any vesicular-arbuscular mycorrhizal fungal species, but the degree of vesicular-arbuscular mycorrhizal infection and its effects can differ with different host–endophyte combinations (Mosse, 1973). Since the early endophyte screening trials of Mosse (1972) it has become more and more obvious that vesicular-arbuscular mycorrhizal fungi differ greatly in their symbiotic effectiveness. Though a particular vesicular-arbuscular mycorrhizal fungus can infect and colonize many host plants, it has a preferred host which exhibits maximum symbiotic response when colonized by that particular vesicular-arbuscular mycorrhizal fungus. This led to the concept of "host preference" in vesicular-arbuscular mycorrhizal fungi. Several recent studies support the notion of host preference by vesicular-arbuscular mycorrhizal fungi (Nemec, 1977; Bagyaraj et al., 1988; Reena and Bagyaraj, 1990a,b; Vinayak and Bagyaraj, 1990).

Attempts have been made to use the genetic variability in fungal efficiency and host response to select vesicular-arbuscular mycorrhizal fungal isolates that are able to improve plant production (Trouvelot et al., 1987; Reena and Bagyaraj, 1990a,b). Variations in the extent and effect of vesicular-arbuscular mycorrhizal colonization have also been linked to the genotype of the host plant (Krishna et al., 1985; Lackie et al., 1988). The recent isolation of myc⁻ plant mutants (Duc et al., 1989) and the discovery that vesicular-arbuscular mycorrhizal colonization is a heritable trait (Mercy et al., 1990) suggest the possibility of tailoring plant–fungus combinations for maximum efficiency. This may lead to developing plants that will form endomycorrhiza specifically with certain vesicular-arbuscular mycorrhizal fungi (Gianinazzi et al., 1989).

## VI.  Inoculation methods

Various methods for the introduction of vesicular-arbuscular mycorrhizal inoculum into field-grown crops have been examined.

### A.  Transplanted crops

Seedlings are raised in sterilized or unsterilized soil supplied with selected vesicular-arbuscular mycorrhizal fungi in small nursery beds or containers and planted out when mycorrhizal colonization is well established. This method has been successfully used by us to produce worthwhile and economic growth responses in agronomically important crops such as chilli, finger millet, tomato and tobacco (Govinda Rao et al., 1983; Sreeramulu and Bagyaraj, 1986), horticultural crops such as citrus, mango, asters and marigold (Bagyaraj *et al*, 1988; Vinayak and Bagyaraj, 1990), and forest tree species such as *Leucaena* spp., *Tamarindus indica, Acacia nilotica* and *Calliandra calothyrsus* (Bagyaraj *et al.*, 1989; Reena and Bagyaraj, 1990a,b).

This method can be used in many other transplanted crops important in agriculture, horticulture and forestry. The possibility of introducing efficient vesicular-arbuscular mycorrhizal fungi to cereals through forest tree species in alley cropping needs investigation.

## B.   Field-sown crops

### 1.   Coating seeds with vesicular-arbuscular mycorrhizal inoculum

Coating seeds with mycorrhizal inoculum using methods similar to those used with *Rhizobium* would be the easiest method of inoculating plants with vesicular-arbuscular mycorrhizal fungi if it provided consistently good infecton. Seeds are coated with an adhesive, e.g. methyl cellulose, to which inoculum is expected to stick. Unfortunately, because of their large size, it is much more difficult to attach vesicular-arbuscular mycorrhizal propagules than bacteria in this way. However, this approach has proved satisfactory for large-seeded crops such as citrus in field nurseries (Hattingh and Gerdemann, 1975).

### 2.   Mycorrhizal pellets

Rather than coating seeds with vesicular-arbuscular mycorrhizal inoculum, it may be technically more feasible to incorporate seeds into the inoculum to form multiseeded pellets. These pellets, about 1 cm in diameter, consist of soil inoculum from pot cultures stabilized with clay (Powell, 1979; Hayman *et al.*, 1981), or peat inoculum. Good infection has been achieved by this means. Hall and Kelson (1981) have described a system for producing some 5000 vesicular-arbuscular mycorrhiza-infected soil pellets per man-day with seeds attached by gum arabic.

### 3.   Fluid drilling

Seeds and inoculum can be incorporated into a uniform suspension in a viscous fluid, for example 4% (w/v) methyl cellulose, and applied in furrows as a slurry. This procedure has proved successful under field conditions (Hayman *et al.*, 1981). The advantage of this technique is that the inoculum is less bulky and does not increase appreciably the volume of material drilled. It can also be combined with rhizobia for legumes.

### 4. Inoculation in furrows

Placing inoculum under or beside seeds sown in a furrow (Owusu-Bennoah and Mosse, 1979; Hayman et al., 1981) is another possible method of inoculating field-sown crops. In field experiments, mycorrhizal inoculum has been placed below the seeding point. The rates of application of inoculum vary from 0.8 to 167 $t\,ha^{-1}$, rates of 20–30 $t\,ha^{-1}$ being commonly used (see Powell, 1984). Most of these rates would be unacceptable in practical agriculture.

Powell and Bagyaraj (1982) used a tractor-drawn drill (Model Nodet) to handle fresh or granulated vesicular-arbuscular mycorrhizal inoculum for onion and maize crops. Mycorrhizal inoculum, when placed 3 cm below the seed at the rate of 1.9 $t\,ha^{-1}$, increased onion bulb yield by 17%. With modification in the drill, the amount of inoculum used could be reduced and it was found that inoculum used at the rate of 400 $kg\,ha^{-1}$ and placed along with the seed, increased the yield of maize by 15%. These field trials show that existing farm machinery can be modified to accept mycorrhizal inoculum and give biologically and statistically significant increases in crop yield in the field. However, the lowest amount of inoculum (400 $kg\,ha^{-1}$) used in this field trial was too high to be used in practical agriculture.

### 5. Pre-cropping

Populations of vesicular-arbuscular mycorrhizal propagules can be raised in situ by growing strongly mycorrhizal host plants and leaving the infected roots and associated spores in the soil to infect the next crop. Hence, judicious crop rotations involving use of mycorrhizal crop plants and addition of organic amendments to encourage indigenous endophytes might be considered as a means of manipulating the population and inoculum size of vesicular-arbuscular fungi (Bagyaraj, 1990). These techniques could also be used to increase the population of an efficient introduced endophyte. The top soil can also be used as a crude inoculum. Sreenivasa and Bagyaraj (1988c) suggested partially sterilizing microplots by burning straw in furrows (2 kg straw for a plot of 1 m × 1 m with five furrows, each 15 cm deep) and then raising sorghum or Rhodes grass (*Chloris gayana*) inoculated with a desired vesicular-arbuscular mycorrhizal fungus. After 75 days, the top 15 cm of soil along with roots (chopped) can be used as mycorrhizal inoculum. In this way farmers can raise inoculum in their own field from a starter culture obtained from a reliable source.


## VII.  Management of indigenous vesicular-arbuscular mycorrhizal fungi

There are two principal ways of ensuring that benefits in terms of crop production are obtained from mycorrhizal associations:

(1)  by inoculating the crop with selected efficient fungi; and

(2)  by promoting the activity of effective indigenous mycorrhizal fungi by adequate cultural practices (Sieverding, 1986).

Mycorrhiza workers have shown more interest in the process of inoculating plants with efficient vesicular-arbuscular mycorrhizal fungi. In order to promote mycorrhizal activity, information is required about the indigenous vesicular-arbuscular mycorrhizal fungi which must be managed. It may then be possible to alter the composition of vesicular-arbuscular mycorrhizal fungal species in the soil by certain agronomic practices, in such a way that effective fungi are enhanced and ineffective fungi are depressed. Such attempts have already been made by some workers. It is known that prolonged monoculture with certain agricultural crops, such as wheat, increases vesicular-arbuscular mycorrhizal populations (Strzemska, 1975), while continuous cropping with mustard decreases mycorrhizal populations (Harinikumar and Bagyaraj, 1988). Heavy doses of phosphatic fertilizer depress the size of mycorrhizal fungal populations (Kruckelman, 1975), while organic manure increases their size (Harinikumar and Bagyaraj, 1989). Cultivation of a non-mycorrhizal host like mustard, kale, etc. or leaving the land fallow is known to reduce the number of native vesicular-arbuscular mycorrhizal fungi significantly (Powell, 1982; Harinikumar and Bagyaraj, 1988).

Due to the vast amount of information required and the lack of an easy, rapid methodology to define the status of the mycorrhizal population in the field, the use of this management strategy must at present be considered questionable. However, it can be expected in the years to come that some general recommendations for the management of indigenous vesicular-arbuscular mycorrhizal fungi to improve crop production may become available for each soil and crop.

## VIII. Appropriate vesicular-arbuscular mycorrhizal technology

### A. Transplanted crops

As pointed out earlier, there is an immediate potential for use of vesicular-arbuscular mycorrhizal inoculation for transplanted crops by farmers. Inoculum containing suitably selected strains can be used in the nursery bed. Growers simply need to incorporate inoculum in nursery beds, seedling trays, or polythene bags at the appropriate rate by hand. Seedlings thus raised will be colonized by the introduced fungus and can then be planted out in the field. In developed countries seedlings are usually raised in fumigated soil or potting mix. Inoculation of such substrates with selected vesicular-arbuscular mycorrhizal fungi for the crop is an easy and appropriate technology. In the USA commercial use of vesicular-arbuscular mycorrhizal inoculation by this method is used in the nursery production of citrus (Menge, 1977). Inoculation of the nursery with efficient vesicular-arbuscular mycorrhizal fungi has been successful in the large-scale production of white ash seedlings for reafforestation in Canada (Furlan et al., 1985) and for many horticultural plants in France (Gianinazzi et al., 1989).

In many developing countries, and sometimes in developed countries, seedlings are raised in unsterile soil or substrate. Inoculation with appropriate fungi in seedling trays containing unsterilized potting mix was found to give vigorously growing seedlings of asparagus which performed well when planted out in the field (Powell et al., 1985). We found inoculation of unsterile nursery beds with efficient vesicular-arbuscular mycorrhizal fungi enhanced the growth rate of slow-growing tropical tree species and rootstocks of citrus and mango. This technology, which can be adopted easily by farmers, is now being tried on a large scale.

### B. Field-sown crops

Development of appropriate technology for mycorrhizal inoculation is more difficult for crops grown from seed in the field. Work of Powell and Bagyaraj (1982) showed that farm machinery can be modified to accept mycorrhizal inoculum and to give significantly increased yields of directly field-sown crops like maize. The technology for applying vesicular-arbuscular mycorrhizal inoculum using farm machinery for directly field-sown crops is already available and should be developed further. The main limiting factor would be production of large quantities of inoculum. Vesicular-arbuscular mycorrhizal inoculation by hand, in developing countries, may be feasible for crops with low field densities (because of larger spacing between plants) such as fruit trees and root crops such as yam, cassava, etc.

## IX.  Conclusions

It has now been proved beyond doubt that vesicular-arbuscular mycorrhizal fungi can enhance plant growth. We can therefore use these fungi to improve crop productivity, at least in transplanted crops. Positive responses to vesicular-arbuscular mycorrhizal fungi are most likely in soils of low phosphate content on which mycorrhiza-dependent plants are being grown. It is important to determine the mycorrhizal dependency of the different crop plants grown in a region and to select those plants which are highly mycorrhiza-dependent for application of inoculation techniques. Screening to determine which are the most efficient vesicular-arbuscular mycorrhizal fungi in association with highly mycorrhiza-dependent crop plants should be undertaken.

Initially experiments can be carried out with crops which are mostly raised in nursery, seedling trays or polythene bags and then planted out in the field. As only small quantities of the mycorrhizal inoculum would be required, the technology could be applied immediately to practical farming without much difficulty. Studies to promote the activity of effective indigenous vesicular-arbuscular mycorrhizal fungi, through judicious manipulation of agronomic practices, should be intensified. The possibility of reducing phosphate fertilizer input and of using cheaper fertilizers such as rock phosphate along with vesicular-arbuscular mycorrhizal fungi should be investigated. It appears to be legitimate to regard the use of vesicular-arbuscular mycorrhizal fungi as being an alternative strategy for a more rational and sustainable agriculture.

Concentrated efforts to improve inoculum production technology of vesicular-arbuscular mycorrhizal fungi should be given top priority. More research is also needed on inoculation techniques. In view of the widespread interest in vesicular-arbuscular mycorrhiza during the last decade, it is likely that the agricultural application of this mutualistic symbiosis will continue to increase in the coming years.

## References

Abbott, L. K. and Robson, A.D. (1982). *Austral. J. Agri. Res.*, **23**, 389–408.

Bagyaraj, D. J. (1984). In *VA Mycorrhiza* (C. L. Powell and D. J. Bagyaraj, eds), pp. 131–153. CRC Press, Boca Raton, FL.

Bagyaraj, D. J. (1989). In *Mycorrhiza for Green Asia* (A. Mahadevan, N. Raman and K. Natarajan), pp. 326–328. University of Madras, Madras.

Bagyaraj, D. J. (1990). In *Handbook of Applied Mycology*, Vol. I, *Soil and Plants* (D. K. Arora, B. Rai, K. G. Mukerji and G. Knudsen, eds), pp. 3–34. Marcel Dekker, New York.

Bagyaraj, D. J., Manjunath, A. and Govinda Rao, Y. S. (1988). *J. Soil Biol. Ecol.* **8**, 98–103.

Bagyaraj, D. J., Byra Reddy, M. S. and Nalini, P. A. (1989). *Forest Ecol. Manage.* **27**, 791–801.

Carling, D. E. and Brown, M. F. (1980). *Soi. Sci. Soc. Am. J.* **44**, 528–531.

Chatel, D. L., Greenwood, R. M. and Parker, C. A. (1968). *Proc. 9th Int. Congr. Soil Sci. Trans.* **11**, 65–73.

Crush, J. R. and Pattison, A. C. (1975). In *Endomycorrhizas* (F. E. Sanders, B. Mosse and P. B. Tinker, eds), pp. 485–493. Academic Press, London.

Dehne, H. W. and Backhaus, G. F. (1986). *Z. Pflanzenkr. Pflanzenschutz* **93**, 415–424.

Duc, G., Trouvelot, A., Gianinazzi-Pearson, V. and Gianinazzi, S. (1989). *Plant Sci.* **60**, 215–222.

Ferguson, J. J. and Woodhead, S. H. (1982). In *Methods and Principles of Mycorrhizal Research* (N. C. Schenck, ed.), pp. 47–54. American Phytopathological Society, St Paul, MN.

Furlan, V., Fortin, J. A. and Campagna, J. P. (1985). In *Proceedings of the 6th North American Conference on Mycorrhizae* (R. Molina, ed.), p. 84. Forest Research Laboratory, Oregon State University, Corvallis, OR.

Gerdemann, J. W. (1975). In *The Development and Function of Roots* (J. G. Torrey and D. J.Clarkson, eds), pp. 575–591. Academic Press, New York.

Gerdemann, J. W. and Nicolson, T. H. (1963). *Trans. Br. Mycol. Soc.* **46**, 235–244.

Gianinazzi-Pearson, V., Gianinazzi, S. and Trouvelot, A. (1985). *Can J. Bot.* **63**, 1521–1524.

Gianinazzi, S., Gianinazzi-Pearson, V. and Trouvelot, A. (1989). In *Biotechnology of Fungi for Improving Plant Growth* (J. M. Whipps and R. D. Lumsden, eds), 41–54. Cambridge University Press, Cambridge.

Govinda Rao, Y. S., Bagyaraj, D. J. and Rai, P. V. (1983) *Zentralbl. Mikrobiol.* **138**, 415–419.

Hall, I. R. (1984). *J. Agric. Sci.* **102**, 719–723.

Hall, I. R. and Kelson, A. (1981). *N.Z. J. Agric. Res.* **24**, 221–222.

Harinikumar, K. M. and Bagyaraj, D. J. (1988). *Plant Soil.* **110**, 77–80.

Harinikumar, K. M. and Bagyaraj, D. J. (1989). *Biol. Fert. Soils.* **7**, 173–175.

Hattingh, M. J. and Gerdemann, J. W. (1975). Inoculation of Brazilian sour orange seed with endomycorrhizal fungus. *Phytopathology* **65**, 1013–1016.

Hayman, D. S., Morris, E. J. and Page, R. J. (1981). *Ann. Appl. Biol.* **99**, 247–253.

Hetrick, B. A. D. (1984) In *VA mycorrhiza* (C. L. Powell and D. J. Bagyaraj, eds), pp. 35–55. CRC Press, Boca Raton, FL.

Jeffries, P. (1987). *Crit. Rev. Biotechnol.* **5**, 319–357.

Krishna, K. R., Shetty, K. G., Dart, P. J. and Andrews, D. J. (1985). *Plant Soil* **86**, 113–125.

Kruckelmann, H. W. (1975). In *Endomycorrhizas* (F. E. Sanders, B. Mosse and P. B. Tinker, eds), pp. 511-525. Academic Press, New York.

Lackie, S. M., Bowley, S. R. and Peterson, R. L. (1988). *New Phytol.* **108**, 477-482.

Menge, J. A. (1984). In *VA Mycorrhiza* (C. L. Powell and D. J. Bagyaraj, eds), 187-203. CRC Press, Boca Raton, FL.

Menge, J. A., Lembright, H. and Johnson E. L. V. (1977). *Proc. Int. Soc. Citric.* **1**, 129-132.

Menge, J. A., Johnson, E. L. V. and Platt. R. G. (1978). *New Phytol.* **81**, 553-559.

Mercy, M. A., Shivashankar, G. and Bagyaraj, D. J. (1990). *Plant Soil* **121**, 292-294.

Mosse, B. (1972). *Rev. Ecol. Biol. Soc.* **9**, 529-537.

Mosse, B. (1973). *Ann. Rev. Phytopathol.* **11**, 171-196.

Nemec, S. (1977). *Proc. Am. Phytopath. Soc.* **4**, 227.

Nicolson, T. H. (1967). *Sci. Progr.* **55**, 561-568.

Nopamornbodi, O., Rojanasiriwong, W. and Thamsuakul, S. (1988). In *Mycorrhiza for Green Asia* (A. Mahadevan, N. Raman and K. Natarajan, eds), pp. 315-316. University of Madras, Madras.

Owusu-Bennoah, E. and Mosse, B. (1979). *New Phytol.* **83**, 671-679.

Plenchette, C., Fortin, J. A. and Furlan, V. (1983). *Plant Soil* **70**, 199-209.

Porter, W. M. (1979). *Austral. J. Soil. Res.* **17**, 515-518.

Powell, C. L. (1979). *New Phytol.* **83**, 81-85.

Powell, C. L. (1982). *N.Z. J. Agric. Res.* **25**, 461-464.

Powell, C. L. (1984). In *VA Mycorrhiza* (C. L. Powell and D. J. Bagyaraj, eds), pp. 205-222. CRC Press, Boca Raton, FL.

Powell, C. L. and Bagyaraj, D. J. (1982) *Proc. N. Z. Agron. Soc.* **12**, 85-88.

Powell, C. L., Bagyaraj, D. J., Clark, G. E. and Caldwell, K. J. (1985). *N.Z. J. Agric. Res.* **28**: 293-297.

Reena, J. and Bagyaraj, D. J. (1990a). *World J. Microbiol. Biotech.* **6**, 59-63.

Reena, J. and Bagyaraj, D. J. (1990b). *Arid Soil Res. Rehabil.* **4**, 261-268.

Schubert, A. and Hayman, D. S. (1986). *New Phytol.* **103**, 79-80.

Sieverding, E. (1986). In *Physiological and Genetical Aspects of Mycorrhizae* (V. Gianinazzi-Pearson and S. Gianinazzi, eds), pp. 475-478. INRA, Paris.

Sreenivasa, M. N. and Bagyaraj, D. J. (1988a). *Plant Soil.* **106**, 289-290.

Sreenivasa, M. N. and Bagyaraj, D. J. (1988b). *Plant Soil* **109**, 125-127.

Sreenivasa, M. N. and Bagyaraj, D. J. (1988c). *Indian Farmers' Digest* **21**, 27-29.

Sreeramulu, K. R. and Bagyaraj, D. J. (1986). *Plant Soil* **93**, 299-302.

Strzemska, J. (1975). In *Endomycorrhizae* (F. E. Sanders, B. Mosse and P. B. Tinker, eds), pp. 537-543. Academic Press, New York.

Tommerup, I. C. and Kidby, D. K. (1979). *Appl. Environ. Microbiol.* **37**, 831-835.

Trouvelot, A., Gianinazzi, S. and Gianinazzi-Pearson, V. (1987). In *Mycorrhizae in the Next Decade. Practical Applications and Research Priorities* (D. M. Sylvia, L. L. Hung and J. H. Graham, eds), p. 39, University of Florida, Gainsville, FL.

Vinayak, K. and Bagyaraj, D. J. (1990). *Biol. Agric. Hort.* **6**, 305–311.
Wood, T. (1985). In *Proceedings of 6th North American Conference on Mycorrhizae* (R. Molina, ed.), p. 84. Forest Research Laboratory, Oregon State University, Corvallis, OR.

# 20

# Vesicular-arbuscular Mycorrhizal Fungi in Nitrogen-fixing Legumes: Problems and Prospects

GABOR J. BETHLENFALVAY

*United States Department of Agriculture, Agricultural Research Service, Western Regional Research Center, Albany, CA 94710, USA*

## I. Introduction

The "rationale of mycorrhiza formation" for plants is the uptake of mineral nutrients, declared Stahl (1900) 90 years ago. But (so argued Stahl, all in one sentence), "though one must not infer from the occurrence of both bacteriotrophy and mycotrophy in the same plant that the meaning of the two symbioses for the host is different, this assumption is still probable, since otherwise it would be unlikely that two so strange and in their end effect equivalent mechanisms should have been acquired by the Papilionaceae in their struggle for survival".

Nonetheless, he concluded, it seems most likely that the rationale is valid for the legumes also, and this is the prevailing view on the subject to this day.

What Stahl did not know, and what Jones (1924) and Asai (1944) pointed out later, was the special, nutritional relationship between the two microsymbionts: the high phosphorus requirement of the nitrogen-fixing root nodule, the high nitrogen requirement of the chitin-walled vesicular-arbuscular mycorrhizal fungus, and the high carbon requirement of both. Since each symbiont can supply the other's needs in excess, the endophytes can bring about synergistic growth responses in the host plant when the association is grown in nutrient-deficient soil (Schenck and Hinson, 1973; Mosse, 1977). Such observations led to a view of the legume microsymbionts as biological substitutes for fertilizers (Azcón-Aguilar *et al.*, 1979).

But how do the symbionts of the tripartite legume association relate to each other when the growth medium is not deficient, such as in moderately fertilized field soils, a condition often found in modern agriculture? Or under experimental conditions, where scientific insight into the biology of plant symbioses rather than productivity is of interest? One of the reasons why this question is difficult to answer experimentally, is the lack of adequate, non-symbiotic controls in the quest for the nature of "symbiotic response". Without controls, one is reduced to using what appears to be the most appropriate comparison plant (another treatment) for whatever the objective of the experiment happens to be. For the legume association, this comparison treatment is necessarily a double approximation. What follows is an attempt to illustrate the problems encountered in such whole-plant experiments with mycorrhiza in nodulated legumes.

## II.   Different growth effects, same symbiosis

Organisms that have come to depend on one another (symbiosis) show this dependence by a range of symbiotic responses. When both symbionts benefit, we call the symbiosis mutualistic; when one benefits at the expense of the other, the former is known as a parasite. When neither seems to benefit, the commensal partners may still provide hidden advantages to one another, or to outsiders. The vesicular-arbuscular mycorrhizal fungus, an obligate biotroph (Warner and Mosse, 1980), always benefits, and it is the cost–benefit ratio to the host plant that determines its role in the partnership.

## A. Mutualism

Many workers in mycorrhiza think of the vesicular-arbuscular mycorrhizal symbiosis exclusively as mutualistic. This is not the case. Growth depression, though transient, has also been observed in vesicular-arbuscular mycorrhizal plants (Tinker, 1978; Stribley et al., 1980). Although the response of a plant to colonization by vesicular-arbuscular mycorrhizal fungi depends on many biotic and environmental factors (Barea and Azcón-Aguilar, 1983; Abbott and Robson, 1984; Hayman, 1987), the most important is phosphorus availability (Hayman, 1987). Host plant growth is enhanced when the benefits of increased phosphorus nutrition to the host outweigh the disadvantage of carbohydrate loss to the endophyte (Harley, 1969). Experimentally, one can follow the development of the mutualistic effect in the plant by relating changes in available phosphorus in the growth medium with changes in the vesicular-arbuscular mycorrhizal plant relative to non-vesicular-arbuscular mycorrhizal comparison grown under the same phosphorus regime (Bethlenfalvay et al., 1982b). This can best be done in a synthetic, phosphorus-free medium, using nutrient solutions or slow-release phosphorus sources, such as hydroxyapatite. Determinations of the development of both macro- and microsymbiont are possible if sequential, time-course measurements are made (Bethlenfalvay et al., 1982a).

A number of methods are available to determine both the internal and external portion of vesicular-arbuscular mycorrhizal fungal mycelium (Bethlenfalvay and Ames, 1987; Bethlenfalvay et al., 1981; Giovanetti and Mosse, 1980). All of these methods involve estimates, to some extent. Recently, techniques have been developed to estimate also the viability of the external (extraradical) hyphae of vesicular-arbuscular mycorrhizal fungi (Sylvia, 1988), adding little to the cumbersomeness, but a great deal to the precision and validity of mycorrhizal work. The difficulty of quantifying vesicular-arbuscular mycorrhizal soil hyphae generally prevents workers from including statements in their reports on this important organ of the fungus-root, even though the tenuous nature of the relationship between root colonization and the growth effect is well-known (Graham et al., 1982).

## B. Parasitism

Growth depression in vesicular-arbuscular mycorrhizal plants is different from pathogenic parasitism, in that neither irreversible tissue damage nor necrosis is observed as a result of vesicular-arbuscular mycorrhizal colonization. In the vesicular-arbuscular mycorrhizal association, where the host plant and its fungal endophyte live together in an intimate, balanced relationship, symptoms of symbiotic (non-pathogenic) parasitism occur when the balance is disturbed (Gerdeman, 1974). The effects of disturbance can be dramatic (Bethlenfalvay et al., 1982b). Experimentally, the conditions imposed to produce the parasitic response involve a manipulation of phosphorus levels in the medium (Bethlenfalvay et al., 1982d, 1983). When phosphorus is extremely limiting, competition for it by the symbionts results in growth depression of the host. If phosphorus availability is increased slightly, vesicular-arbuscular mycorrhizal roots gain an advantage over non-vesicular-arbuscular mycorrhizal roots in exploiting this limiting resource, due to the phosphorus-uptake efficiency of the vesicular-arbuscular mycorrhizal soil mycelium (Sanders et al., 1977). This is the stage of mycotrophic growth, which is limited to a range of suboptimal soil phosphorus concentrations (Ross, 1971; Bethlenfalvay et al., 1983) and is evaluated as growth enhancement relative to non-vesicular-arbuscular mycorrhizal plants suffering from the same limitations. Growth inhibition also occurs at levels of phosphorus availability above those which result in mycotrophy. At such levels, there is sufficient phosphorus for the plant to supply its needs without its (now parasitic) symbiont, but not enough to inhibit colonization. Thus, the fungus can proliferate, and may divert significant amounts of carbon from the host (Buwalda and Goh, 1982). Eventually, with increasing phosphorus additions, colonization is eliminated.

Responses to phosphorus availability modified by vesicular-arbuscular mycorrhizal fungi appear to be aggravated in nodulated soybean plants (Bethlenfalvay et al., 1982e), where inhibition of host-plant growth is more pronounced, apparently as a result of increased competition for carbohydrates by the nodules and the high energy requirement of nitrogen fixation.

### III.  Is there a best endophyte?

Co-evolution of vascular plants with vesicular-arbuscular mycorrhizal fungi over geological time is likely to have produced combinations best adapted for survival within the biotic communities of their origin (Trappe, 1987; Pirozynski and Dalpé, 1989). An exotic crop plant is not likely to encounter a symbiotic mycoflora best suited for its requirements, when introduced to a new area (Hall, 1987). This deficiency is of particular interest for the highly mycotrophic legume plant, and led to research evaluating the merit of native versus introduced vesicular-arbuscular mycorrhizal fungi in the field (Hall, 1987).

Root systems are typically colonized by more than one vesicular-arbuscular mycorrhizal fungus. Mutual exclusion is not observed, but success in occupancy varies and is not necessarily related to host response (Daft, 1983; Lopez-Aguillon and Mosse, 1987). Host response can differ with fungal species (Graw et al., 1979; Wilson, 1988), and its variability may be due to seasonal development (Daft et al., 1981) or to varying uptake or exclusion capabilities of different fungi for different elements (Sieverding and Toro, 1988; Bethlenfalvay and Franson, 1989). Multiple colonization by mixed inocula containing fungi with different host effects and environmental adaptations might therefore be more consistently beneficial to the host plant (Daft, 1983) than one (ideal, but non-existent) endophyte. Disturbance, agrochemicals, management practices and cultural and environmental stress conditions are likely to influence the composition of the vesicular-arbuscular mycorrhizal mycoflora, whose components may possess different levels of tolerance for conditions imposed by unsustainable agriculture (see Trappe, 1982; Trappe et al., 1984; Jeffries, 1987). Utilization of the endophytes in agricultural situations (Dagoberto et al., 1986; Hayman, 1987) therefore not only poses the problem of selecting one "best" organism, but a collection of organisms which are compatible with each other, with the native microflora, with new host plants and with the new host soil.

## A. Selection of the endophytes

As major characteristics for the selection of vesicular-arbuscular mycorrhizal fungi, Abbott and Robson (1985) listed the ability to infect rapidly and extensively, to form an effective soil mycelium for the uptake of phosphorus, and to produce a large number of propagules. Their scheme included criteria for the collection and culture of the fungi under different edaphic conditions, and for response evaluation. Recently, they further extended their discussion of the field management of vesicular-arbuscular mycorrhizal fungi and proposed new methods to evaluate their effects in terms of plant growth and phosphorus nutrition (Abbott and Robson, 1990).

Selecting the right organism to help solve the right problem is further complicated by differences in plant responses not only to vesicular-arbuscular mycorrhizal fungal species but also to their geographical isolates ("edaphotypes", sensu Bethlenfalvay et al., 1989). Furthermore, host effects are not restricted to those of phosphorus nutrition, but are pervasive in both plant form and function. Therefore, utilization of vesicular-arbuscular mycorrhizal fungi in the field on more than a hit-or-miss basis will necessitate a complex screening and selection phase. Thus, problem identification is an important first step prior to launching the screening process. The uses for vesicular-arbuscular mycorrhizal fungi are manifold in terms of desirable host plant or host soil responses: drought resistance, tolerance to metal toxicity, mineral nutrient uptake from the soil, organic nutrient transport to the soil biota, effective nitrogen fixation, and others. With the objectives identified, isolates collected from specific sites can be tested for a specific requirement. Site selection for collecting the fungi is important, because useful adaptations are presumably site-specific. Alternatively, starting with a given set of available inocula, one may test for a wide range of their possible capabilities.

The "common garden" technique long employed in experimental botany (Heslop-Harrison, 1964) to equalize conditions for response testing, is a good approach for pre-screening vesicular-arbuscular mycorrhizal fungi in the greenhouse or growth chamber. It imposes uniform conditions (same new soil, same new host) for the screening of fungi that hail from diverse places. However, to make the comparisons between host responses to different fungal species or edaphotypes meaningful, the infectivity of the inocula must be carefully determined (Franson and Bethlenfalvay, 1989). This is a time-consuming procedure, and a step that is often omitted.

Recently the Diagnosis and Recommendation Integrated System (DRIS), long used as a diagnostic tool for fertilizer application (Walworth and Sumner, 1987), has been tested for its applicability to the ranking of vesicular-arbuscular mycorrhizal effects on plant nutrition (Bethlenfalvay et al., 1990). Analysis by DRIS is based on nutrient balance ratios, and its use minimizes morphogenic and genotypic effects on the accuracy of deficiency diagnoses. It also predicts which nutrient is most limiting to yield (Hallmark et al., 1987). Thus, plant nutritional responses to vesicular-arbuscular mycorrhizal fungi in a given soil may be evaluated for each nutrient. Since vesicular-arbuscular mycorrhizal fungi may affect the uptake of each nutrient differently, resulting in both growth enhancement and inhibition (see Ames and Bethlenfalvay, 1987), DRIS promises to become a particularly valuable tool in vesicular-arbuscular mycorrhizal screening efforts.

## B.  Selection of endophyte combinations

The dependence of nodulation on vesicular-arbuscular mycorrhizal colonization when phosphorus is limiting is well-documented (Mosse, 1977; Barea and Azcón-Aguilar, 1983; Barea et al., 1987; Smith and Gianinazzi-Pearson, 1988). Effects of root nodules on the development of vesicular-arbuscular mycorrhizal fungi, on the other hand, are little-known (Bethlenfalvay et al., 1985a). The management of the bacterial and fungal endophytes of the tripartite legume association is important for the production of these crops (Daft et al., 1985). Compatibility of such endophyte combination should therefore be determined, because it affects co-endophyte effectiveness (Bayne and Bethlenfalvay, 1987). Determinations should be made utilizing nutrient-sufficient plants, otherwise nutrient-deficiency effects will mask symbiotic effects. Before evaluating tripartite relationships, one must first determine the effects of each endophyte on the host individually, in the simpler, bipartite associations. This is necessary, since the evaluation of one of the two endophytes of the tripartite association must be made against the bipartite association colonized by the co-endophyte and supplemented nutritionally with nitrogen or phosphorus (Bayne and Bethvenfalvay, 1987).

Generally, factors which are easily determined in characterizing compatibility and effectiveness of all three symbionts are their developmental characteristics (extent of vesicular-arbuscular mycorrhizal colonization, nodule mass, and plant mass). When nutrient-sufficient comparison plants are employed in such studies, antagonistic (Bethlenfalvay et al., 1985a), as well as stimulatory relationships are revealed between the microsymbionts. Plant and endophyte activities can also be determined, by leaf gas-exchange (Brown and Bethlenfalvay, 1987), nitrogenase activity (Brown et al., 1988) and vesicular-arbuscular mycorrhizal hyphal viability (Sylvia, 1988) measurements, and by nutritional analysis (Jarrel and Beverly, 1981). Stable isotope ($^{13}$C, $^{15}$N) labelling (Rundel et al., 1988) can be a useful tool, but is as yet little used in studies to determine the complex source–sink relationships (Bayne et al., 1984) between the symbionts.

## IV. The vesicular-arbuscular mycorrhizal legume associated with non-legumes

Nutrient transfer between the roots of associated plants is enhanced when the roots are colonized by vesicular-arbuscular mycorrhizal fungi (Francis and Read, 1984; van Kessel et al., 1985). This has profound implications for our understanding of plant community structure (Fitter,

1985). The concept of resource distribution optimized by the movement of nutrients along concentration gradients between vesicular-arbuscular mycorrhizal donor and receiver plants (Read et al., 1985) is of particular interest in plant associations including legumes, because of the role of legumes as sources of nitrogen. Practices, such as soil disturbance (Evans and Miller, 1988) or fallowing (Thompson, 1987), which directly or indirectly disrupt the integrity of the vesicular-arbuscular mycorrhizal soil mycelium, have been shown to be detrimental to plants. A continuous hyphal system, on the other hand, can enhance seedling establishment (Read and Birch, 1988) and nutrient uptake (Fairchild and Miller, 1988). While quantitatively important, vesicular-arbuscular mycorrhizal-mediated nutrient fluxes between plants in the field have not yet been reported (Newman, 1988), significant nutrient transfer between soybean and corn plants has been demonstrated under controlled conditions (G. J. Bethlenfalvay et al., unpubl. res.).

For an unambiguous demonstration of nutrient transfer by vesicular-arbuscular mycorrhizal hyphae, a root-free zone must be created between the root systems of the donor and receiver plants (Camel et al., 1991). This can be achieved by delimiting a soil bridge between donor and receiver plants by means of fine screens permeable by hyphae, but not by roots. Such an arrangement eliminates nutrient transfer by the "soil pool pathway" (such as exudation, diffusion of ions, root anastomosis, and access to biologically bound nutrients following mineralization, see Newman and Ritz, 1986). The direction of lateral fluxes of water between the soils of the bridge and adjacent root compartments can be controlled by watering schedules which keep soil water content slightly lower in the donor-plant soil. With water flux, if any, moving from receiver to donor, nutrient-ion counterfluxes become unlikely. In studies utilizing nitrogen fertilizers enriched or depleted in $^{15}N$, the soils of donor, bridge and receiver compartments can be tested for $^{15}N$ abundance to check for nitrogen diffusion.

Application of $^{13}C$ as $CO_2$ is also a useful tool in vesicular-arbuscular mycorrhizal nutrient-transfer studies. Exposure of corn plants to labelled $CO_2$ resulted in a rapid movement of $^{13}C$ to associated soybean plants, where it was shown to enhance nitrogenase activity significantly. Comparative work with nodulated and unnodulated soybean showed that the nodules are a strong sink for phosphorus, which they may obtain from their vesicular-arbuscular mycorrhizal companion plant by mycorrhizal hyphal transport (G. J. Bethlenfalvay et al., unpubl. res.). It appears that mycorrhiza-mediated nutrient fluxes are controlled by source–sink relationships and tend to balance nutrient supply and demand between interdependent plants.

## V. Growing experimental tripartite legume associations

Perhaps the most difficult aspect of experimentation with the legume symbiosis is the lack of adequate controls. The well-known nutritional effects of the microsymbionts are most conspicuous when the plants are grown under conditions limiting in nitrogen, or phosphorus, or both. However, the non-symbiotic plants grown under such nutrient regimes do not provide meaningful comparisons in applied or basic studies, since such legumes are unlikely to occur or survive long in the field, while they are too stunted to serve as physiological "controls" in experimentation (Bethlenfalvay et al., 1987).

### A. The symbiotic test plant

It is possible to grow well-nodulated soybean plants with root colonization by vesicular-arbuscular mycorrhizal fungi in excess of 50% within three weeks after planting. Rapid establishment of the symbiotic condition is important in experiments with plants grown in pot cultures, since the vesicular-arbuscular mycorrhizal effect is dependent on root density and tends to diminish or disappear as the roots become pot-bound (Baath and Hayman, 1984). However, the factors contributing to the growth of the association must be favourable to attain rapid symbiotic development. Some of these factors are temperature (Sieverding, 1988), light intensity (Diederichs, 1982a), day length (Diederichs, 1982b), soil conditions (Bethlenfalvay et al., 1985b), viability (Sylvia, 1988), and dormancy (Tommerup, 1983) of the vesicular-arbuscular mycorrhizal inoculum, effectiveness of the Rhizobia (Barea and Azcón-Aguilar 1983), intersymbiont compatibilities, and the little-known effects of the syndrome known as "suppressive soil" (Wilson et al., 1988). The agents of the latter may be introduced into the usually sterilized growth medium with the non-sterile vesicular-arbuscular mycorrhizal soil inoculum, and may be responsible, to a large extent, for variation between vesicular-arbuscular mycorrhizal experiments.

Depending on the nature of the medium utilized, the available and bound nutrient content of which should be known to the experimenter, nutrients added in soluble or slow-release capsule form should be calculated to maximize the information to be gained from the experiment. For instance, if the uptake of one specific nutrient ion is of interest, none of the other nutrients should be limiting. In the tripartite association, there is a delicate balance in the demand (Rhizobium) and tolerance (vesicular-arbuscular mycorrhizal fungus) of the microsymbionts for phosphorus. This balance shifts during the ontogeny of the association (Bethlenfalvay et al., 1982c). Adequate phosphorus supplies are essential for nodulation in the early stages of plant and nodule development, but excess phosphorus will inhibit vesicular-arbuscular mycorrhizal colonization (Barea and Azcón-Aguilar, 1983). In soybean, grown in soil severely deficient in available phosphorus, an initial concentration of 0.2 mM $KH_2PO_4$ in nutrient solution applied as a drip twice a week until nodules became functional, generally produced a rapid rate of vesicular-arbuscular mycorrhizal colonization. With the establishment of colonization, phosphorus may be withdrawn since the fungi then assume the role of supplying the nodulated plant with its phosphorus requirement. An initial nitrogen supplement ($\approx 1$ mM $NH_4NO_3$) helps the nodules to develop rapidly, and prevents transient growth depression of the host during nodule development (Bethlenfalvay et al., 1978).

## B.  The non-symbiotic comparison plant

Different aspects of a plant's form and function react differently to nutrients supplied by its symbiotic partners or by fertilization (Bethlenfalvay et al., 1989). Success in growing symbiotic test plants and non-symbiotic comparison plants that are equivalent in more than one physiological or morphological parameter is therefore unlikely. Hence, the experimenter must decide which of the many possible growth responses he will seek to equalize. If leaf gas exchange is of interest, for example, adjustments of nitrogen and phosphorus regimes in the comparison plant may produce leaf areas equivalent to those of the test plant. However, the root–shoot ratios and leaf nutrient concentrations, factors which also have an impact on water fluxes through the plant, will be different (Brown et al., 1988).

Attempts to adjust nutrient regimes to make the comparison plants comparable to the symbiotic test plants are not always successful. Thus, it is possible, for instance, that decreasing the nitrogen supply to a non-nodulating legume to slow its growth to that of its nodulated counterpart, will achieve only a yellowing of its leaves instead. similar scenarios occur when phosphorus input by fertilization versus vesicular-arbuscular mycorrhizal colonization are to be compared. It is therefore advisable, in working with such comparisons, to learn to grow both sets of plants satisfactorily, prior to starting time-consuming, large experiments.

## VI. Vesicular-arbuscular mycorrhizal legumes in soil nutrition

Agriculture has been production-oriented since its early beginnings. Currently, however, a shift in emphasis is taking place to resource conservation, motivated by global erosion and desertification of agricultural lands (Beatty, 1982; Gibbs and Carlson, 1985; Healy et al., 1986). Sustainable agriculture (Mosse, 1986; Sprent, 1986) utilizes practices which are equally concerned with crop productivity and soil conservation. Vesicular-arbuscular mycorrhizal legumes have a central role in this effort: since the soil mycelium of vesicular-arbuscular mycorrhizal fungi represents a significant portion of the soil microbial biomass (Hayman, 1978) and since the cell walls of its hyphae are composed mainly of the amino-sugar chitin (Weijman and Meuzelaar, 1987), the soil mycelium may be one of the most important vehicles for nitrogen and carbon input into the soil. The contribution of vesicular-arbuscular mycorrhizal legumes to soil organic matter, and as a consequence, to soil aggregate stability (Lynch and Bragg, 1984; Oades, 1884), therefore promises to become a new and exciting experimental field for research with the legume association.

The literature on the effects of vesicular-arbuscular mycorrhiza on soil aggregation has been reviewed by Thomas et al. (1986), and was brought into agro-ecological context by Bethlenfalvay and Newton (1990). The difficulties and challenges of experimentation with a system as complex as the soil microbiota were highlighted by Linderman (1988). Others (Tisdall and Oades, 1982; Gupta and Germinda, 1988) pointed out the role of micro-organisms and their organic products in the formation of aggregates at different levels of size and organization. An important new direction in soil-oriented, rather than plant-centred, vesicular-arbuscular mycorrhizal research involves the close association of vesicular-arbuscular mycorrhizal fungi and chitin-decomposing actinomycetes (Ames, 1989; Ames et al., 1989) which are important for the formation of small aggregates. Experimentally, the effects of mycorrhizal roots, non-mycorrhizal roots, and mycorrhizal hyphae can be segregated by using split-root systems, inoculated or not inoculated with the fungi. The two parts of the root system can then be further separated by the use of fine screens, which permit passage only for hyphae ($\approx 5\text{--}20\,\mu\text{m}$ diameter) to adjacent soil compartments. The effect of each component of the fungus-root, with and without root nodules, can then be evaluated for aggregate formation.

## VII.  Conclusions

A fundamental role is ascribed to vesicular-arbuscular mycorrhizal fungi in the ecophysiology of nodulated legumes. The concept of cost–benefit relationships in plant symbioses recognizes the influence of changing conditions, such as nutrient or water availability, on the balance of nutrient supply and demand between the symbionts. Such changes may bring about transient shifts in the symbiotic condition, causing it to alternate between the mutualistic and parasitic modes. The carbon–nitrogen–phosphorus supply–demand relationship is a fundamental expression of the three symbionts as sources and sinks for one another's products within the closed system of an isolated plant. In the open system of inter-plant relationships within the plant–soil community, a major role of the legume association is envisioned as a nitrogen source, functioning not as a self-contained unit but as one involved in the fluxes of nutrients between the members of the community. The complexity of such a system is a challenge for each new experiment designed not only to gain basic insights into its workings, but also to put the findings to practical use.

## References

Abbott, L. K. and Robson, A. D. (1984). In *VA Mycorrhiza* (L. C. Powell and D. J. Bagyaraj, eds), pp. 113–130. CRC Press, Boca Raton, FL.

Abbott, L. K. and Robson, A. D. (1985). In *Proceedings of the 6th North American Conference on Mycorrhizae* (R. Molina, ed.), pp. 89–91. Forest Research Laboratory, Oregon State University, Corvallis, OR.

Abbott, L. K. and Robson, A. D. (1990). In *The Rhizosphere and Plant Growth*, Beltsville Symposium XIV (P. B. Cregan and D. L. Keister, eds). USDA-ARS, Beltsville, MD (in press).

Ames, R. N. (1989). *New Phytol.* **112**, 423–427.

Ames, R. N. and Bethlenfalvay, G. J. (1987). *J. Plant Nutr.* **10**, 1313–1321.

Ames, R. N., Mihara, K. L. and Bayne, H. G. (1989). *New Phytol.* **111**, 67–71.

Asai, T. (1944). *Jpn. J. Bot.* **13**, 463–485.

Azcón-Aguilar, C., Azcón, R. and Barea, J. M. (1979). *Nature* **279**, 325–327.

Baath, E. and Hayman, D. S. (1984). *Plant Soil* **77**, 373–376.

Barea, J. M. and Azcón-Aguilar, c. (1983). *Adv. Agron.* **36**, 1–54.

Barea, J. M., Azcón-Aguilar, C. and Azcón, R. (1987). *New Phytol.* **106**, 717–725.

Bayne, H. G. and Bethlenfalvay, G. J. (1987). *Plant Cell Environ.* **10**, 607–612.

Bayne, H. G., Brown, M. S. and Bethlenfalvay, G. J. (1984). *Physiol. Plant.* **62**, 576–580.

Beatty, M. T. (1982). *Soil Erosion: its Agricultural, Environmental, and Socioeconomic Implications*, Council for Agricultural Science and Technology, Report No. 92, Ames, Iowa.

Bethlenfalvay, G. J. and Ames, R. N. (1987). *Soil Sci. Soc. Am. J.* **51**, 834–837.
Bethlenfalvay, G. J. and Franson, R. L. (1989). *J. Plant Nutr.* **12**, 953–970.
Bethlenfalvay, G. J. and Newton, W. E. (1990). In *The Rhizosphere and Plant Growth*, Beltsville Symposium XIV (D. L. Keister, ed.) USDA-ARS, Beltsville, MD (in press).
Bethlenfalvay, G. J., Abu-Shakra, S. S. and Phillips, D. A. (1978). *Plant Physiol.* **62**, 127–130.
Bethlenfalvay, G. J., Pacovsky, R. S. and Brown, M. S. (1981). *Soil Sci. Soc. Am. J.* **45**, 871–875.
Bethlenfalvay, G. J., Brown, M. S. and Pacovsky, R. S. (1982a). *New Phytol.* **90**, 537–545.
Bethlenfalvay, G. J., Brown, M. S. and Pacovsky, R. S. (1982b). *Phytopathology* **72**, 889–893.
Bethlenfalvay, G. J., Pacovsky, R. S., Brown, M. S. and Fuller, G. (1982c). *Plant Soil* **68**, 43–54.
Bethlenfalvay, G. J., Pacovsky, R. S. and Brown, M. S. (1982d). *Phytopathology* **72**, 894–897.
Bethlenfalvay, G. J., Pacovsky, R. S., Bayne, H. G. and Stafford, A. E. (1982e). *Plant Physiol.* **70**, 446–450.
Bethlenfalvay, G. J., Barne, H. G. and Pacovsky, R. S. (1983). *Physiol. Plant.* **57**, 543–548.
Bethlenfalvay, G. J., Brown, M. S. and Stafford, A. E. (1985a). *Plant Physiol.* **79**, 1054–1058.
Bethlenfalvay, G. J., Ulrich, J. M. and Brown, M. S. (1985b). *Soil Sci. Soc. Am. J.* **49**, 1164–1168.
Bethlenfalvay, G. J., Brown, M. S. and Newton, W. E. (1987). In *Proceedings of the 7th North American Conference on Mycorrhizae* (D. M. Sylvia, L. L. Hung, and J. H. Graham, eds), pp. 231–233. University of Florida, Gainesville, FL.
Bethlenfalvay, G. J., Franson, R. L. Brown, M. S. and Mihara, K. L. (1989). *Physiol. Plant.* **76**, 226–232.
Bethlenfalvay, G. J., Franson, R. L. and Brown, M. S. (1990). *Agron. J.* **82**, 302–304.
Brown, M. S. and Bethlenfalvay, G. J. (1987). *Plant Physiol.* **85**, 120–123.
Brown, M. S., Thamsurakul, S. and Bethlenfalvay, G. J. (1988). *Physiol. Plant.* **74**, 159–163.
Buwalda, J. G. and Goh, K. M. (1982). *Soil Biol. Biochem.* **14**, 103–106.
Camel, S. B., Reyes-Solis, M. G. Ferrera-Cerrato, R., Franson, R. L., Brown, M. S. and Bethlenfalvay, G. J. (1991). *Soil Sci. Soc. Am. J.* **55**, 389–393.
Daft, M. J. (1983). *Plant Soil* **71**, 331–337.
Daft, M. J., Chilvers, M. T. and Nicholson, T. H. (1981). *New Phytol.* **85**, 181–189.
Daft, M. J., Clelland, D. M. and Gardner, I. C. (1985). *Proc. Roy. Soc. Edinburgh* **89B**, 283–298.
Dagoberto, A., Bojórquez, A., Ferrera-Cerrato, R., Trinidad-Santos, A. and Volke-Haller, V. (1986). *Agrociencia* **65**, 141–160.
Diederichs, C. (1982a). *Angew. Bot.* **57**, 45–53.
Diederichs, C. (1982b). *Angew. Bot.* **57**, 55–67.
Evans, D. G. and Miller, M. H. (1988). *New Phytol.* **110**, 67–75.
Fairchild, G. L. and Miller, M. H. (1988). *New Phytol.* **110**, 75–84.

Fitter, A. H. (1985). *New Phytol.* **99**, 257–265.

Francis, R. and Read, D. J. (1984). *Nature* **307**, 53–56.

Franson, R. L. and Bethlenfalvay, G. J. (1989). *Soil Sci. Soc. Am. J.* **53**, 754–756.

Gerdeman, J. W. (1974). In *The Plant Root and its Environment* (E. W. Carson, ed.), pp. 205–217. University Press of Virginia, Charlottesville, VA.

Gibbs, M. and Carlson, C. (eds) (1985). *Crop Productivity—Research Imperatives Revisited*, an international conference held at Boyne Highlands Inn, 13–18 October 1985.

Giovanetti, M. and Mosse, B. (1980). *New Phytol.* **86**, 131–144.

Graham, J. H., Linderman, R. G. and Menge, J. A. (1982). *New Phytol.* **91**, 183–189.

Graw, D., Moawad, M. and Rehm, S. (1979). *Z. Acker. Pflanzenbau* **148**, 85–98.

Gupta, V. V. S. R. and Germida, J. J. (1988). *Soil Biol. Biochem.* **20**, 777–786.

Hall, I. R. (1987). *Angew. Bot.* **61**, 127–134.

Hallmark, W. B., Walworth, J. L. Sumner, M. E., deMooy, C. J., Pesek, J. and Shao, K. P. (1987). *J. Plant Nutr.* **10**, 1381–1390.

Harley, J. L. (1969). *The Biology of Mycorrhiza*, 2nd edn, p. 334. Leonard Hill, London.

Hayman, D. S. (1978). In *Interactions between Non-pathogenic Soil Microorganisms and Plants* (Y. R. Dommergues and S. V. Krupta, eds), pp. 401–442. Elsevier, Amsterdam.

Hayman, D. S. (1987). In *Ecophysiology of VA Mycorrhizal Plants* (G. R. Safir, ed.), pp. 171–192. CRC Press, Boca Raton, FL.

Healy, R. G., Waddell, T. E. and Cook, K. A. (1986). *Agriculture and Environment in a Changing World Economy*. The Conservation Foundation, Washington, District of Columbia.

Heslop-Harrison, J. (1964). *Adv. Ecol. Res.* **2**, 159–247.

Jarrell, W. M. and Beverly, R. B. (1981). *Adv. Agron.* **34**, 197–224.

Jeffries, P. (1987). *CRC Crit. Rev. Biotechnol.* **4**, 319–357.

Jones, F. R. (1924). *J. Agric. Res.* **29**, 459–470.

Linderman, R. G. (1988). *Phytopathology* **78**, 366–371.

Lopez-Aguillon, R. and Mosse , B. (1987). *Plant Soil* **97**, 155–170.

Lynch, J. M. and Bragg, E. (1985). *Adv. Soil Sci.* **2**, 133–171.

Mosse, B. (1977). In *Exploiting the Legume–Rhizobium Symbiosis in Tropical Agriculture* (J. M. Vincent, A. S. Whitney and J. Bose, eds), pp. 275–292. University College of Tropical Agriculture Miscellaneous Publication 145, University of Hawaii, Honolulu, Hawaii.

Mosse, B. (1986). *Biol. Agric. Hort.* **3**, 191–209.

Newman, E. I. (1988). *Adv. Ecol. Res.* **18**, 243–270.

Newman, E. I. and Ritz, K. (1986). *New Phytol.* **104**, 77–87.

Oades, J. M. (1984). In *Biological Processes and Soil Fertility* (J. Tinsley and J. F. Darbyshire, eds), pp. 319–337. Martinus Nijhoff/Dr W. Junk, The Hague.

Pironzynski, K. A. and Dalpé, Y. (1989). *Symbiosis* **7**, 1–36.

Read, D. J. and Birch, C. P. D. (1988). *Proc. Roy. Soc. Edinburgh* **94B**, 13–24.

Read, D. J., Francis, R. and Finlay, R. D. (1985). In *Ecological Interactions in Soil* (A. H. Fitter, ed.), pp. 193–217. Blackwell Scientific, Oxford.

Ross, J. P. (1971). *Phytopathology* **61**, 1400–1403.

Rundel, P. W., Ehleringer, J. A. and Nagy, K. A. (1988). *Table Isotopes in Ecological Research*, Ecological Studies, Vol. 68. Springer–Verlag, New York.

Sanders, F. E., Tinker, P. B., Black, R. L. B. and Palmerly, S. M. (1977). *New Phytol.* **78**, 257–268.

Schenck, N. C. and Hinson, K. (1973). *Agron. J.* **65**, 849–850.

Sieverding, E. (1988). *Angew. Bot.* **62**, 295–300.

Sieverding, E. and Toro, S. (1988). *J. Agron. Crop Sci.* **161**, 322–332.

Smith, S. E. and Gianinazzi-Pearson, V. (1988). *Ann. Rev. Plant Physiol.* **39**, 221–244.

Sprent, J. (1986). *Biol. Agric. Hort.* **3**, 153–165.

Stahl, E. (1900). *Jahrb. Wiss. Bot.* **34**, 540–668.

Stribley, D. P., Tinker, P. B. and Rayner, J. H. (1980). *New Phytol.* **86**, 261–266.

Sylvia, D. M. (1988). *Soil Biol. Biochem.* **20**, 39–43.

Thomas, R. S., Dakessian, S., Ames, R. N., Brown, M. S. and Bethlenfalvay, G. J. (1986). *Soil Sci. Soc. Am. J.* **50**, 1494–1499.

Thomspon, B. D. (1987). *Austral. J. Argic. Res.* **38**, 847–867.

Tinker, P. B. (1978). *Physiol. Vég.* **16**, 743–751.

Tisdall, J. M. and Oades, J. M. (1982). *J. Soil Sci.* **33**, 141–163.

Tommerup, I. C. (1983). *Trans. Br. Mycol. Soc.* **81**, 37–45.

Trappe, J. M. (1982). In *Advances in Food-Producing Systems in Arid and Semiarid Lands* (J. T. Menassah and E. J. Briskey, eds), pp. 581–599. Academic Press, New York.

Trappe, J. M. (1987). In *Ecology of VA Mycorrhizal Plants* (G. A. Safir, ed.), pp. 5–25. CRC Press, Boca Raton, FL.

Trappe, J. M., Molina, R. and Castellano, M. (1984). *Ann. Rev. Phytopathol.* **22**, 331–359.

van Kessel, C., Singleton, P. W. and Hoben, H. J. (1985). *Plant Physiol.* **79**, 562–563.

Walworth, J. L. and Sumner, M. E. (1987). *Adv. Soil Sci.* **6**, 149–188.

Warner, A. and Mosse, B. (1980). *Trans. Br. Mycol. Soc.* **74**, 407–410.

Weijman, A. C. and Meuzelaar, H. L. (1978). *Can. J. Bot.* **57**, 284–291.

Wilson, D. O. (1988). *Plant Soil* **110**, 69–75.

Wilson, G. W. T., Daniels Hetrick, B. A. and Gerschefske Kitt, D. (1988). *Mycologia* **80**, 338–343.

# 21

# Vesicular-arbuscular Mycorrhizal Fungi in Nitrogen-fixing Systems

JOSE MIGUEL BAREA, ROSARIO AZCON and CONCEPCION AZCÓN-AGUILAR

*Departamento de Microbiología, Estación Experimental del Zaidín, Profesor Albareda 1; 18008 Granada, Spain*

## I. Introduction

### A. Microbial interactions in the rhizosphere

The presence of roots and the resulting supply of certain plant products, which form microbial substrates, to the root–soil interface, stimulate the development of soil microbiota in the rhizosphere (Bowen, 1989; Foster and Bowen, 1982). Current methods for studying soil and rhizosphere micro-organisms include the use of electron microscopy, direct and indirect antibody staining techniques, radiotracers or enzyme-linked

METHODS IN MICROBIOLOGY
VOLUME 24  ISBN 0-12-521524-X

Copyright © 1992 by Academic Press Limited
All rights of reproduction in any form reserved

immunosorbants; respiration and ATP measurements; analysis of bio-chemical (enzymatic) activity; DNA probes, etc. These methods have been recently discussed in several textbooks (Lynch, 1983; Curl and Truelove, 1986; Paul and Clark, 1989). The analysis of the methodologies used to study microbial dynamics in the rhizosphere continues to be a topic of interest (Bowen, 1990), and descriptions of techniques for screening the root-colonizing ability of soil micro-organisms are common in the literature (Misaghi, 1990).

Both microbe–plant and microbe–microbe interactions have been the focus of a variety of experimental approaches. Assays to demonstrate the role of soil micro-organisms in changing the availability status of plant nutrients and to evaluate the production of growth substances or enzymes have been reviewed in the above-mentioned textbooks. It is known that these microbial activities are able to modify the geometry, distribution and size of the root system (Bowen, 1980).

Micro-organisms in the rhizosphere live in discrete microhabitats where they interact. Antagonism, competition and synergism have been demonstrated by *in vitro* studies (Stotzky, 1972). However, our knowledge of microbial interactions *in situ* is still fragmentary. Efforts are being made to remedy this since the information is vital if we are to try to manage rhizosphere populations with respect to plant nutrition and health (Linderman, 1986; Newman, 1978; Suslow, 1982; Campbell, 1985).

## B.  Interactions involving vesicular-arbuscular mycorrhizal fungi as components of rhizosphere microbiota

The fungus–plant mutualistic symbioses termed mycorrhiza are critical components of the root–soil interface. They function as the absorptive organs of the plant, but they are also members of the soil population. The spores and extramatrical mycelia of vesicular-arbuscular mycorrhizal fungi thus constitute a fraction of the soil-borne fungal biomass (Harley and Smith, 1983). Both qualitative and quantitative analyses of vesicular-arbuscular mycorrhizal fungi have been carried out and these studies are reported elsewhere in this volume (Chapter 2 by Tommerup and Chapter 3 by Sylvia).

Important to our understanding of rhizosphere biology is the study of the interactions between vesicular-arbuscular mycorrhizal fungi and other soil micro-organisms. A number of experiments, as reviewed by Barea and Azcón-Aguilar (1982a) and Azcón-Aguilar and Barea (1991), describe the effects of soil microbiota on vesicular-arbuscular mycorrhiza formation and function and, conversely, the influence of vesicular-arbuscular mycorrhiza on rhizosphere populations. Some selected assays will be briefly discussed below, as an introduction to the main topic of this chapter.

## 1. *Effect of soil micro-organisms on vesicular-arbuscular mycorrhiza formation*

Only a fraction of a given vesicular-arbuscular mycorrhizal fungus spore population inoculated into unsterilized soils is able to germinate (Powell, 1976). This is proof of the occurrence of a fungistasis against spores of vesicular-arbuscular mycorrhizal fungi that can, on occasion, prevent germination completely (Tommerup, 1985). *In vitro* assays, however, demonstrate both antagonistic and beneficial effects of soil micro-organisms on spores of vesicular-arbuscular mycorrhizal fungi. Actinomycetes usually appear as antagonists but Mugnier and Mosse (1987) found that some isolates stimulated spore germination.

Water–agar has been employed as the basal medium in a number of studies on spore germination and subsequent hyphal growth; for example, to assess the influence of soil extracts. Soil extracts have been tested in both the unsterilized and autoclaved conditions, and even as dialysates (Azcón-Aguilar and Barea, 1991). Soil micro-organisms were able to stimulate the germination of spores in co-cultures in Petri dishes before any physical contact occurred (Azcón-Aguilar *et al.*, 1986a; Azcón, 1987), and even when the vesicular-arbuscular mycorrhizal fungus and the stimulatory micro-organisms grew in separate compartments of the plate (Mugnier and Mosse, 1987). This indicates that water-soluble and/or volatile substances are involved in the interaction.

Hyphal development from germinating surface-sterilized spores was stimulated in water–agar, as demonstrated by methods described by Mayo *et al.* (1986) and Azcón-Aguilar *et al.* (1986a). It has also been suggested that soil micro-organisms can improve the "saprophytic" growth of *Glomus mosseae* in soil. The method (Azcón-Aguilar and Barea, 1985) consisted of incubating the spores of the vesicular-arbuscular mycorrhizal fungus for 2 weeks, either axenically or together with a mixed population of micro-organisms, in the absence of any host plant. Then, mycotrophic seedlings were used to evaluate the microbial effects on the vesicular-arbuscular mycorrhizal fungus in terms of initiation of infection. Methods to determine biologically active substances (plant hormones, amino acids, vitamins, etc.), which are possibly responsible for the microbial influence on vesicular-arbuscular mycorrhizal fungi, are in widespread use in soil biotechnology (Lynch, 1983).

## 2. Effect of vesicular-arbuscular mycorrhiza on microbial populations

The development of a mycorrhizosphere is the consequence of changes induced by vesicular-arbuscular mycorrhiza formation in the mineral composition of both the root tissues and the rhizosphere, and in the rate of production and composition of root exudates (Linderman, 1988). Scanning electron microscopy confirms the development of bacteria on the surface of external mycelia of vesicular-arbuscular mycorrhiza (Vancura *et al.*, 1988).

Assays for particular functional groups of micro-organisms show that vesicular-arbuscular mycorrhiza specifically affect some of these groups (Meyer and Linderman, 1986a; Azcón, 1989). The stimulatory effect was proved to be also dependent on the particular vesicular-arbuscular mycorrhizal fungus involved (Secilia and Bagyaraj, 1987). Ames *et al.* (1984b) found in their pot experiments a positive correlation between particular bacterial populations and the mycorrhizal status and recommended that the results be expressed as colony-forming units per g of dry root.

## II.  Vesicular-arbuscular mycorrhiza and nitrogen-fixing bacteria associations

It is well known that a range of soil bacteria possess the ability to cycle nitrogen from the atmosphere to the biosphere by means of the so-called nitrogen-fixing process. All of these bacteria (with the exception of *Sesbania* sp., which also nodulates on stems) live either in the endorhizosphere (forming root nodules), in intimate association with the root surface or rhizoplane, or in the rhizosphere. Therefore, they must coexist with vesicular-arbuscular mycorrhizal fungi and/or vesicular-arbuscular mycorrhiza in the ecosystem. Coexistence commonly involves interaction and there is evidence that such interactions occur, either at the colonization and/or at the functional and nutritional levels (Azcón-Aguilar and Barea, 1991).

### A.  *Rhizobium*

Leguminous plant species in the subfamilies Papilionoideae and Mimosoideae, which usually develop nodules in association with *Rhizobium* (this generic name includes *Bradyrhizobium* and *Azorhizobium*, and it is used here to simplify presentation), are also characteristically colonized by vesicular-arbuscular fungi (Hayman, 1986). Both microsymbionts are known to interact with one another and with the host legume (Barea and Azcón-Aguilar, 1983). Moreover, the genetic analysis carried out by Duc *et al.* (1989), which led to the isolation of non-mycorrhizal legume plant mutants (myc⁻), demonstrated that the expression of the myc⁻ character was associated with the nod⁻ (non-nodulation) character. These findings indicate a close relationship between vesicular-arbuscular mycorrhiza and nodule formation processes at the level of a common genetic control mechanism exerted by the host plant. Ecophysiological and biochemical approaches have been developed to elucidate the basis of the vesicular-arbuscular mycorrhiza (or vesicular-arbuscular mycorrhizal fungus) × *Rhizobium* interactions. Such interactions have been found to occur: (1) at the formation of the symbiosis; (2) at the level of the nutritional status of the plant; and (3) at the level of development and activity of the tripartite symbiosis, which is host-mediated. Selected experiments concerning these interactions will now be analysed systematically.

*1. Interactions concerning vesicular-arbuscular mycorrhiza and nodule formation*

*Rhizobium* and vesicular-arbuscular mycorrhizal fungi can interact at the pre-colonization stages, or during the early development of vesicular-arbuscular mycorrhiza and root nodules. Direct evidence for a microbe–microbe interaction in the rhizosphere is lacking but experiments carried out *in vitro* (Gonzalez, 1988) showed that *Rhizobium* sp. produced an enhancement of the development of the mycelium emerging from surface-sterilized spores of the vesicular-arbuscular mycorrhizal fungus *G. mosseae* in co-culture.

Azcón-Aguilar *et al.* (1980) obtained cell-free extracellular polysaccharides (EPS) from *Rhizobium meliloti* and applied these products in a defined medium (soil–sand, 1:1) to assess their influence on vesicular-arbuscular mycorrhiza formation between *Medicago sativa* seedlings and *G. mosseae*, under controlled conditions. The EPS compounds known to increase root exudation rates in the legume (Olivares *et al.*, 1977), were able to enhance the level of the vesicular-arbuscular mycorrhiza (number of "entry points" and growth of "infection units") in *M. sativa* roots. These results are evidence of a positive influence of rhizobial product on vesicular-arbuscular mycorrhizal fungus, which, since it involves exudation from roots, may be at least partly mediated by the plant (Barea, 1986).

Specific bioassays allowed the detection and quantification of the presence of the plant hormones auxins, gibberellins and cytokinins in cell-free supernatants of *Rhizobium meliloti* cultures (Azcón *et al.*, 1978). The addition of such supernatants to pot cultures of *M. sativa* and *G. mosseae* resulted in an improvement of vesicular-arbuscular mycorrhiza formation at a level similar to that induced by a mixture of these hormones, as pure substances, applied at concentrations equivalent to those bioassayed in cell-free supernatants of rhizobial cultures (Azcón *et al.*, 1978; Azcón-Aguilar and Barea, 1978). Since nodulation was not apparent, the effects of *Rhizobium* were not obviously due to nitrogen fixation.

It has been shown that vesicular-arbuscular mycorrhizal fungi and *Rhizobium* do not compete for infection sites and that both endophytes colonize simultaneously (Smith and Bowen, 1979), except in cases where the photosynthetic rate is limiting (Bethlenfalvay *et al.*, 1985) (see Section II.A.3). However, under certain conditions, vesicular-arbuscular mycorrhiza can change the pattern of distribution of nodules along the root system (Patterson *et al.*, 1990). Plant growth promoting rhizobacteria (PGPR) can influence the effect of vesicular-arbuscular mycorrhiza on nodulation (Meyer and Linderman, 1986b). This is the basis for the development of experimental approaches in the general context of rhizosphere biotechnology.

An important point in rhizobial ecology is to ascertain whether vesicular-arbuscular mycorrhiza can improve the ability of a given efficient strain of *Rhizobium* to compete with other less efficient but more competitive strains. The pot experiment by Ames and Bethlenfalvay (1987) however did not support the conclusion that vesicular-arbuscular mycorrhiza altered the competitiveness for nodule sites within a rhizobial population. Moreover, the amount of vesicular-arbuscular mycorrhizal colonization did not affect the relative competitive abilities of two *Rhizobium* strains (Hicks and Loynachan, 1987) under either controlled or field conditions. In contrast, Nambiar and Anjaiah (1989) found that vesicular-arbuscular mycorrhizal fungi influenced the competitiveness of *Rhizobium* strains and, conversely, that these affected the amount of vesicular-arbuscular mycorrhizal colonization.

The suggestion that vesicular-arbuscular mycorrhiza could reduce root hair formation, thereby affecting nodulation (Linderman, 1988), must be tested experimentally.

### 2.    Interactions concerning the nutritional status of the plant

Between the pioneering work by Asai (1944) and very recent publications, for example those of Azcón *et al.* (1991), a great number of

papers have described the positive effects of vesicular-arbuscular mycorrhiza on nodulation and nitrogen fixation by *Rhizobium*–legume
associations. The conclusions of these studies have recently been reviewed (Azcón-Aguilar and Barea, 1991). In summary, the available
information indicates that, in spite of the occurrence of a range of
responses, the main reason for the positive interaction observed in the
tripartite symbiosis is that the vesicular-arbuscular mycorrhizal infection
satisfies the high phosphorus demand of the nitrogen fixation processes.
The vesicular-arbuscular mycorrhiza has been shown to have a generalized influence on plant nutrition but more localized effects, exerted at
the root, nodule or bacteroid levels, have also been described. These
localized effects can be explained by the ranking of phosphorus-depen-
dency in the processes involved which is as follows: nitrogenase activity
> nodulation > plant growth.

Experiments have been carried out under growth chamber, greenhouse and field conditions. The commonest are pot experiments in the
greenhouse. The soils used for both greenhouse and field experiments
are usually low-phosphorus soils and the natural population of microsymbionts is, or is not, present, according to the aims of the study. The
main legume species tested are those recorded by Hayman (1986).

The control treatments commonly used include: (1) unamended non-
inoculated soils; (2) additions of nitrogen (N) and/or phosphorus (P) to
match inoculation effects; (3) use of P-response curves; (4) split root
systems. In such experiments harvesting is usually carried out as (1) one
final harvest; (2) two to four harvests from the same plant following
regrowth; or (3) a time-course with harvests of different replicate pots
(or plots) at each scheduled plant growth stage.

The plant parts harvested for analysis include: (1) shoots (or stems
and leaves separately); (2) roots; (3) nodules; (4) pods; and (5) mature
fruits.

The commonest parameters determined are: (1) dry matter yield; (2)
N and P concentration and content; (3) K, Ca and Mg content; (4)
micronutrients; (5) extent of nodulation and vesicular-arbuscular mycorrhizal development; (6) photosynthesis-related parameters ($CO_2$
exchange rate, leaf area, starch concentration, N or P use efficiency of
$CO_2$ fixation, etc.); (7) other ecophysiological parameters (for example
those related to water relationships); and (8) $N_2$-fixation estimates.

Determination of $N_2$-fixation have been achieved by several methods:
(1) assessment of nodule number and biomass (indicative); (2) measurement of leghaemoglobin concentration in the nodules (estimative); (3)
measurement of "difference in N yield" (indirect, estimative); (4)
acetylene reduction assays (indirect, for relative comparisons); and (5)

isotope ($^{15}$N)-aided methodologies (the only ones allowing direct qualitative and quantitative measurement).

Among the published reports, which have supplied both methodological and/or new information on the topic, some experimental approaches are selected for a brief description. These can be grouped as follows:

*(a) Experiments concerning general effects.* A number of greenhouse experiments have demonstrated that vesicular-arbuscular mycorrhizal inoculation improves nodulation and nodule activity, and that the effects, which appear to be P-mediated, seem to be closely related to plant growth and nutritional responses. Single and/or dual inoculations and a final harvest were the basis of these assays. Some early and recent examples are the reports by Crush (1974), Daft and El-Giahmi (1974), Badr El-Dim and Moawad (1988) and Patterson *et al.* (1990). Studies with woody legumes will be discussed in Section II.A.4.

Measurements of acetylene reduction activity and the amount of $H_2$ evolved have enabled the "specific nodule activity" to be calculated, and have allowed the effects of vesicular-arbuscular mycorrhiza on nodule biomass to be distinguished from those on nodule activity (Pacovsky *et al.*, 1986).

*(b) Time-course experiments.* Since the nodules usually contain two to three times more P than the root on which they are formed (Mosse, 1977), the idea that nodules and rhizobial bacteroids actually have a "special demand" for P prompted experiments to ascertain if vesicular-arbuscular mycorrhiza exert a localized preferential P supply to nodule tissues and whether this precedes any effect on plant growth. Accordingly, time-course experiments have been carried out by a number of workers. Successive harvests corroborate that, early in their development, the nodules have first call on P resources, probably as a consequence of their higher P-dependency relative to that of other plant organs (Smith and Daft, 1977; Smith *et al.*, 1979; Asimi *et al.*, 1980; Gueye *et al.*, 1987).

*(c) Experiments based on "P-response curves".* The construction of "P-response curves" for both vesicular-arbuscular mycorrhizal and non-mycorrhizal plants allows some physiological vesicular-arbuscular mycorrhiza-mediated effects on plant development other than those directly P-mediated to be studied (Abbott and Robson, 1984). Such methodology has been used to establish the causes of some vesicular-arbuscular mycorrhiza–*Rhizobium* interactions on nodulation and $N_2$ fixation, and

also to assess the influence of the vesicular-arbuscular mycorrhizal supply of nutrients, other than P, on such interactions. The use of N- and P-matched non-symbiotic plants is useful for such purposes.

Early experiments on this topic (Smith and Daft, 1977; Waidyanatha *et al.*, 1979; Asimi *et al.*, 1980) confirmed that preferential P-supply arose in vesicular-arbuscular mycorrhiza-infected plants and that this satisfied the demands of nitrogenase activity. More recent reports (Mosse, 1986; Fernandes *et al.*, 1987; Habte and Manjunath, 1987; Bell *et al.*, 1989; Morton *et al.*, 1990) have investigated the critical P concentrations in the soil solution for nodulation and $N_2$ fixation using different experimental designs. These concentrations depend on the ecophysiological conditions and on the symbionts involved.

Other studies concerning the use of "P-response curves" or comparisons involving non-symbiotic legumes given N and P will be discussed in other sections of this chapter.

*(d) Experiments using $^{15}N$.* It is known that the addition of a small amount of $^{15}N$-enriched inorganic fertilizer and the use of an appropriate "non-fixing" reference crop is the basis for a direct method which allows the relative contribution of the N sources for a fixing crop, i.e. soil, atmosphere and fertilizer, to be distinguished (Danso, 1988). Consequently, $^{15}N$-based methodologies also offer the possibility of assessing the influence of any treatment on $N_2$ fixation distinct from soil or fertilizer N supply. This isotope assay can be used to measure the effect of vesicular-arbuscular mycorrhiza on $N_2$ fixation.

The development of the technique and the equations to calculate the amount of N derived from fixation, and other associated parameters, were adapted from the specialized literature for use in vesicular-arbuscular mycorrhiza inoculation studies by Barea *et al.* (1987, 1989a,b). The experiments described in these papers, together with those reported by Subba Rao *et al.* (1986), Kucey and Bonetti (1988) and Shivaram *et al.* (1988), give data on the evaluation of the amount of N actually fixed by a legume–*Rhizobium* association as affected by vesicular-arbuscular mycorrhiza inoculation, under a number of ecophysiological conditions.

When only qualitative estimates of the $N_2$ fixation level are needed, for example to rank the effectiveness of several vesicular-arbuscular mycorrhizal fungi a given legume–*Rhizobium* association, the $^{15}N$-aided methods do not need a reference crop. This minimizes errors and saves $^{15}N$-labelled material. A small amount of the nitrogenous tracer (for example, $2\,mg\,kg^{-1}$ of soil, i.e. $5\,kg\,N\,ha^{-1}$, as $(NH_4)_2SO_4$ with 10% $^{15}N$ atomic excess) is added to each pot (or plot) for all treatments. After growing the plant, the lower the $^{15}N/^{14}N$ ratio in the plant tissue, the better the treatment is in enhancing $N_2$ fixation (Danso, 1988). Such a qualitative procedure has been used by Azcón *et al.* (1988, 1991) for several purposes (see Sections II.A.4 and III).

By calculating the $A_N$ value (the apparent size of the pool of soil N) for a crop it has been possible to demonstrate that vesicular-arbuscular mycorrhiza do increase such $A_N$ values (Ames *et al.*, 1984a; Barea *et al.*, 1989b), indicating that mycorrhizal plants can either use available N forms more efficiently, or derive N from sources less available to non-mycorrhizal plants. This is important in mixed cropping where legumes are usually involved. Since vesicular-arbuscular mycorrhizal mycelia can link different plant species and thus make common the pool of available nutrients for the intercropped plant species (Newman, 1988), the N released into the overlapping mycorrhizospheres by legume root exudation, or by nodule decay, can result in N becoming available for non-fixing plants. With appropriate experimental designs using $^{15}N$, the role of vesicular-arbuscular mycorrhiza can be assessed (Kessel *et al.*, 1985; Haystead *et al.*, 1988; Barea *et al.*, 1989a,b).

*(e) Implication for nutrients other than P.* As previously discussed (Hayman, 1986), vesicular-arbuscular mycorrhiza can help nodulation and $N_2$ fixation by supplying the plants with nutrients, in addition to P, that are known to be involved in these processes. Particularly the uptake of Cu, Zn, and sometimes Fe, have been described as being improved in vesicular-arbuscular mycorrhizal legumes (Pacovsky, 1986; Kucey and Janzen, 1987; Rai, 1988). A recommended experimental approach to investigate such a role of vesicular-arbuscular mycorrhiza would be that followed by Pacovsky (1986), which combines sequential harvests with the use of appropriate P-treated non-vesicular-arbuscular mycorrhizal plants (soybeans), as controls for the inoculated ones. He used non-vesicular-arbuscular mycorrhizal plants of similar weight, growth stage, and P status, which are very useful for assessing micronutrient uptake.

Other key studies on vesicular-arbuscular mycorrhizal fungus × *Rhizobium* regarding nutritional interactions include those related to effects in stress situations (Section II.A.4) and those concerning the manipulation of the tripartite symbiosis, whether under field or controlled conditions (Section III).

## 3.   Host-mediated effects on symbiotic interactions

The use of N- and/or P-sufficient ("fertilizer-compensated") non-nodulated and/or non-vesicular-arbuscular mycorrhizal legume plants allows several aspects of the interactions between *Rhizobium* and vesicular-arbuscular mycorrhizal fungi to be studied. Since the experimental details are presented elsewhere (see Bethlenfalvay, Chapter 20, this volume) only brief descriptions will be given, as follows:

*(a) Carbon economy of the tripartite association.* Since they are hetero-trophic micro-organisms, vesicular-arbuscular mycorrhizal fungi and *Rhizobium* obtain C compounds from their autotrophic hosts by means of biotrophic transfer across the living membranes of the partners involved. Thus, the interest in assessing the energy cost of the tripartite symbiosis has prompted several experiments. Pang and Paul (1980) used both $^{14}CO_2$ and $^{15}N_2$ to follow the distribution of these nutrients as affected by microsymbiont inoculation in faba beans. Similarly, Kucey and Paul (1982) studied C allocation in the tripartite symbiosis and found that the $CO_2$ fixation rate, expressed as $g C g^{-1}$ shoot dry matter $h^{-1}$, increased in symbiotic plants. This is a mechanism that enhances photosynthesis and so compensates for the cost of the symbiosis.

A further development of the technique, also based on determining the distribution of $^{14}CO_2$, is that reported by Harris *et al.* (1985). These authors measured the "specific leaf area" in soybean and found that it increased in symbiotic plants. This may explain, at least partially, the increase they detected in rates of $^{14}CO_2$ fixation in leaves of vesicular-arbuscular mycorrhizal and nodulated soybeans. This, together with measurements of increased starch mobilization and of raised leaf P concentration, suggests that the host becomes adapted to support its endophytes.

*(b) Localized versus systematic effect of vesicular-arbuscular mycorrhiza on nodulation.* A model experiment in this area is that of Ames and Bethlenfalvay (1987). They used a split-root system, one side of which (Side 1) they either inoculated with a vesicular-arbuscular mycorrhizal fungus or gave increasing amounts of assimilable P. The other side (Side 2) was given the lowest P dose applied on Side 1. Both sides received a rhizobial inoculum. Analysis of N and P content, and of nodule activity by acetylene reduction, on both sides of the split-root system, showed a localized, non-systemic and non-P-mediated influence of vesicular-arbus-cular mycorrhiza on $N_2$-fixation.

Results such as these suggest that complex interactions could be involved (Barea and Azcón-Aguilar, 1982b).

*(c) Photosynthetic nutrient-use efficiency in symbiotic legumes.* Source (chloroplast) to sink (the endophytes) relationships are important in the regulation of the symbiotic status in vesicular-arbuscular mycorrhiza +

nodulated legumes. In a series of experiments Brown and Bethlenfalvay (1988) and Brown *et al.* (1988) studied some of the influencing factors. The process of photosynthesis, the C source, is in turn a sink for N and P, therefore parameters such as carbon dioxide exchange rate (CER), leaf area and dry weight, and leaf N, P and starch concentration, are critical for ascertaining the photosynthetic nutrient-use efficiency (CER per unit leaf nutrient). Brown and Bethlenfalvay (1988) demonstrated by this approach that both types of endophytes enhance CER at N and/or P concentrations lower than those of the non-symbiotic soybeans.

Using soybeans, Brown *et al.* (1988) determined P-use efficiency of $CO_2$ fixation in vesicular-arbuscular mycorrhizal nodulated plants, and found that it was higher than in non-mycorrhizal plants receiving a high level of P (HP). However, the P-use efficiency of nitrogenase activity was higher in the HP plants. Brown *et al.* (1988) also measured the ratio of nodule to chloroplast activity, expressed as mol $C_2:H_2$ reduced (mol $CO_2$ fixed)$^{-1}$ and found it higher in HP than in mycorrhizal soybeans. Under the ecophysiological conditions of this experiment, vesicular-arbuscular mycorrhizal plants had a lower nodule activity than HP plants. In spite of this, vesicular-arbuscular mycorrhiza increased both plant biomass and N concentration relative to HP. These facts confirm that vesicular-arbuscular mycorrhiza elicit host-mediated effects.

All in all it is important to recognize that because of the complexity of host-mediated microsymbiont interactions, and because of the diversity of environmental conditions to be tested, individual experiments are only one step in the research process to find the causes of such interactions.

## 4. Interactions under stress conditions

A number of situations of different origin, adverse to plant development, are known to influence the interaction between vesicular-arbuscular mycorrhizal fungi and *Rhizobium* (Mosse, 1986). In particular, the effect of vesicular-arbuscular mycorrhizal fungi in helping nodulation and $N_2$-fixation are of great relevance, and these have been recently reviewed (Barea, 1990). Some selected papers describing experiments on the topic are described, the emphasis being on those using controls adequately supplied with P.

*(a) Water stress.* Experiments using $^{15}N$ have confirmed that drought stress affects $N_2$-fixation (Kirda *et al.*, 1989). Since vesicular-arbuscular mycorrhiza can help plants to cope with such stress, the role of the mycorrhizal symbiosis in legumes is particularly interesting. Features

such as leaf water potential, photosynthesis rate, stomatal conductance and hormonal patterns have been examined in association with measurements of nodule activity (Cooper, 1984). Peña et al. (1988) compared the effects of vesicular-arbuscular mycorrhiza and P addition on the Medicago sativa–Rhizobium meliloti symbiosis. They stressed the plants by cyclically withholding water and rewatering to different soil water potentials. Positive effects of vesicular-arbuscular mycorrhiza on nodule activity (acetylene reduction) were observed and it was suggested that non-P-mediated mycorrhizal effects were involved. Azcón et al. (1988) investigated the effects of drought and mycorrhizal infection on $N_2$-fixing activity ($^{15}N$) using a simple experimental design consisting of increasing P levels, or a vesicular-arbuscular mycorrhizal inoculum, and differing soil water contents. They found that vesicular-arbuscular mycorrhizal infection was the most effective treatment in improving nodulation and $N_2$ fixation at the lower levels of water potential.

(b) Salinity. Negative effects of salinity on nodulation and $N_2$ fixation can be compensated by vesicular-arbuscular mycorrhiza. El-Atrach et al. (1989) compared vesicular-arbuscular mycorrhiza inoculation with a series of assimilable P levels, combined with several levels of induced salinity. A $^{15}N$ tracer was added to all pots. A positive effect of vesicular-arbuscular mycorrhiza on $N_2$ fixation was demonstrated, particularly at the higher level of salinity, where vesicular-arbuscular mycorrhizal inoculation provided the greatest improvement of nodulation and $N_2$ fixation.

(c) Heavy metals and "nutrients in excess". There is published information supporting the role of vesicular-arbuscular mycorrhiza in decreasing the acquisition by plants of certain elements when they are present at superoptimal, mostly toxic, concentrations in soil (Mosse et al., 1981). In legumes the role of vesicular-arbuscular mycorrhiza appears critical also for $N_2$ fixation. Heggo et al. (1990) discuss available experimental information and give an example for the design of greenhouse assays. Vesicular-arbuscular mycorrhiza exert a "buffer effect", decreasing Zn, Cd and Mn concentrations in soybeans to physiological levels, to the benefit of nodulation. This is important in polluted environments (Barea, 1990).

The role of vesicular-arbuscular mycorrhiza in decreasing the uptake of a nutrient when it is present in superoptimal concentrations in soil is a topic of current interest. For example, acid soils can contain an excess of available Mn, and calcareous soils an excess of assimilable Ca. Using red clover as test legume, Arines et al. (1989) demonstrated a benefit of vesicular-arbuscular mycorrhizal fungi in lowering the acquisition of Mn in acid soils where the nutrient was present at toxic levels. They used an approach based on the calculation of the nutrient: phosphate ratios in plant tissues to estimate the relative lowering in the uptake of a given nutrient (Arines and Vilariño, 1989). A similar situation was found by Azcón *et al.* (1991) regarding the role of vesicular-arbuscular mycorrhiza in *Rhizobium* activity ($^{15}$N) in calcareous soils containing superoptimal concentrations of assimilable Ca. They found that vesicular-arbuscular mycorrhiza had a beneficial "buffering effect" in decreasing Ca acquisition from these soils.

*(d) Revegetation of eroded and desertified soils.* A working methodology to study the tripartite interactions vesicular-arbuscular mycorrhizal fungus–legume–*Rhizobium* to exploit the benefit of $N_2$ fixation in revegetation programmes will be discussed in Section III.B.

*(e) Others (plant pathogens).* Evidence is accumulating to show that vesicular-arbuscular mycorrhiza could be used as possible biocontrol agents (Bagyaraj, 1984). Literature concerning legumes has been reviewed (Hayman, 1986). Experimental design to evaluate the significance of vesicular-arbuscular mycorrhiza in protecting nodulation from plant pathogen attacks must include testing of interactions among particular legume species and vesicular-arbuscular mycorrhizal fungus–*Rhizobium*–pathogen species/strains.

## B. Root-nodulating actinomycetes

*Frankia* spp. (actinomycete) are able to form $N_2$-fixing nodules in the root of certain woody, non-leguminous, plants. These plant usually form mycorrhiza, and vesicular-arbuscular mycorrhiza have often been described (Rose, 1980; Daft *et al.*, 1985; Gardner, 1986; Trappe, 1986). The presence of both types of endophytes enable "actinorrhizal" plants to thrive in nutrient-poor soils and increase their potential in forestry (Gauthier *et al.*, 1983). The physiological aspects and ecological significance of this symbiosis are discussed elsewhere (Cervantes and Rodríguez-Barrueco, Chapter 22, this volume) and only brief mention is appropriate here.

Interactions between vesicular-arbuscular mycorrhizal fungi and *Frankia* are essentially similar to those with *Rhizobium*. However "communication" between the coexisting endophytes seems easer in actinorrhizal symbiosis because vesicular-arbuscular mycorrhizal fungi are able to colonize the nodular tissues (a feature not common in legumes). Nevertheless, vesicular-arbuscular mycorrhizal fungi do not penetrate the zone of the actinorrhizal nodules containing the *Frankia* endophyte (see Gardner, 1986).

The isotope $^{15}$N was used to evaluate N$_2$ fixation by *Casuarina* sp. inoculated with *Frankia* isolates using 1 m$^3$ containers (Gauthier *et al.*, 1985). It is important to use isotopic techniques to ascertain the actual levels of N$_2$ fixation and to evaluate the impact of vesicular-arbuscular mycorrhiza inoculation on N$_2$ fixation, even though such improvement is known to occur (Gauthier *et al.*, 1985; Gardner *et al.*, 1984; Russo, 1989). Ectomycorrhiza coexist with vesicular-arbuscular mycorrhiza in some cases and experiments on the appropriate time of inoculation of the different mutualists involved are being designed to investigate the physiological relationships in these tetrapartite symbioses (Chatarpul *et al.*, 1989).

## C.   Other symbiotic nitrogen-fixing associations

According to Jeffrey (1987) nine genera of the gymnosperm order Cycadales can be nodulated by the cyanobacteria *Nostoc* or *Anabaena*. Such symbiotic N$_2$-fixing associations are helped by vesicular-arbuscular mycorrhizal fungi (Trappe, 1986). Ecophysiological studies on these systems would be of interest.

Some species of the genus *Sesbania* form stem nodules, in addition to those that typically develop in the root. The endophyte is a specific *Rhizobium* sp. Using $^{15}$N, Ndoye and Dreyfus (1988) gave experimental evidence of N$_2$-fixing activity in the stem nodulation. The enhancement of such activity by vesicular-arbuscular mycorrhiza inoculation is a topic deserving future research.

*Parasponia* (Ulmaceae) is the only non-leguminous genus able to be nodulated by *Rhizobium* sp. The N$_2$-fixing root nodules coexist with vesicular-arbuscular mycorrhizal fungi (Trappe, 1986), and endophyte interactions could be similar to those in legume species.

## D.   Nitrogen-fixing free-living micro-organisms

The colonization of the rhizosphere soil or of the rhizoplane by free-living bacteria possessing nitrogenase activity is a research topic of considerable interest.

The inoculation of seeds or seedlings with cultures of these micro-organisms produces changes in the rooting pattern and shoot morphology which resemble those induced by plant hormones. Hormonal substances have indeed been demonstrated in these cultures (Barea and Brown, 1974; Azcón and Barea, 1975). However, the ability of these procaryotes to fix $N_2$ in situ and thus to enhance the flow of ammonium from the microbial cells to the host is limited. The use of $^{15}N$, either as gas ($^{15}N_2$), or by the "isotope dilution technique" ($^{15}N$-labelled ammonium sulphate) suggests that restricted supply of N may occur from fixation (Giller et al., 1984; Rennie and Thomas, 1987; Malik et al., 1988).

Since both the involvement of growth substances and N supply can benefit vesicular-arbuscular mycorrhiza formation and function (Azcón et al., 1978; Bagyaraj, 1984; Harari et al., 1988) a number of experiments have been aimed at studying the interactions between vesicular-arbuscular mycorrhiza and diazotrophic bacteria. Most of the assays used very simple designs consisting of the inoculation of either vesicular-arbuscular mycorrhizal fungi, the diazotroph, or both together. Control plants added with N and/or P were sometimes used to compensate for the input of nutrient due to microbial colonization. Plant species from the family Gramineae were the usual hosts, but some horticultural crops were also assayed. Representative experiments on vesicular-arbuscular mycorrhiza × diazotrophic free-living bacteria are:

## 1. Azotobacter

Greenhouse experiments indicate positive interactions as estimated by growth and nutritional parameters. Beneficial effects of vesicular-arbuscular mycorrhiza formation on Azotobacter establishment have been reported (Barea et al., 1973; Bagyaraj and Menge, 1978; Brown and Carr, 1984; Ho, 1988). A field experiment using tomato as test plant described additive vesicular-arbuscular mycorrhizal fungus × Azotobacter effects on fruit yield (Mohandas, 1987).

## 2. Azospirillum

Studies by Barea et al. (1983), Subba Rao et al. (1985a,b), Pacovsky et al. (1985) and Pakovsky (1988, 1989) have described increases, either synergistic or additive, on plant growth due to the interaction between vesicular-arbuscular mycorrhizal fungi and Azospirillum. The occurrence of Azospirillum in cultures of surface-sterilized spores of vesicular-arbuscular mycorrhizal fungi, which appear to show nitrogenase activity (acetylene reduction), is of interest (Tilak et al., 1989).

### 3.  Other diazotrophs

The ability of bacteria belonging to the genera *Bacillus* (Rennie and Thomas, 1987), *Klebsiella* and *Beijerinckia* (Malic *et al.*, 1988) to fix and transfer N to the plant have been examined using the isotopic method ($^{15}N$ dilution). Their interaction with vesicular-arbuscular mycorrhizal fungi offers a research situation similar to that for the other free-living organisms.


### III.  Manipulation of vesicular-arbuscular mycorrhiza–nitrogen-fixing bacteria

Most of the experiments on this topic concern symbiotic systems largely involving *Rhizobium*. As is well known (see Bagyaraj, Chapter 19, this volume), problems associated with production of inocula and the development of inoculation techniques limit our ability to manipulate vesicular-arbuscular mycorrhiza. However, biotechnological and ecophysiological studies now provide the potential to harness the vesicular-arbuscular mycorrhiza–*Rhizobium* interactions so that their possible benefits in agriculture, horticulture, and forestry can be realized. Accordingly the experimental approaches will be analysed in two groups: (1) those aimed at assessing "functional compatibility" between symbionts, and (2) those developed under natural conditions, with a practical aim, mostly using field assays.


### A.  Functional compatibility

Gianinazzi-Pearson (1984) introduced the concept of "functional compatibility" which associates "compatibility" with "symbiotic effectivess" in vesicular-arbuscular mycorrhiza. Obviously this could be applied also to tripartite symbioses involving $N_2$-fixing bacteria. The symbiotic responses in a given vesicular-arbuscular mycorrhizal fungus–legume–*Rhizobium* relationship would thus depend on the expression of the genotypic make-up of the three interacting partners, these in turn being modulated by environmental influences.


### 1.  Host genotype influences

Isotopic techniques ($^{15}N$) have been appplied to confirm the genetic variability for $N_2$ fixation in legume–*Rhizobium* symbioses (Danso *et al.*, 1987; Duc *et al.*, 1988); actinorrhiza (Sanginga *et al.*, 1990); and in associations involving free-living bacteria (Giller *et al.* 1986). This effect is conditioned by the host genotype. It is also well known that the level of mycotrophy of the host plant influences the vesicular-arbuscular mycorrhizal response (Barea and Azcón, 1989). The host genotype has been found to affect interactions between vesicular-arbuscular mycorrhizal fungus and *Rhizobium* (Duc *et al.*, 1989; Mercy *et al.*, 1990; Rao *et al.*, 1990).

## 2. *Microbial selection*

When a legume is inoculated either singly or with a dual combination with different *Rhizobium* strains and/or vesicular-arbuscular mycorrhizal fungi, the extent of symbiotic response greatly depends on the species/strains of the microsymbionts involved (Azcón-Aguilar *et al.*, 1986a,c; Cardoso, 1986; Cruz *et al.*, 1988; Louis and Lim, 1988). Double antibiotic-marking for *Rhizobium*, and the application of either morphological studies (López-Aguillon and Mosse, 1987) or iso-enzymatic pattern analysis (Hepper *et al.*, 1988), can be used to distinguish endophytes co-colonizing a common root system.

Specific relationships between endophytes (*Rhizobium* and vesicular-arbuscular mycorrhizal fungi) have been recently described (Bayne and Bethlenfalvay, 1987; Ianson and Linderman, 1989; Von Alten *et al.*, 1989). Moreover, by using $^{15}N$, Azcón *et al.* (1991) confirmed these selective interactions between different species of vesicular-arbuscular mycorrhizal fungi and strains of *R. meliloti* as deduced from determinations of $N_2$ fixation rates.

## 3. *Soil treatments and experimental conditions*

Soil characteristics, fertilizer applications, and other treatments, known to influence plant and/or microbial development (Mosse *et al.*, 1981) are able to modulate the phenotypic response to different vesicular-arbuscular mycorrhizal fungus–$N_2$-fixing bacteria species/strains combination (Barea and Azcón-Aguilar, 1983; Hayman, 1986; Barea, 1990). The experiments, carried out by applying either single of multiple variates (induced environmental conditions, or influencing factors) usually follow very simple designs. These have been reviewed in the papers mentioned above. Some recent experiments are described by Siqueira and Paula (1986), Hirata *et al.* (1988), Michelsen and Rosendahl (1990) and Nair *et al.* (1990) and provide complementary information on the effects of environmental factors and/or conditions on vesicular-arbuscular mycorrhizal fungus–diazotrophic bacterial interactions.

## B.  Field studies

Interactions between *Rhizobium* and vesicular-arbuscular mycorrhizal fungi have been studied under natural conditions (Barea and Azcón-Aguilar, 1983; Hayman, 1986).

### 1.  Experimental designs

Preliminary to a field assay itself, it is recommended that the effectivity of native vesicular-arbuscular mycorrhizal fungi in the given environment be assessed. Evaluation of their vesicular-arbuscular mycorrhizal effects on nodulation or plant growth (Barea *et al.*, 1980; Howeler *et al.*, 1987) and measurements of P-inflow values (McGonigal and Fitter, 1988) have been carried out. Interactions of native and introduced vesicular-arbuscular mycorrhizal fungi have been studied and their effects on nodulation described (Azcón-Aguilar *et al.*, 1986b). The management of natural populations by cultural practices has been discussed (Howeler *et al.*, 1987).

Limitations imposed by the difficulty in producing vesicular-arbuscular mycorrhizal inocula imply that field experiments must be carried out in small plots. Some examples of such experiments are as follows: Azcón-Aguilar *et al.* (1989) carried out a study using only growth and nutritional parameters to compare single or dual inoculation of vesicular-arbuscular mycorrhizal fungus–*Rhizobium*. Rai (1988) developed another field experiment which provided information on nodulation and nitrogenase activity (acetylene reduction) as affected by mycorrhizal infection effect. Barea *et al.* (1987, 1989a) evaluated these interactions using [15]N-aided technology.

### 2.  Trends and future development

In spite of the fact that the biotechnology of inocula production needs to be improved (see Stribley, 1989; Gianinazzi *et al.*, 1990; Barea, 1990), currently available inocula allow investigation of the interactions of vesicular-arbuscular mycorrhizal fungi and diazotrophic bacteria under field conditions. In this context some horticultural plants may be benefited by the dual inoculation of diazotrophic bacteria and vesicular-arbuscular mycorrhizal fungi, since the amounts of mycorrhizal inocula required in horticulture are often less than in extensive agriculture and required amounts can be produced by current techniques.

The use of dual inoculation in woody plants, either legumes or actinorrhizal species, is likely to become more prevalent, in particular for revegetation of eroded, desertified or degraded ecosystems. Such inoculation involves the use of vesicular-arbuscular mycorrhiza together with either *Rhizobium* for woody legumes (Roskoski *et al.*, 1986; Barea *et al.* 1990), or *Frankia* for actinorrhizal species (Gardner, 1986).

There are research programmes aimed at taking advantage of dual inoculation as a biotechnological tool for the management of $N_2$-fixing trees in restoring and maintaining soil fertility. Among these studies, those of Barea *et al.* (1990) involving use of woody legumes, vesicular-arbuscular mycorrhizal fungi and *Rhizobium* in eroded (desertified) soils in arid zones provide useful examples.

Studies of the impact of microsymbiont inoculation on tree establishment, development and $N_2$ fixation are a main part of such programmes. Methodologies using $^{32}P$ and $^{15}N$ are being applied to ascertain the role of mycorrhiza in improving $N_2$ fixation and nutrient and water use efficiency by woody legumes. The role of mycorrhiza in the transfer of nutrients to associated non-fixing plants and in restoring the N and organic matter status of the soil is particularly important.

The work plan includes:

(1) Survey of woody legumes growing in the arid, commonly eroded, soil in the zone selected.
(2) Isolation of rhizobia and mycorrhiza fungi from the rhizosphere of the selected native plant species.
(3) Development of methods, including tissue culture, to produce appropriate plant material.
(4) Selection of microsymbionts by following criteria of compatibility + efficiency with the soil–plant test system.
(5) Production of appropriate inocula.
(6) Development of a procedure to obtain healthy plantlets in optimized nodulated and mycorrhizal status, using the test soil as substrate.
(7) Study of plant establishment in the chosen life zone sites as affected by microbial inoculation.
(8) Use of isotope techniques to evaluate $N_2$ fixation, N-transfer and nutrient turnover in the ecosystem to benefit non-fixing associated crops.
(9) Measurement of the impact on soil fertility. Evaluation of the impacts of the biotechnologies listed above on the stabilization of an agroforestry system.

## IV.   Conclusions and perspectives

The microbial interactions concerning vesicular-arbuscular mycorrhiza or vesicular-arbuscular mycorrhizal fungi and nitrogen-fixing bacteria are of relevance because they can improve plant establishment, development and nutrient acquisition. Taking this into account, two main groups of experimental approaches can be distinguished: (1) studies concerning the establishment of micro-organisms in the rhizosphere (or mycorrhizosphere); and (2) manipulation of these microbial association as a biotechnological tool to improve plant growth. Regarding the first group of topics, several related to the dynamics of microbial colonization of the root–soil interface are the subject of current research. The use of methodologies such as electron microscopy, antibiotic-marking, immunofluorescence and isotopic techniques, ELISA, genetic approaches, etc., are contributing to the understanding of these interactions and to the establishment of the molecular basis of vesicular-arbuscular mycorrhizal fungus–plant–diazotrophic bacteria relationships.

The other group of topics include experiments to assess the determinants of symbiotic functional compatibility for particular fungus–plant–bacteria–environment combinations. This will allow the development of biotests to determine a suitable and feasible microbial inoculation pattern. Additionally, the use of isotopic techniques is crucial to ascertaining the actual significance of the microbial interactions in nutrient cycling in the ecosystem. Biotechnology related to microbial inoculant production and inoculation techniques is a keystone deserving present and future attention in research.

## References

Abbott, L. K. and Robson, A. D. (1984). In *VA Mycorrhiza* (C. L. Powell and D. J. Bagyaraj, eds), pp. 113–130. CRC Press, Boca Raton, FL.

Ames, R. N. and Bethlenfalvay, G. J. (1987). *New Phytol.* **106**, 207–215.

Ames, R. N., Porter, L. K., St John, T. V. and Reid, C. P. P. (1984a). *New Phytol.* **97**, 269–276.

Ames, R. N., Reid, C. P. P. and Inghan, E. R. (1984b). *New Phytol.* **96**, 555–563.

Arines, J. and Vilariño, A. (1989). *Biol. Fert. Soils* **8**, 293–297.

Arines, J., Vilaniño, A. and Sainz, M. (1989). *New Phytol.* **112**, 215–219.

Asai, T. (1944). *Jap. J. Bot.* **12**, 359–408.

Asimi, S., Gianinazzi-Pearson, V. and Gianinazzi, S. (1980). *Can. J. Bot.* **58**, 2200–2206.

Azcón, R. (1987). *Soil Biol. Biochem.* **19**, 417–419.

Azcón, R. (1989). *Soil Biol. Biochem.* **21**, 639–644.

Azcón, R. and Barea, J. M. (1975). *Plant Soil* **43**, 609–619.
Azcón, R., Azcón-Aguilar, C. and Barea, J. M. (1978). *New Phytol.* **80**, 359–369.
Azcón, R., El-Atrach, F. and Barea J. M. (1988). *Biol. Fert. Soils* **7**, 28–31.
Azcón, R., Rubio, R. and Barea, J. M. (1991). *New Phytol.* **117**, 399–404.
Azcón-Aguilar, C. and Barea, J. M. (1978). *Can. J. Microbiol.* **24**, 520–524.
Azcón-Aguilar, C. and Barea, J. M. (1991). In *Mycorrhizal Functioning* (M. F. Allen, ed.). Chapman & Hall, London (in press).
Azcón-Aguilar, C. and Barea, J. M. (1985). *Trans. Br. Mycol. Soc.* **84**, 536–537.
Azcón-Aguilar, C., Azcón, R. and Barea, J. M. (1989). *Nature* **279**, 325–327.
Azcón-Aguilar, C., Barea, J. M. and Olivares, J. (1980). *Second International Symposium on Microbial Ecology*, University of Warwick, Coventry, UK, Abstr. No. 187.
Azcón-Aguilar, C., Díaz-Rodriguez, R. M. and Barea, J. M. (1986a). *Trans. Br. Mycol. Soc.* **86**, 337–340.
Azcón-Aguilar, C., Barea, J. M., Azcón, R. and Díaz-Rodriguez, R. M. (1986b). *Agric. Ecosyst. Environ.* **15**, 241–252.
Azcón-Aguilar, C., Gianinazzi-Pearson, V., Fardeau, J. C. and Gianinazzi, S. (1986c). *Plant Soil* **96**, 3–15.
Badr-El-Din, S. M. S. and Moawad, H. (1988). *Plant Soil* **108**, 117–124.
Bagyaraj, D. J. (1984). In *VA Mycorrhiza* (C. Ll. Powell, and D. J. Bagyaraj, eds), pp. 131–153. CRC Press, Boca Raton, Fl.
Bagyaraj, D. J. and Menge, J. A. (1978). *New Phytol.* **80**, 567–573.
Barea, J. M. (1986). In *Physiological and Genetical Aspects of Mycorrhizae* (V. Gianinazzi-Pearson and S. Gianinazzi, eds), pp. 177–187. INRA, Paris.
Barea, J. M. (1990). In *Advances in Soil Science* (B. A. Stewart, ed.), Vol. 15, pp. 1–40, Springer-Verlag, New York.
Barea, J. M. and Azcón, R. (1989). In *Frontiers in Applied Microbiology* (K. G. Mukerji, V. P. Singh and K. L. Garg, eds), Vol. 3, pp. 135–166. Rastogy and Company, Subhash Bazar, Meerut, India.
Barea, J. M. and Azcón-Aguilar, C. (1982a). In *Les Mycorrhizes: Biologie et Utilization* (S. Gianinazzi, V. Gianinazzi-Pearson and A. Trouvelot, eds), pp. 181–193. INRA, Paris.
Barea, J. M. and Azcón-Aguilar, C. (1982b). *Appl. Environ. Microbiol.* **43**, 810–813.
Barea, J. M. and Azcón-Aguilar, C. (1983). In *Advances in Agronomy* (N. C. Brady, ed.), Vol. 36, pp. 1–54. Academic Press, New York.
Barea, J. M. and Brown, M. E. (1974). *J. Appl. Bacteriol.* **37**, 583–593.
Barea, J. M., Brown, M. E. and Mosse, B. (1973). *Rothamsted Ann. Rep.*, 81–82.
Barea, J. M., Escudero, J. L. and Azcón-Aguilar (1980). *Plant Soil*, **54**, 283–296.
Barea, J. M., De Bonis, A. F. and Olivares, J. (1983). *Soil Biol. Biochem.* **15**, 705–709.
Barea, J. M., Azcón-Aguilar, C. and Azcón, R. (1987). *New Phytol.* **106**, 717–725.
Barea, J. M., Azcón, R. and Azcón-Aguilar, C. (1989a). *New Phytol.* **112**, 399–404.
Barea, J. M., El-Atrach, F. and Azcón R. (1989b). *Soil. Biol. Biochem.* **21**,

581–589.
Barea, J. M., Salamanca, C. P. and Herrera, M. A. (1990). In *Fast Growing Trees and Nitrogen Fixing Trees* (D. Werner, and P. Müller, eds), pp. 303–311. Gustav Fischer-Verlag, Stuttgart.
Bayne, H. G. and Bethlenfalvay, G. J. (1987). *Plant Cell Environ.* 10, 607–612.
Bell, M. J., Middleton, K. J. and Thompson, J. P. (1989). *Plant Soil.* 117, 49–57.
Bethlenfalvay, G. J., Brown, M. S. and Stafford., A. E. (1985). *Plant Physiol.* 79, 1054–1058.
Bowen, G. D. (1980). In *Contemporary Microbial Ecology* (D. C. Ellwood, M. J. Latham, J. N. Hedger, J. N. Lynch and J. M. Slater, eds), pp. 283–304. Academic Press, London.
Bowen, G. D. (1989). In *The Rhizosphere and Plant Growth*, p. 21. Beltsville Symposium XIV, MD.
Brown, M. E. and Carr, G. R. (1984). *J. Appl. Bacteriol.* 56, 429–437.
Brown, M. S. and Bethlenfalvay, G. J. (1988). *Plant Physiol.* 86, 1292–1297.
Brown, M. S., Thamsurakul, S. and Bethlenfalvay, G. J. (1988). *Physiol. Plant.* 74, 159–163.
Campbell, R. (1985). *Plant Microbiology.* Edward Arnold, London.
Cardoso, E. J. B. N. (1986). *R. bras. Ci. Solo* 10, 17–23.
Chatarpaul, L., Chakravarty, P. and Subramaniam, P. (1989). *Plant Soil* 118, 145–150.
Cooper, K. M. (1984). In *VA Mycorrhiza* (C. Ll. Powell and D. J. Bagyaraj, eds), pp. 156–186. CRC Press, Boca Raton, FL.
Crush, J. R. (1974). *New Phytol.* 73, 743–752.
Cruz, R. E. De La, Manalo, M. Q., Aggangan, N. S. and Tambalo, J. D. (1988). *Plant Soil* 108, 111–115.
Curl, E. A. and Truelove, B. (1986). *The Rhizosphere.* Springer-Verlag, Berlin.
Daft, M. J. and El-Giahmi A. A. (1974). *New Phytol.* 73, 1139–1147.
Daft, M. J., Clelland, D. M. and Gardner, I. C. (1985). *Proc. Roy. Soc. Edinburgh* 85, 283–298.
Danso, S. K. A. (1988). In *Nitrogen Fixation by Legumes in Mediterranean Agriculture* (D. P. Beck and L. A. Materon, eds), pp. 245–358. ICARDA Martinus Nijhoff, Dordrecht.
Danso, S. K. A., Hera, C. and Douka, C. (1987). *Plant Soil* 99, 163–174.
Duc, G., Mariotti, A. and Amarger, N. (1988). *Plant Soil* 106, 269–276.
Duc, G., Trouvelot, A., Gianinazzi-Pearson, V. and Gianinazzi, S. (1989). *Plant Sci.* 60, 214–222.
El-Atrash, F., Azcón, R. and Barea, J. M. (1989). *International Conference on the Mechanisms of the Relationship between Soil–Plant–Microorganism in the Rhizosphere*, p. 85.
Fernandes, A. B., Siqueira, J. O., Menezes, M. A. L. and Guedes, G. A. A. (1987). *R. bras. Ci. Solo* 11, 101–108.
Foster, R. C. and Bowen, G. D. (1982). In *Phytopathogenic Prokaryotes* (R. Mount and C. Lacey, eds), Vol. 1, pp. 159–185. Academic Press, New York.
Gardner, I. C. (1986). *MIRCEN J.* 2, 147–150.
Gardner, I. C., Clelland, D. M. and Scott, A. (1984) *Plant Soil* 78, 189–199.
Gauthier, D., Diem, H. G. and Dommergues, Y. (1983). In *Casuarina Ecology Management and Utilization* (S. J. Midgeley, J. W. Turnbull and R. D. Johnston, eds), pp. 211–217. CSIRO, Melbourne.

Gauthier, D., Diem, H. G. and Dommergues, Y. (1985) *Soil. Biol. Biochem.* **17**, 375–379.
Gianinazzi-Pearson, V. (1984). In *Genes Involved in Plant–Microbe Interactions* (D. P. S. Verma and T. H. Hohn, eds), pp. 225–253. Springer-Verlag, Wien.
Gianinazzi, S., Trouvelot, A. and Gianinazzi-Pearson, V. (1990) *Agric. Ecosyst. Environ.* **29**, 153–161.
Giller, K. E., Day, J. M., Dart, P. J. and Wani, S. P. (1984). *J. Microbiol. Meth.* **2**, 307–316.
Giller, K. E., Wani, S. P. and Day, J. M. (1986). *Plant Soil* **90**, 255–263.
Gonzalez, S. B. (1988). Ecologia y biotecnologia de micorrhizas en leguminosas (soja y alfalfa). PhD Thesis, University of Granada.
Gueye, M., Diem, H. G. and Dommergues, Y. R. (1987). *MIRCEN J.* **3**, 75–86.
Habte, M. and Manjunath, A. (1987). *Appl. Environ. Microbiol.* **53**, 797–801.
Harari, A., Kigel, J. and Okon, Y. (1988). *Plant Soil* **110**, 275–282.
Harley, J. L. and Smith, S. E. (1983). *Mycorrhizal Symbiosis.* Academic Press, London.
Harris, D., Pakovsky, R. S. and Paul, E. A. (1985). *New Phytol.* **101**, 427–440.
Hayman, D. S. (1986). *MIRCEN J.* **2**, 121–145.
Haystead, A., Malajczuk, N., Crove, T. S. (1988). *New Phytol.* **108**, 417–423.
Heggo, A. and Angle, J. S. (1990). *Soil. Biol. Biochem.* **22**, 865–869.
Hepper, C. M., Azcón-Aguilar, C., Rosendahl, S. and Sen, T. (1988). *New Phytol.* **110**, 207–215.
Hicks, P. M. (1987). *Agron. J.* **79**, 841–844.
Hirata, H., Masunaga, T. and Koiwa, H. (1988). *Soil. Sci. Plant Nutr.* **34**, 441–449.
Ho, I. (1988). *Plant Soil* **105**, 291–293.
Högberg, P. (1990). *New Phytol.* **115**, 483–486.
Howeler, R. H., Sieverding, E. and Saig, S. (1987). *Plant Soil* **100**, 249–283.
Ianson, D. C. and Linderman, R G. (1989). In *The Rhizosphere and Plant Growth*, Beltsville Symposium XIV, MD, p. 61.
Jeffrey, D. W. (1987). *Soil–Plant Relationships.* Timber Press, Portland, OR.
Kaur, S. and Singh, O. S. (1988). *Plant Soil* **112**, 293–295.
Kessel, Ch. Van, Singleston, P. W. and Hoben, H. J. (1985). *Plant Physiol.* **79**, 562–563.
Kirda, C., Danso, S. K. A. and Zapata, F. (1989). *Plant Soil* **120**, 49–56.
Kucey, R. M. N. and Bonetti, R. (1988). *Can. J. Soil Sci.* **68**, 143–149.
Kucey, R. M. N. and Janzen, H. H. (1987). *Plant Soil* **104**, 71–78.
Kucey, R. M. N. and Paul, E. A. (1982). *Soil Biol. Biochem.* **14**, 407–412.
Linderman, R. G. (1986). *HortScience* **21**, 1299–1302.
Linderman, R. G. (1988). *Phytopathology* **78**, 366–371.
Lopez-Aguillon, R. and Mosse, B. (1987). *Plant Soil* **97**, 155–170.
Louis, I. and Lim, G. (1988). *Plant Soil* **112**, 37–43.
Lynch, J. M. (1983). *Soil Biotechnology. Microbiological Factors in Crop Productivity.* Blackwell Scientific Publications, Oxford.
Malik, K. A., Bilal, R., Azam, F. and Sajjad, M. I. (1988). *Plant Soil* **108**, 43–51.
Mayo, K., Davis, R. E. and Motta, J. (1986). *Mycology* **78**, 426–431.
McGonigle, T. P. and Fitter, A. H. (1988). *New Phytol.* **108**, 59–65.
Mercy, M. A., Shivashanker, G. and Bagyaraj, D. J. (1990). *Plant Soil* **112**,

292–294.
Meyer, J. R. and Linderman, R. G. (1986a). *Soil Biol. Biochem.* **18**, 191–196.
Meyer, J. R. and Linderman, R. G. (1986b). *Soil Biol. Biochem.* **18**, 185–190.
Michelsen, A. and Rosendahl, S. (1990). *Plant Soil* **124**, 7–13.
Misaghi, I. J. (1990). *Soil Biol. Biochem.* **22**, 1985–1988.
Mohandas, S. (1987). *Plant Soil* **98**, 295–297.
Mosse, B. (1973). *Ann. Rev. Phytopathol.* **11**, 171–196.
Mosse, B. (1977). In *Exploiting the Legume–Rhizobium Symbiosis in Tropical Agriculture* (J. M. Vincent, A. S. Witney and J. Bose, eds), pp. 275–292. College of Tropical Agriculture, University of Hawaii, Miscellaneous Publication 145.
Mosse, B. (1986). *Biol. Agric. Hort.* **3**, 191–209.
Mosse, B., Stribley, D. P. and Le Tacon, F. (1981). *Adv. Microb. Ecol.* **5**, 137–210.
Morton, J. B., Yarger, J. E. and Wright, S. F. (1990). *Soil Biol. Biochem.* **22**, 127–129.
Mugnier, J. and Mosse, B. (1987). *Phytopathology* **77**, 1045–1050.
Nair, S. K., Peethambaran, D. K., Geetha, D., Nayar, K. and Wilson, K. I. (1990). *Plant Soil* **123**, 153–154.
Nambiar, P. T. C. and Anjaiah, V. (1989). *Biol. Fertil. Soils* **8**, 311–318.
Ndoye, I. and Dreyfus, B. (1988). *Soil Biol. Biochem.* **20**, 209–213.
Newman, E. I. (1978). *Biol. Rev.* **53**, 511–554.
Newman, E. I. (1988). *Adv. Ecol. Res.* **18**, 244–270.
Olivares, J., Montoya, E. and Palomares, A. (1977). In *Recent Developments in Nitrogen Fixation* (W. Newton, J. R. Postgate and C. Rodriguez-Barrueco, eds), pp. 375–376. Academic Press, London.
Pacovsky, R. S. (1986). *Plant Soil* **95**, 379–388.
Pacovsky, R. S. (1988). *Plant Soil* **110**, 283–287.
Pacovsky, R. S. (1989). *Soil Biol. Biochem.* **21**, 953–960.
Pacovsky, R. S., Fuller, G. and Paul, E. A. (1985). *Soil Biol. Biochem.* **17**, 525–531.
Pacovsky, R. S., Fuller, G., Stafford, A. E. and Paul, E. A. (1986). *Plant Soil* **92**, 37–45.
Pang, P. C. and Paul, E. A. (1980). *Can. J. Soil Sci.* **60**, 241–250.
Patterson, N. A., Chet, I. and Kapulnik, Y. (1990). *Symbioses* **8**, 9–30.
Paul, E. A. and Clark, F. E. (1989). *Soil Microbiology and Biochemistry.* Academic Press, San Diego.
Peña, J. I., Sanchez-Díaz, M., Aguirrolea, J. and Recana, M. (1988). *J. Plant Physiol.* **133**, 79–83.
Powell, C. Ll. (1976). *Trans. Br. Mycol. Soc.* **66**, 439–445.
Rai, R. (1988). *J. Plant Nutr.* **11**, 863–869.
Rao, P. S. K., Tilak, K. V. B. R. and Arunachalam, V. (1990). *Plant Soil* **122**, 137–142.
Rennie, R. J. and Thomas, J. B. (1987). *Plant Soil* **100**, 213–223.
Rose, S. L. (1980). *Can. J. Bot.* **58**, 1449–1454.
Roskoski, J. P., Pepper, I. and Pardo, E. (1986). *For. Ecol. Manag.* **16**, 57–68.
Russo, R. O. (1989). *Plant Soil* **118**, 151–155.
Sanginga, N., Bowen, G. D. and Danso, S. K. A. (1990). *Soil Biol. Biochem.* **22**, 539–547.
Secilia, J. and Bagyaraj, D. J. (1987). *Can. J. Microbiol.* **33**, 1067–1073.

Shivaram, S., Rai, P. V. and Hegde, S. V. (1988). *Plant Soil* **111**, 11–16.
Siqueira, J. O. and Paula, M. A. (1986). *R. bras. Ci. Solo.* **10**, 97–102.
Smith, S. E. and Bowen, G. D. (1979). *Soil Biol. Biochem.* **11**, 469–473.
Smith, S. E. and Daft, M. J. (1977). *Austral. J. Plant Physiol.* **4**, 403–413.
Smith, S. E., Nicholas, D. J. D. and Smith, F. A. (1979). *Austral. J. Plant Physiol.* **6**, 305–316.
Stotzky, G. (1972). *Crit. Rev. Microbiol.* **2**, 59–137.
Stribley, D. P. (1989). In *Microbial Inoculation of Crop Plants* (R. Campbell, and R. M. Macdonald, eds), pp. 49–65. IRL Press, Oxford University Press, Oxford.
Subba Rao, N. S., Tilak, K. V. B. R. and Singh, C. S. (1985a). *Plant Soil* **84**, 283–286.
Subba Rao, N. S., Tilak, K. V. B. R. and Singh, C. S. (1985b). *Soil Biol. Biochem.* **17**, 119–121.
Subba Rao, N. S., Tilak, K. V. B. R. and Singh, C. S. (1986). *Plant Soil* **95**, 351–359.
Suslow, T. V. (1982). In *Phytopathogenic Prokaryotes* (R. Mount and C. Lacey, eds), Vol. 1, pp. 187–223. Academic Press, New York.
Tilak, K. V. B. R., Li, C. Y. and Ho I. (1989). *Plant Soil* **116**, 286–288.
Tommerup, I. C. (1985). *Trans. Br. Mycol. Soc.* **85**, 267–278.
Trappe, J. M. (1986). In *Ecophysiology of VA Mycorrhizal Plants* (G. R. Safir, ed.), pp. 5–25. CRC Press, Boca Raton, FL.
Vancura, V., Orozco, M. O., Prikryl, Z. and Grauov, O. (1988). *Second European Symposium on Mycorrhizae*, Prague, Czechoslovakia. 108 pp.
Von Alten, H., Tanneberg, A. and Schonbeck, F. (1989). In *The Rhizosphere and Plant Growth*, p. 65. Beltsville Symposium XIV, MD.
Waidyanatha, U. P. S., Yogaratnam, N. and Ariyaratne, W. A. (1979). *New Phytol.* **82**, 147–152.

# 22

# Relationships between the Mycorrhizal and Actinorhizal Symbioses in Non-legumes

E. CERVANTES and C. RODRÍGUEZ-BARRUECO

*IRNA-CSIC, Apartado 257, 37071 Salamanca, Spain*

## I. General introduction

Some non-leguminous, nitrogen-fixing plants carry dual symbiotic infection involving both mycorrhizal fungi and actinorhizal actinomycetes. Since it is likely that the efficient function of one of the symbioses is to some extent dependent upon resources supplied by the other, it is appropriate in a volume dealing with mycorrhizal research to pay some attention to the co-occurring symbioses.

Indeed, the nitrogen-fixing process carried out in the plant nodules is energy demanding and therefore liable to require phosphorus, both for ATP synthesis and for the production of photosynthetic enzymes involved in the provision of carbon compounds. Consequently, the mycor-

METHODS IN MICROBIOLOGY
VOLUME 24  ISBN 0-12-521524-X

Copyright © 1992 by Academic Press Limited
All rights of reproduction in any form reserved

rhiza of actinorhizal plants may be essential in the field to obtain higher yields, especially when the plants are growing in phosphorus-deficient soils.

The characterization of the genus *Frankia*, the microsimbiont in the actinorhizal symbiosis, is approaching maturity but the optimal application of methods already developed requires the consideration of the real contribution of mycorrhiza associated with the actinorhizal symbioses.

The dramatic increase in population growth throughout the world in the present century has resulted in ecological problems, amongst which desertification is one of the most important. Areas with high risk of desertification include broad regions in North and South America, Africa, Asia, Australia and Southern Europe and the rate of growth of population may exceed the support capacity of the local agriculture in large regions, especially in Southern Asia and Africa (Clark, 1989). Possible strategies aimed at diminishing such destructive tendencies, have to be developed by the scientific community and local administrations and must have natural conservation as a main objective.

Since the early 1950s, increasing attention has been focused on the possibility of using actinomycete-nodulated plants in forestry management, land reclamation and soil improvement. Atmospheric dinitrogen fixation rates in these plants are of the same order of magnitude as those observed in legumes. Thus, rates over $100 \, \text{kg} \, \text{N} \, \text{ha}^{-1} \, \text{yr}^{-1}$, which is approximately the amount of nitrogen added as fertilizer in intensive forest exploitations, are reported. At least 23 genera belonging to phylogenetically divergent families are known to be able to form the actinorhizal symbiosis (Table I). Of these, *Casuarina* is the only genus extending to tropical areas: its ability to grow under extremely difficult conditions makes it an interesting candidate for reforestation programmes. In temperate climates, *Alnus* is intensively used in interplantings and as an alternate crop in forest timber plantations providing a biological source of fixed nitrogen. Species of *Alnus*, *Myrica* and *Elaeagnus* have been planted for many decades, mixed with other species for erosion control and soil improvement. Experiments with mixed plantings have involved different mixtures (e.g. poplar–alder for pulp and total biomass production). In all of them, significant increases in soil nitrogen content have been obtained, this leading to higher yields of companion or subsequent crops. Other uses of actinorhizal plants have included reclaiming strip mine areas and revegetating and sustaining road sides. The broadening and optimization of the utilization of actinorhiza are priority objectives for scientists working in this field. In this contribution, a brief account of the present knowledge of the symbiosis is outlined. If a wider practical approach to use of actinorhizal

**TABLE I**

Genera of actinorhizal plants indicating from which the endophyte has been isolated in pure culture (partly taken from Lechevalier, 1986)

| Family | Genus | Total species/ nodulated species | Endophyte isolated |
|--------|-------|----------------------------------|--------------------|
| Betulaceae | *Alnus* | 35/33 | + |
| Casuarinaceae | *Casuarina* | 45/24 | + |
| | *Allocasuarina* | | + |
| | *Gymnostoma* | | + |
| Coriariaceae | *Coriaria* | 15/13 | − |
| Datiscaceae | *Datisca* | 2/2 | + |
| Elaeagnaceae | *Elaeagnus* | 45/16 | + |
| | *Hippophae* | 3/1 | + |
| | *Shepherdia* | 3/3 | + |
| Myricaceae | *Myrica* | 35/26 | + |
| | *Comptonia* | 1/1 | + |
| Rhamnaceae | *Ceanothus* | 55/31 | + |
| | *Colletia* | 17/1 | + |
| | *Discaria* | 10/2 | − |
| | *Kentrothamnus* | 1/1 | − |
| | *Trevoa* | 1/1 | − |
| Rosaceae | *Cercocarpus* | 20/4 | + |
| | *Chaemabatia* | 20/1 | − |
| | *Cowania* | 4/1 | + |
| | *Dryas* | 4/3 | − |
| | *Purshia* | 2/2 | + |
| | *Rubus* | 200/? | − |
| | *Talquenea* | | |

plants is sought, the reader is referred to Rodríguez-Barrueco and Moiroud (1990).

## II. Mycorrhiza in actinorhizal plants

Mycorrhizal species have been found in most actinorhizal plant genera so far examined in both ectomycorrhizal or vesicular-arbuscular mycorrhizal symbioses. The presence of a mycorrhizal symbiosis in a plant species depends largely on many soil characteristics. Thus, infection of *Myrica gale* by both ecto- and vesicular-arbuscular mycorrhiza is restricted to well-drained soils, being absent in wet soils (Harley and Harley, 1987). However, sometimes even in well-drained conditions, *M. gale* fails to form vesicular-arbuscular mycorrhizal associations (Berliner and Torrey, 1989).

## A.  Ectomycorrhizal associations

Among the ectomycorrhizal fungi present in *Alnus* are members of the
Cortinariaceae (*Alnicola, Inocybe, Phlegmacium*), the Hygrophoraceae
(*Hygrocybe*), the Russulaceae (*Russula, Lactarius*), the Boletaceae
(*Gyrodon*) and the Fungi Imperfecti (*Cenococcum*) (Trappe, 1962;
Horak, 1963). *Alpova diplophloeus* (Melanogastraceae) forms well-deve-
loped ectomycorrhiza with many *Alnus* species (Trappe, 1975; Clemencon, 1977; Godbout and Fortin, 1983), both in North America and in
Europe. It was suggested from these findings that the low diversity
shown by fungal sporocarps detected under *Alnus* was an indication of a
marked specificity of the fungal associate towards that actinorhizal
genus. Furthermore, the inoculation of five different species of *Alnus*
with a wide variety of enriched preparations of ectomycorrhizal fungi led
to fungus–plant associations with only three of the fungi tested (Molina,
1979, 1981), mainly *Alpova diplophloeus, Astraeus pteridis* and *Paxillus
involutus*. Profuse mantle and Hartig net formations were obvious in
*Alpova*, while those features were only weakly represented when
*Paxillus* was the inoculum. The latter association was also accompanied
by an accumulation of phenolic substances in the root cell walls,
indicating a strong reaction on the part of the host plant. The longer
time required for penetration and developement of the Hartig net by
most ectomycorrhizal fungi other than *Alpova*, suggests a certain
specialization of the latter towards *Alnus*. There are contradictory
reports of the status of other fungi, e.g. *Cenococcum geophylum*,
following tests carried out under laboratory conditions (Molina, 1979,
1981). These have been attributed to a diminished viability of the fungus
through long storage (Marx and Daniel, 1976). The nutrient regime of
the plants might also have an adverse effect on the establishment of the
fungal associate. In all these matters, the nutritional influence of both
fungus and *Frankia* is still to be assessed, and to that end *Alpova*
appears to be a valuable tool to make progress in the understanding of
the physiology and biochemistry of the host–fungus specificity relations.

The following genera of ectomycorrhizal fungi have been reported to
form mycorrhiza with *Casuarina*: *Cenococcum, Pisolithus, Hymeno-
gaster, Thelephora, Rhizopogon* and *Amanita*. Rose (1980) and
Gardner (1986) suggest that the role of *Casuarina* as a pioneering plant
may be assisted by the presence of mycorrhiza on the root system; this
may also be helped by the presence of proteoid roots in members of the
Casuarinaceae, tightly packed rows of rootlets probably structured to
increase the ability of the plant to absorb nutrients (Diem and Dommergues, 1990).

The ectomycorrhizal association has been shown to enhance phosphate uptake in the case of *Alnus viridis* subsp. *viridis* (Mejstrik and Benecke, 1969).

## B. Endomycorrhizal associations

Studies of vesicular-arbuscular fungi associated with actinorhizal plants indicates that species of the genus *Glomus*, *Gigaspora* and *Acaulaspora* are primarily involved. Vesicular-arbuscular mycorrhizal associations with roots of actinorhizal plants have been shown in several cases to yield enhanced phosphorus uptake as well as an increased nodulation and nitrogen fixation (Rose and Youngberg, 1981; Gauthier *et al.*, 1983; Gardner *et al.*, 1984). Diem and Gauthier (1982) found that double inoculation of *C. equisetifolia* with *Glomus mosseae* and *Frankia* (as crushed nodules), significantly improved plant growth and nodulation. After six months of growth, mycorrhizal nodulated plants contained twice as much nitrogen as seedlings that had only been inoculated with *Frankia* (Diem and Dommergues, 1990). These results may be dependent on the nutritional status of the plants. Thus, inoculation of *Alnus acuminata* with *Glomus intraradices* in addition to *Frankia* shifted the exogenous P level for maximum acetylene reduction from 50 down to 10 ppm, suggesting that this vesicular-arbuscular mycorrhizal infection may enhance the activity of nitrogenase in seedlings grown at low P level (Russo, 1989).

Sometimes, both ecto- and vesicular-arbuscular mycorrhiza have been found in the same plant, as is the case in *Cercocarpus* and *Purshia* (Williams, 1979), *Alnus glutinosa* (Hall *et al.*, 1979) and other actinorhizal species (Rose, 1980), resulting in a tetrapartite symbiosis that can support a better rate of growth of the plant (Rose and Youngberg, 1981; Gauthier *et al.*, 1983; Gardner, 1986; Chatarpaul *et al.*, 1989).

In the development of these multiple symbioses, mycorrhizal fungi may compete with *Frankia* for sites on the roots, and once established around the nodule lobes, the ectomycorrhiza may inhibit the spread of *Frankia* into the new lobe (Gardner, 1986). Vesicular-arbuscular hyphae are often found in the nodular tissues. Here they may be surrounded by ectomycorrhizal hyphae which penetrate the outer cortex of the *Alnus* nodule or the adjoining root. However, such hyphae are never found in the zone of the nodule containing the bacteroids (Daft, 1983) (see Figs 1–5).

Other reports indicate that vesicular-arbuscular mycorrhizal infection may improve drought tolerance of plants (Safir *et al.*, 1971, 1972) and resistance to fungal root rots and other pathogens (Trappe, 1972). Also,

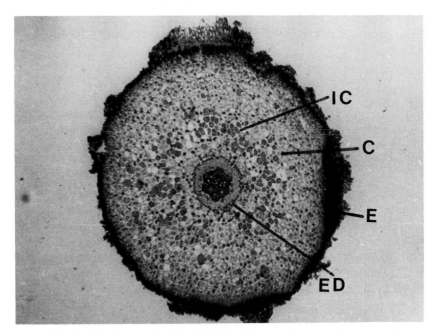

**Fig. 1.**   Cross-section of a root nodule lobe of *Alnus glutinosa*. E, epidermis; C, cortex; ED, endodermis; IC, infected cell with *Frankia*. Magnification, 40×.

it is interesting to consider that mycorrhizal mycelium can form a network in the soils (Newman, 1988) which is able to interconnect actinorhiza with other non nitrogen-fixing plants.

### III.   The actinorhizal symbioses

In this section, some recent developments in the study of actinorhizal symbiosis will be described with emphasis on practical applications. For a more detailed account of most of the particular topics reviewed here, the reader is referred to *The Biology of* Frankia *and Actinorhizal Plants* (Shwintzer and Tjepkema, 1990).

   Research on the actinorhizal symbiosis has been hampered by the difficulty in obtaining pure cultures of *Frankia* from nodules. The first isolation of *Frankia* from nodules, including re-infection with the micro-organism in pure culture and its re-isolation from the nodules formed, was reported by Callaham, Del Tredici and Torrey in 1978. Since then, the actinorhizal endophyte has been isolated from many host

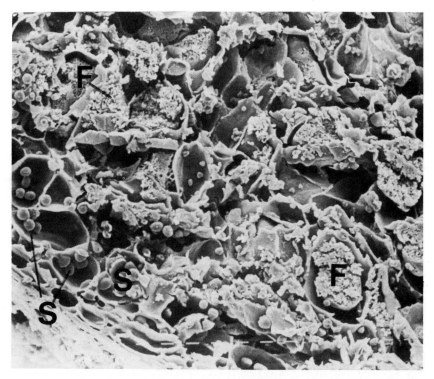

**Fig. 2.**  Scanning electron micrograph of the nodule cortex of *Casuarina cunninghamiana* showing *Frankia* hyphal masses (F) within cells and starch granules (S). Magnification, 425×.

plants (Table I) but isolation remains to be achieved with many others.

The genus *Frankia* is now relatively well described as a symbiotic, filamentous, Gram-positive, sporangium- and vesicle-forming actinomycete. The vesicles are the sites of nitrogen fixation and of localization of the nitrogenase enzyme complex (Meesters *et al.*, 1987). In the symbiotic state, two types of *Alnus* nodules exist depending on the sporulating capacity of *Frankia in situ*: spore-positive and spore-negative. The capacity to sporulate in the nodule is considered to be a characteristic genetically determined by the endophyte independently from the host plant (Van Dijk, 1984; Van den Bosch and Torrey, 1983) and it is interesting to note that there appear to be no strains present in culture that consistently induce spore positive nodules (Torrey, 1987). Spore formation by *Frankia in vitro* can be affected by the composition of the culture medium. Root nodules of *Hippophae rhamnoides* and *Casuarina* spp. have never been shown to produce sporangia, but

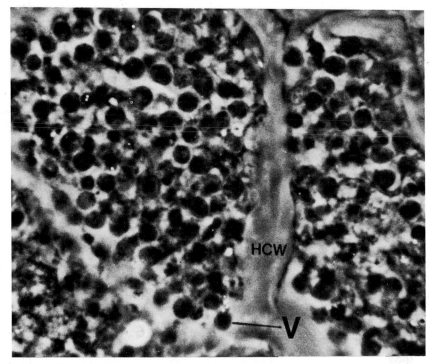

**Fig. 3.**  Vesicular structures (v) of *Frankia* in cortical cells of a root nodule of *Alnus glutinosa*. HCW, host cell wall. Magnification, 1700×.

cultures of *Frankia* isolated from them sporulate abundantly during growth in an appropriate medium. Spore-negative nodules have been reported in other actinorhizal plants. However, *Frankia* isolates from these nodules formed spores *in vitro* in an appropriate medium. Thus, the expression of genes involved in sporulation may also be regulated through environmental conditions.

Vesicle formation may also be environmentally regulated. Thus, *Frankia* strains growing in the nodules of *Casuarina* sometimes show neither sporangia nor vesicles, but in pure cultures develop spherical vesicles under appropriate conditions.

Apart from these morphological characters, *Frankia* strains can be identified by their different capacity to synthesize nitrogenase, the presence of specific cell wall components, the protein profile and serological properties. Growth requirements are also highly variable. Many strains grow well with propionic acid as a carbon source and can also utilize fatty acids; others may also utilize organic acids, sugars and other compounds.

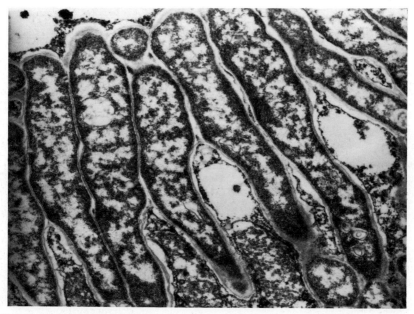

**Fig. 4.** Transmission electron micrograph of *Frankia* terminal hyphae in tissue of a root nodule of *Coriaria myrtifolia*. Magnification, 12 000×.

The genome size of *Frankia* has been measured giving values similar to *Streptomyces*, and approximately twice the size of that of *Escherichia coli* (An, 1983). The DNA composition is in the range of 66–75% C+G, which is in agreement with the high C+G content of actinomycetes. Several plasmids have been found in *Frankia* strains and sometimes they carry symbiotic functions like the structural nitrogenase genes (Simonet *et al.*, 1986).

Systems for the transfer of genetic material between *Frankia* strains are lacking, and no conjugative plasmids or transduction phages have been reported. Effective transformation systems are currently being investigated, based mainly on protoplast preparation (Normand *et al.*, 1987).

The classification of the symbiotic nitrogen-fixing bacteria belonging to the genus *Rhizobium* has been classically based on the existence of cross-inoculation groups. Such groups also exist in the actinorhizal symbioses and were traditionally investigated in experiments using crushed nodule inocula (Rodriguez-Barrueco and Subramaniam, 1988). Host plant–endophyte specificity studies with the increasing number of *Frankia* isolates available indicate at least four cross-inoculation groups (Baker, 1987):

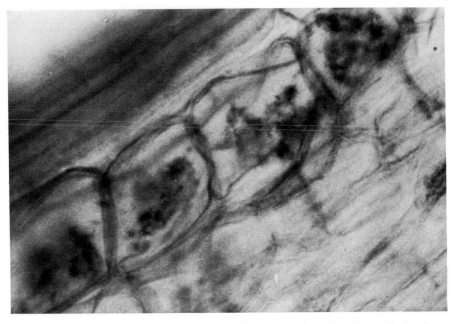

**Fig. 5.**   Arbuscular structures formed by *Glomus* sp. infecting the cortical cells
of *Casuarina equisetifolia*.

- Strains nodulating *Alnus* and *Myrica*.
- Strains nodulating *Casuarina* and *Myrica*.
- Strains nodulating the Elaeagnaceae and *Myrica*.
- Strains nodulating only the Elaeagnaceae.

Studies on deoxyribonucleic acid relatedness correlate well with this classification, showing the existence of various genomic species in each of these cross-inoculation groups (Fernández *et al.*, 1989).

In order to use this symbiotic association for practical purposes, the methodologies required include:

- Isolation of *Frankia* strains.
- Culture, characterization and collection of strains from different sources.
- Production of inocula.
- Searching for better host–endophyte combinations.

as described below.

## A. Isolation of *Frankia* strains from actinorhizal nodules

The coralloid structure of the nodules in many plant species facilitates the adherence of soil particles and supports an abundant microflora which has to be eliminated before the endophyte can be isolated.

The woody nature of the plant tissue allows the utilization of potent sterilizing agents, which are able to penetrate the outer cortical spaces, and eliminate undesired micro-organisms. Excessive exposure to the sterilizing agent may have a destructive effect on the endosymbiont located in the cortical parenchyma. Suspension of the nodules in a 3% aqueous solution of osmium tetroxide for 5 min, after thorough washing in tap water has been widely used as the method for sterilization. Before using such methods, one must consider the anatomy of the nodule. For example, in *Myrica faya* (Miguel and Rodríguez-Barrueco, 1974) and *Rubus ellipticus* (Becking, 1984), the endosymbiont is located in a circular band near the epidermis, thus being less protected from such treatments as the endophyte of *Alnus*, *Myrica gale*, *Elaeagnus*, and *Casuarina* with a more generalized location. Indeed, no strain of *Frankia* has been succesfully isolated from *Myrica faya* nor *Rubus ellipticus*, although this may also be related to their restricted geographical distribution (the first in Macronesia, the second in Indonesia); also, in the case of *Rubus*, further evidence will be required confirming the nodulation of this genus by *Frankia*.

Other sterilizing methods may involve the use of antibiotics or other chemical agents such as hydrogen peroxide. Filtration and differential centrifugation have also been used to obtain purified preparations of *Frankia* (Baker *et al.*, 1979; Benson, 1982).

All these methods require the release of the micro-organism from the plant cells for direct contact with the isolation medium. For this purpose it may be useful to employ lytic enzymes. Alternatively, release may be achieved through mechanical disruption of the plant cell walls with mortar or homogenizer. This may produce a greater risk of contamination and the oxidation of nodule organic compounds may also reduce *Frankia* viability, so it is advisable to cut the nodule tissue into small pieces with a sterile razor blade. If required, a concentrate of *Frankia* propagules may be obtained by differential centrifugation of nodule homogenates (i.e. the endophyte will precipitate between 45% and 60% sucrose in differential centrifugation: Baker *et al.*, 1979).

## B. Culture of isolated *Frankia*

From 1866, when Woronin first described the microbial nature of the endophyte of actinorhiza, to 1959, the year in which Pommer reported the first isolation of a micro-organism from *Alnus glutinosa* nodules (Pommer, 1959), there was a period of discontinuous efforts to obtain the micro-organism in pure culture and of uncertainty about its real nature. Since the isolation of *Frankia* by Callaham *et al.* (1978), succesful isolations and subsequent cultivations have been reported using relatively simple culture media. A typical medium, for example BS (described by Burggraaf and Shipton, 1982), contains glucose and propionate as carbon sources, casein hydrolysate as a nitrogen source, trace elements and vitamins and several salts such as $K_2HPO_4$, $NaH_2PO_4$, $MgSO_4$, $CaCl_2$ and FeNaEDTA. Other media often used contain also a lipid supplement (i.e. Tween 80 in the BuCT medium), yeast extract and peptone as in Qmod medium or different combinations of mineral salts. The addition of phospholipids and the agitation of the medium improves significantly the growth of *Frankia* strains isolated from *Casuarina* (J. Schwencke, pers. commun). The addition of lipid extracts from roots may improve the results, and dipterocarpol has been shown to be important for the growth of several *Frankia* strains (Quispel *et al.*, 1989). The combination of solid and semisolid media has given good results in the first steps of isolation when small nodule fragments are placed between two agar slides in a Petri dish.

A special difficulty arises with the isolation of *Frankia* from nodules containing abundant sporangia in the tissue. This emphasizes the fact that it is not known exactly which structures are responsible for the initiation of colonies in the medium. *Frankia* is able to develop hyphal structures in culture, which are called "reproductive torulose hyphae". These are able to multiply (Diem and Dommergues, 1984), although they have been interpreted as being structures produced in response to stress. An exclusively hyphal culture has been shown to yield a good infective capacity in the case of *Alnus glutinosa*. Several plant compounds may affect the pattern of development of *Frankia*, thus *p*-coumaric acid inhibits sporogenesis and *p*-hydroxybenzoic acid stimulates vesicle production. These molecules thus act as chemical mediators controlling development (Perradin *et al.*, 1983). Tannin production by the plant may help to inhibit succesive infections through the root hairs after the initial infection is established. The required oxygen tension of the culture medium seems to vary from strain to strain, there being strict microaerophily in some and full aerophily in others.

## C.  Production of inocula

Frequently, inoculation of actinorhizal plants is made with soil or crushed nodules. This technique, used commonly because of its effectiveness and also because of the lack of pure cultures of *Frankia* in the past, has several disadvantages. First, it is difficult to standarize as it is impossible to control every chemical or micro-organism present in soil or in crushed nodules; second, seedlings may become contaminated with undesirable agents such as pseudomonads or nematodes.

Several recent reports concern the utilization of pure cultures of *Frankia*. These have been used as inocula for *Casuarina equisetifolia* in Senegal (Dreyfuss *et al.*, 1988) and *C. cunninghamiana* in Australia and Zimbabwe (Reddell *et al.*, 1988).

Inoculation of trees raised in containers is carried out by mixing the soil or substrate with the inoculum because of the lack of motility of the micro-organism in the soil. Research on methods to increase the biomass of pure cultures to be used as inocula will give results of practical importance. Experiments so far with *Casuarina* are promising (Dreyfus *et al.*, 1988).

## D. Searching for better host–endophyte combinations

Dreyfuss *et al.* (1988) define the nitrogen-fixing potential (NFP) as the amount of nitrogen fixed by a symbiotic system in a constraint-free environment. The amount of nitrogen actually fixed is called the actual nitrogen fixation (ANF). The improvement of these parameters is a priority in nitrogen-fixation research. Once the more actively nitrogen-fixing individuals in a population are identified, the superior genotypes must be vegetatively propagated (Sougoufara *et al.*, 1989).

The first report of *in vitro* propagation of alder came from Garton *et al.* (1981). The method involves rooting stem or internode cuttings and has been developed for most species of *Alnus*. In the case of *Casuarina equisetifolia*, immature female inflorescences were used as starting stock (Sougoufara *et al.*, 1989). The conditions for such cultures are discussed in detail by Tremblay *et al.* (1986).

Leaf protoplast culture has been reported for *Alnus glutinosa*, *A. incana* and *A. crispa*, and in future the development of such methods may be useful for the creation of somatic hybrids and eventually for the transfer of desired characteristics between species.

## IV. Perspectives

The different endophyte–plant genotype combinations that may be engineered by the techniques described above will not have the same efficiency in different soils and habitats; thus, to check the efficiency of a symbiotic association, field studies are required. In such studies, the effect of inoculation with *Frankia* will be dependent on the environmental conditions (Dixon and Wheeler, 1983) including nutrients, soil pH, salinity, water and temperature. Ideally, these studies must include inoculation with ecto- and vesicular-arbuscular mycorrhiza. Before these applied goals may be achieved, more research is required concerning the molecular basis and the physiology of the establishment of the mycorrhizal interaction, the real nature of the interaction and its physiological details in stress environments and the role played by specific environmental factors through the development of the association.

More studies are required to determine the maximum potential not only from these associations, but also from other associative symbionts which may contribute to plant nutrition and growth. There are reports, for example, of *Azospirillum* enhancing root growth of *Casuarina* seedlings (Rodríguez-Barrueco *et al.*, 1991).

## References

An, C. S. (1983). Relationships of *Frankia* among themselves and to other actinomycetes based on DNA homology studies. PhD Thesis, University of Tennessee, Knoxville, TE. 97 pp.

Baker, D. D. (1987). *Physiol. Plant.* **70**, 245–248.

Baker, D. D., Torrey, J. G. and Kidd, G. H. (1979). *Nature*, **281**, 76–78.

Becking, J. H. (1984). *Plant Soil* **78**, 105–128.

Benson, D. R. (1982). *Appl. Env. Microbiol.* **44**, 461-465.

Berliner, R. and Torrey, J. G. (1989). *Can. J. Bot.* **67**, 1708–1712.

Burggraaf, A. J. P. and Shipton, W. A. (1982). *Plant Soil* **61**, 135–147.

Callaham, D., Del Tredici, P. and Torrey, J. G. (1978). *Science* **199**, 899–902.

Chatarpaul, L., Chakravarty, P. and Subramaniam, P. (1989). *Plant Soil* **118**, 145–150.

Clark, W. C. (1989). *Scient. Am.* **261**, 19–26.

Clemencon, H. (1977). *Schweiz. Z. Pelzkunde* **55**, 107–113.

Daft, M. J. (1983). *Plant Soil* **71**, 331–337.

Diem, H. G. and Dommergues, Y. R. (1984). In *Current Research on* Frankia *and Actinorhizal Plants*, Proceedings of the International Symposium, Laval University.

Diem, H. G. and Dommergues, Y. R. (1990). In *The Biology of* Frankia *and Actinorhizal Plants* (C. R. Schwintzer and J. D. Tjepkema, eds), Ch. 16. Academic Press, San Diego.

Diem, H. G. and Gauthier, D. (1982). *C. R. Hebd. Seances Acad. Sci., Ser. C* **294**, 215–218.

Dixon, R. O. D. and Wheeler, C. T. (1983). In *Nitrogen Fixation Research Progress* (J. C. Gordon and C. T. Wheeler, eds), pp. 107–171. Martinus

Nijhoff/Dr. W. Junk Publishers, The Hague, Netherlands.

Dreyfuss, B. L., Diem, H. G. and Dommergues, Y. R. (1988). *Plant Soil* **108**, 191–199.

Fernández, M. P., Meugnier, H., Grimont, P. A. and Bardin, R. (1989). *Int. J. Syst. Bacteriol.* **39**, 424–429.

Gardner, I. C. (1986). *MIRCEN J.* **2**, 147–160.

Gardner, I. C., Clelland, D. M. and Scott, A. (1984). *Plant Soil* **78**, 89–200.

Garton, S., Hosier, M. A., Read, P. E. and Farnham, R. S. (1981). *HortScience* **16**, 758–759.

Gauthier, D., Diem, H. G. and Dommergues, Y. (1983). In *Casuarina: Ecology, Management and Utilization* (S. J. Midgeley, J. W. Turnbull and R. D. Johnston, eds), pp. 211–217. CSIRO, Melbourne.

Godbout, C. and Fortin, J. A. (1983). *New Phytol.* **94**, 256–262.

Hall, I. R., McNab, H. S., Maynard, C. A. and Green, T. L. (1979). *Bot. Gaz.* **140**, 120–126.

Harley, J. L. and Harley, E. L. (1987). *New Phytol.* **105** (Suppl.), 1–102.

Harley, J. L. and Smith, S. E. (1983). *Mycorrhizal Symbiosis*. Academic Press, London.

Horak, V. E. (1963). *Mitteil. Schweiz. Anstalt Fortsl. Versuchs.* **39**, 1–112.

Lechevalier, M. P. (1986). *Nitrogen Fixing Actinomycetes of the Genus* Frankia. In *Proceedings of the 4th International Symposium on Microbial Ecology* (F. Megusar and M. Gantar, eds). Slovene Society for Microbiology, Ljubljana.

Marx, D. H. and Daniel, W. J. (1976). *Can. J. Microbiol.* **22**, 338–341.

Meesters, T. M., van Vliet, V. M. and Akkermans, A. D. L. (1987). *Physiol. Plant.* **70**, 267–271.

Mejstrick, V. and Bennecke, U. (1969). *New Phytol.* **68**, 141–149.

Miguel, C. and Rodríguez-Barrueco, C. (1974). *Plant Soil* **41**, 521–526.

Molina, R. (1979). *Can. J. Bot.* **57**, 1223–1228.

Molina, R. (1981). *Can. J. Bot.* **59**, 325–334.

Newman, E. I. (1988). *Adv. Ecol. Res.* **18**, 243–270.

Normand, P., Simonet, P., Prin, Y. and Moiroud, A. (1987). *Physiol. Plant.* **70**, 259–266.

Perradin, Y., Mottet, M. J. and Lalonde, M. (1983). *Can. J. Bot.* **61**, 2807–2814.

Pommer, E. H. (1959). *Ber. Deutsch. Bot. Gesell.* **72**, 138–150.

Quispel, A., Baerheim Svendsen, A., Shripsema, J., Baas, W. J., Erkelens, C. and Lungtenburg, J. (1989). *Mol. Plant Microbe Intract.* **2**(3), 107–112.

Reddell, P., Rosbrook, P. A., Bowen, G. D. and Gwaze, D. (1988). *Plant Soil* **108**, 79–86.

Rodríguez-Barrueco, C. and Subramaniam, P. (1988). In *Biological Nitrogen Fixation, Recent Developments* (N. S. Subba Rao, ed.), Ch. 11, pp. 283–310. Oxford & IBH Publishers, New Delhi, India.

Rodríguez-Barrueco, C., Cervantes, E., Subba Rao, N. S. and Rodríguez-Cáceres, E. (1991). *Plant Soil* **135**, 121–124.

Rodríguez-Barrueco, C. and Moiroud, A. (1990). In *Biofertilizers* (L. L. Somani *et al.*, eds), pp. 243–270. Pawan Kumar Science Publishers, Jodhpur, India.

Rose, S. L. (1980). *Can. J. Bot.* **58**, 1449–1454.

Rose, S. L. and Youngberg, C. F. (1981). *Can. J. Bot.* **59**, 34–39.

Russo, R. O. (1989). *Plant Soil* **118**, 151–155.

Safir, G. R. Boyer, J. S. and Gerdemann, J. W. (1971). *Science* **172**, 581–583.

Safir, G. R., Boyer, J. S. and Gerdemann, J. W. (1972). *Plant Physiol.* **49**, 700–703.

Schwintzer, C. R. and Tjepkema, J. D. (1990). *The Biology of* Frankia *and Actinorhizal Plants.* Academic Press, London. 408 pp.

Simonet, P., Haurat, J., Normand, P., Bardin, R. and Moiroud, A. (1986). *Mol. Gen. Genet.* **204**, 492–495.

Sougoufara, B., Diem, H. G. and Dommergues, Y. R. (1989). *Plant Soil* **118**, 133–137.

Torrey, J. G. (1987). *Physiol. Plant.* **70**, 279–288.

Trappe, J. M. (1962). *Bot. Rev.* **28**, 538–606.

Trappe, J. M. (1972). *Oregon State Univ. For. Symp.* **3**, 35–51.

Trappe, J. M. (1975). *Nova Hedwigia* **51**, 279–309.

Tremblay, F. M., Perinet, P. and Lalonde, M. (1986). In *Biotechnology in Agriculture and Forestry*, Vol. 1: *Trees* (Y. P. S. Bajaj, ed.), pp. 87–100. Springer-Verlag, Berlin and Heidelberg.

Van Den Bosch, K. A. and Torrey, J. G. (1983). *Can. J. Bot.* **61**, 2898–2909.

Van Dijk, C. (1984). Ecological aspects of spore formation in the *Frankia–Alnus* symbioses. PhD Thesis. Rijkuniversiteit Leiden, The Netherlands. 157 pp.

Williams, S. E. (1979). *Bot. Gaz.* **140**, S115–S119.

# Index